Goodbye to the Flush Toilet

Goodbye to the Flush Toilet

Water-Saving Alternatives to Cesspools, Septic Tanks, and Sewers

Edited by Carol Hupping Stoner

Special Consultant Patricia M. Nesbitt

Illustrations Jerry O'Brien

 **Rodale Press** Emmaus, PA

Library of Congress Cataloging in Publication Data

Main entry under title:

Goodbye to the flush toilet.

Bibliography: p.
Includes index.
1. Water-closets. 2. Water conservation.
I. Stoner, Carol.
TH6498.G66 333.9'14 77-11052
ISBN 0-87857-192-2 paperback

6 8 10 9 7 paperback

Printed in the United States of America on recycled paper

Contents

Acknowledgments vi

Introduction vii

Chapter 1 How We Got Where We Are, or The Why and Wherefrom of Sewers 1
Joel A. Tarr

Chapter 2 How We're Handling Our Wastewater Now, and Alternatives for the Future 21
Carol Hupping Stoner

Chapter 3 A Short Lesson on the Principles of Composting ... 43
Carol Hupping Stoner

Chapter 4 Composting Privies 51
Carol Hupping Stoner

Chapter 5 Commercial and Owner-Built Composting Toilets .. 97
Carol Hupping Stoner

Chapter 6 Dealing with the Greywater 161
Patricia M. Nesbitt

Chapter 7 What's in Store for Flush Toilet and Greywater Alternatives? 205
Patricia M. Nesbitt

Chapter 8 Every Little Bit Counts—Saving Water 235
Carol Hupping Stoner

Hardware Listings 257

Bibliography 271

Index 279

Acknowledgments

Putting this book together has been a fun and satisfying project, due in no small part to the people who helped me with it. Many whose names appear nowhere in this book deserve credit for their assistance, and I'd like to thank them here and now.

Jerry Goldstein, Ray Wolf, and Steve Smyser provided steady inspiration and a healthy flow of contacts and ideas. Carole Turko was a great help with the interviewing and with many last-minute details; Rill Goldstein labored over the appendix, and Julie Ruhe was invaluable in manuscript preparation.

Abby Rockefeller and Carl Lindstrom, from Clivus Multrum, USA; and David Delporto, from Ecos Inc. helped me with the technical information and put me in touch with many composting toilet owners.

My gratitude to Harold Leich who introduced me to many flush toilet alternatives, and to Alex Hershaft who gave me much information on our present sewer systems. Also to Clarence Golueke of the University of California's Sanitary Engineering Research Laboratory; John T. Winneberger, Research Director of Studies on On-site Sewage Management in the California Governor's Office of Appropriate Technology; Rein Laak of the Civil Engineering Department at the University of Connecticut; and Dan Dindal, soil microbiologist with SUNY College of Environmental Science and Forestry in Syracuse, New York—all for their expert advice.

And, finally, special thanks to the many people who shared with me their first-hand experiences—good, bad, and mixed—of living without the flush toilet.

Carol Hupping Stoner
June 1977
Emmaus, Pennsylvania

Introduction

For one person, the typical five-gallon flush contaminates each year about 13,000 gallons of fresh water to move a mere 165 gallons of body waste. What this means is that we're taking a valuable, clean resource —water—and a potentially valuable resource—human excrement—and mixing them together to pollute the water and make the fertilizer potential of body wastes just about useless. And then we pay dearly to separate them again. Surely there must be a better way.

Actually, there are several "better ways," and this book explores many of them. It examines such alternatives as vacuum and flush toilets that either use no water at all or else use only about one quart of water with each flush. It looks at the potential of the self-contained chemical and incinerating toilets, and also at aerobic tanks designed to replace failing septic tanks, at pressure sewers that replace conventional, costly gravity sewers, and at total household water recycling systems.

But the book really focuses on composting toilets and compatible greywater systems. These alternatives, as they are today, are very new. They are still being tested and refined. And they are being scorned by some, nay, by many because innovation, especially when it has to do with human wastes, is difficult for most people to accept. Such systems, however, are good in concept because they work on simple, natural processes. Not only do they not pollute our clean water, they also, when working as they should, render our own wastes useful and safe as a soil amendment and fertilizer by recycling them at the best place possible—the spot that is both the point of disposal and the point of reuse.

Those of us who worked on this book are not naive enough to think that cities that have invested millions or perhaps billions of dollars in sewer systems are going to turn around and encourage their residents to forsake their porcelain water closets for flushless toilets, even if the expense of "sewering-up" is putting an enormous strain on their budgets and the sludge left over after centralized treatment is giving them disposal headaches.

But we do think that there are many places where composting toilets would be effective, safe, and convenient. Sanitary engineers should begin to look at the alternatives proposed here for new housing, as a way to take an increasing burden off already overtaxed sewage treatment plants. Towns whose populations have grown so dense that they have reached their limit as far as cesspools and septic systems are concerned may find that composting toilets and alternative greywater systems are going to save them considerable amounts of money and prevent them from landfill overloads and water pollution worries. Some rural communities have already and will most likely continue to look at composting toilets as viable alternatives. Because these toilets are self-contained, their environmental impact is practically nil, and because they reduce the amount of wastewater each household discharges, homeowners can put in smaller septic systems and leachfields or install smaller and simpler greywater systems.

Outhouses are still big business in the United States, representing some 20 manufacturers and almost $10 million in sales each year. The U.S. Forest Service, which is responsible for 40,000 privy seats in campgrounds throughout the country, recently spent about $60,000 to develop a better privy, and both the Bureau of Land Management and the National Park Service are also looking at ways to improve the one- and two-holers under their jurisdictions. Composting privies and toilets, as well as biological and even experimental solar-assisted oil-flush toilets, are already taking the rough edges off of "roughing it" in some recreational and wilderness sites. Everything indicates that they'll continue to slowly replace more of the adequate but hardly aesthetically pleasing old-fashioned outhouses.

Surprising as it may be, the majority of the world doesn't have sewers or septic systems. The World Health Organization in 1972 estimated that 70 percent of the world population still doesn't have access to piped water at all. A portion, albeit small, of that 70 percent resides in North America. Composting privies and toilets certainly would be an improvement for them, and one that is excessive neither in cost nor technological sophistication.

In this time of resource-consciousness, it behooves us to look at the many ways we have to harness and wisely use all our natural resources. Just as we are beginning to look at a variety of ways for meeting our energy needs—solar, wood, wind, geothermal, along with nuclear and fossil fuels—we need to find a number of effective means for recycling our wastes. What we propose in this book is that all of us consider the many alternatives to the handling of our wastewaters and find the one or ones that in the long run will best serve the particular situation.

CHS

Chapter 1

How We Got Where We Are, or The Why and Wherefrom of Sewers

> *There is no guano comparable in fertility to the detritus of a capital, and a large city is the strongest of stercovaries. To employ the town in manuring the plains would be certain success; for if the gold be dung, on the other hand our dung is gold.*
>
> *What is done with this golden dung? It is swept into the gulf. We send at a great expense fleets of ships to collect at the Southern Pole the guano of petrels and penguins, and cast into the sea the incalculable element of wealth which we have under our hand. All the human and animal manure which the world loses, if returned to the land instead of being thrown into the sea, would suffice to nourish the world.*
>
> Victor Hugo

In 1894 Solomon Schindler, a well-known Boston reformer and follower of Edward Bellamy, published *Young West,* a sequel to Bellamy's *Looking Backward.* Young West is the son of Julian West, the hero of Bellamy's book, and he lives in the twenty-first century. Young West is a volunteer worker in the sewer division of his native city, Atlantis—a city whose sewerage system had "reached a very high degree of perfection," with all "foul matter" sterilized and carried through large tunnels out into the sea. Young West becomes head of the sewer division and begins to question the waste of the valuable materials in the sewage. Convinced that "the productive forces of the earth must become exhausted unless we return to the land . . . as much as we draw from it," he invents a system to make wastes into fertilizer by deodorizing them and combining the waste with a chemical to form fertilizing bricks. The bricks make barren land bloom, and the public applauds Young West as the man who "has caused three blades to grow in place of one." Riding a crest of popularity, he is elected President and applies his system throughout the country. At the completion of his

Joel A. Tarr is Professor of History, Technology, and Urban Affairs at Carnegie-Mellon University, Pittsburgh.

term, as he expressed it, "the country bloomed like a garden, it yielded fruit in abundance, and the people blessed me for it."

The overtones of today's ecology movement in this utopian novel are far from fantasy. For much of the nineteenth century and for centuries before, agriculturists, sanitarians, and others concerned with public health maintained that the logical way to dispose of urban wastes was to return them to the soil where they would be reincorporated into nature's cycle. Actually, while the use of sewage (human wastes in water carriage) in agriculture dates back to about 1800, the application of human wastes directly to the land has a much longer history. The Romans used human wastes as fertilizer, and in the Middle Ages farmers in Flanders purchased "night soil" in the cities to use on the land. Edicts dating back to the seventeenth century in Flanders compelled settlers in the peat-marsh colonies to manure their fields with urban wastes.

This practice of waste recycling continued in many European nations throughout the nineteenth century. In Paris, for instance, the existing sewers were intended only for storm and wastewater. Human excrement was disposed of in cesspits—large underground tanks usually built of some impermeable material. In 1842 there were approximately 50,000 cesspits in Paris serving about 900,000 people. A crew of between 200 and 250 *vidangeurs* cleaned the cesspits by hand and pump and deposited the wastes in a dump on the outskirts of the city. There the liquids and solids were separated and the solid matter dried. Once dried it was transported to rural areas to be used as fertilizer.

In England the so-called "pail system" was extensively used as a form of prewater-carriage waste removal. In Rochdale (population about 70,000) and Manchester (population about 400,000) in the latter half of the nineteenth century, Health Department regulations required construction of a "pail closet" for most households. This pail closet was constructed on a raised flagstone platform and consisted of a hinged seat set over a removal pail or tub. At the side of the closet was an ash sifter which deposited a layer of ashes on the human wastes. The pails or tubs were collected once a week and taken to a depot where their contents were mixed with more ashes, coal dust, and gypsum. After drying, the mixture was sold as fertilizer. In Rochdale in 1873 the sale of the fertilizer supposedly paid for 80 percent of the cost of collecting and preparing it.

The most widespread recycling of human wastes, however, has been in China, Japan, and Korea, where it has been practiced for many centuries. Writing in 1850, an admiring American agriculturist noted that the Chinese were "the most admirable gardeners and trainers of plants" and attributed their success in agriculture to their careful use of human urine and night soil from the cities. This material was collected daily and either used immediately or baked with clay to form dry bricks called *tafeu*. At the beginning of the twentieth century,

In the Far East, for more than thirty centuries, these enormous wastes have been religiously saved, and today the four hundred million of adult population send back to their fields annually 150,000 tons of phosphorus, 376,000 tons of potassium, and 1,158,000 tons of nitrogen comprised in a gross weight exceeding 182 million tons, gathered from every home, from the country villages and from the great cities. . . .

Man is the most extravagant accelerator of waste the world has ever endured. His withering blight has fallen upon every living thing within his reach, himself not excepted; and his besom of destruction in the uncontrolled hands of a generation has swept into the sea soil fertility which only centuries of life could accumulate, and yet this fertility is the substratum of all that is living. It must be recognized that the phosphate deposits which we are beginning to return to our fields are but measures of fertility lost from older soils, and indices of processes still in progress. The rivers of North America are estimated to carry to the sea more than 500 tons of phosphorus with each cubic mile of water. To such loss modern civilization is adding that of hydraulic sewage disposal through which the waste of five hundred millions of people might be more than 194,300 tons of phosphorus annually, which could not be replaced by 1,295,000 tons of rock phosphate, 75 percent pure. The Mongolian races, with a population now approaching the figure named; occupying an area little more than one-half that of the United States; tilling less than 800,000 square miles of land, and much of this during twenty, thirty or perhaps forty centuries; unable to avail themselves of mineral fertilizers, could not survive and tolerate such waste. Compelled to solve the problem of avoiding such wastes, and exercising the faculty which is characteristic of the race, they "cast down their buckets where they were."

Not even in great cities like Canton, built in the meshes of tideswept rivers and canals; like Hankow on the banks of one of the largest rivers in the world; nor yet in modern Shanghai, Yokohama or Tokyo, is such waste permitted. To them such a practice has meant race suicide and they have resisted the temptation so long that it has ceased to exist.

F. H. King, *Farmers of Forty Centuries*

182 million tons of human wastes were annually applied to the land in the above Asian nations; the figures for Japan alone were nearly 24 million tons in 1908.

In America during the eighteenth and most of the nineteenth century, as in Asia and Europe, the wastes of urban populations were often disposed of on the land

A tank wagon for night soil. This is the least objectionable method of cleaning out privies, since both the pumping and hauling are odorless, and the night soil is not touched by human hands.
(From *Survey*, September 2, 1911.)

for fertilizing purposes. As late as 1910, 62 percent of the population lived in places without water-carriage removal of human wastes—that is, without sanitary sewers. Some American cities, such as Boston and Philadelphia, had sewers as early as the eighteenth century, but these were sewers for the removal of surface waters only. Their main function was drainage rather than waste removal. In fact, municipalities often had laws forbidding the depositing of human excrement in the sewers. Urbanites placed their wastes in cesspools or privy vaults (often with removable "ordure tubs"), while household slops were thrown into the yard or the street. Cesspools and vaults were often used until they had filled up and then new ones dug. Some cities required that waste receptacles be regularly emptied and cleaned by scavengers, usually at night (hence the name, "night soil"). The scavengers disposed of the night soil by dumping it into neighboring water courses, burying it in the fields, selling it to processing plants to be made into fertilizer, or selling it to farmers. In

A primitive method of emptying privies by means of dipper and bucket, barrel and cart. (From *Survey,* September 2, 1911.)

Manually pumping the night soil from privy to tank wagon.
(From *Survey,* September 2, 1911.)

some towns and cities farmers paid for the privilege of cleaning cesspools as well as the streets, where they collected valuable horse manure.

American farmers used urban wastes far more than has been realized by historians. In 1880 the U.S. Census published a two-volume set called *Social Statistics of Cities,* compiled by Colonel George E. Waring, Jr., a leading sanitary specialist of the time. Waring collected a great deal of information in these volumes concerning urban waste disposal, sewers, and drainage. In 103 of the 222 cities included in *Social Statistics,* scavengers or farmers collected the night soil and either deposited it directly on the land, composted it with earth and other materials, and then applied

Next to thorough draining, the great lack in American farming is a proper economy and application of manures and fertilizers. By manures, we mean that produced on the farm; and by fertilizers, guano, phosphates, and the like. And no farmer should buy any fertilizers until he saves and applies his manures. From extensive observation, we conclude that not one farmer in one hundred makes the most of his manures. The urine of the cow is as valuable as her dung; and not one farmer in one hundred saves it. The urine and excrement of each member of the family is as valuable as that of the cow; and yet it is not cared for.

Such waste of valuable food for crops cannot be too strongly condemned.

Our subject then, in this chapter, will be to show the farmer how to save and apply manure. And we begin where there is the most general and inexcusable waste—in the privy. The urine and excrement of each member of the family is abundantly sufficient to fertilize a half acre of land yearly. The simplest way to save this, where the vault can be opened, is to cover it with five or six times its bulk of peat or muck once a week. But a much better way, is to have a shallow vault, with a cemented or tight board bottom, sloping to one corner, from whence there should be an ample drain leading into a cesspool at convenient distance from the house. Into the upper corner of the privy vault should run the drain from the sink, not only to save the washings from the sink, but also to keep the vault washed out, and to dilute the urine, which renders it more valuable. Of course, a brick or stone cesspool is the most durable, but an oil butt, or hogshead, sunk in the ground, forms an economical substitute. The place may be hidden from public view by a row of dwarf trees, pines or spruces. Near it should be hauled peat, muck, leaves, straw—any kind of vegetable matter—and the contents of the cesspool poured on to it. For this purpose, a long-handled dipper may be constructed of a keg or firkin. When this heap is thoroughly saturated, fork it over, haul it away, and bring new material. Peat will absorb more ammonia than any other soil, and is therefore the most valuable for this purpose. The manure thus made will be worth more than the same amount of the best barnyard manure. Don't pay a dollar for fertilizers till you have made the most of this valuable matter right at your elbow. Proceed about it at once, for it is money wasting every hour before your eyes.

Charles W. Dickerman, *How To Make the Farm Pay; or The Farmer's Book of Practical Information on Agriculture Stock Raising, Fruit Culture, Special Crops, Domestic Economy and Family Medicine,* 1869 the mixture to the land or sold it to processing plants to be manufactured into fertilizer. Of all the sections of the country, cities in New England and the middle Atlantic and upper southern states made the most extensive agricultural use of urban wastes. Farmers utilized the wastes of 43 out of 55 New England cities, 31 out of 49 Middle Atlantic cities, and 7 out of the 8 cities in the upper South.

In most of these cities scavengers or farmers emptied the cesspools by hand receptacles and buckets, although 11 cities reported using "odorless evacuators." These devices were first adopted in America in the 1860s and had the advantage of providing for speedier and more sanitary cleaning of privies and cesspools. Boston described its odorless evacuator in the following manner: "These machines, which are airtight, have the air pumped out of them and suck the contents of the vault for any distance up to 125 feet through strong hose. A small charcoal furnace, connected with the air pump, destroys any gases as they are pumped out. The work can be done in the daytime without any offense to sight or smell." In some cities, however, like New York in 1872, the organized scavengers opposed the introduction of this new technology because of the fear that it would drive them out of their jobs.

In eight cities, including New York, Baltimore, Cleveland, and Washington, D.C., night soil was sold to processors who made it into fertilizer. The most important company in this regard was the Lodi Manufacturing Company of New Jersey, which was incorporated in 1840. This company used New York City night soil, mixing it with "vegetable fibrous substances and chemical compounds" to make a "New and Improved Poudrette." "Night soil," said a brochure published by the company, "is the most powerful manure at the command of farmers at the present day, and should be preserved by them with scrupulous care if they wish to preserve and increase the fertility of their soil." Poudrette had an advantage over night soil because of ease of conveyance and handling—advertisement published by the company boasted of its effectiveness on lawns, garden vegetables, corn, potatoes, and tobacco.

Most night soil, however, was applied directly to the land. It is difficult to esti-

> *Science knows now that the most fertilizing and effective manure is the human manure. . . . Do you know what these piles of ordure are, those carts of mud carried off at night from the streets, the frightful barrels of the nightman, and the fetid streams of subterranean mud which the pavement conceals from you? All this is a flowering field, it is green grass, it is the mint and thyme and sage, it is game, it is cattle, it is the satisfied lowing of heavy kine, it is perfumed hay, it is gilded wheat, it is bread on your table, it is warm blood in your veins.*
>
> Victor Hugo, *Les Miserables,* 1862

mate the amount of human wastes actually removed from the cities and how much of this was used as fertilizer, but it was far from an insignificant amount. In 1880 Brooklyn reported that 20,000 cubic feet of night soil was taken each year from the city's 25,000 privy vaults and applied to "farms and gardens outside the city." Philadelphia in the same year estimated that the city's 20 "odorless" vault-emptying companies removed about 22,000 tons of fluid matter per year, and that the "matters removed are largely used by farmers and market gardeners of the vicinity. . . ."

In Baltimore the practice of applying night soil to crops continued into the beginning of the twentieth century. The city was the last major municipality to construct a system of sewers for the removal of human wastes. Until approximately 1912, the population depended on 70,000 cesspools and privy vaults to dispose of its excrement. The vaults were emptied by "night soil men" using either odorless evacuators or dippers and buckets. In the 1870s the city itself manufactured a fertilizer from its night soil, but by the turn of the century this practice seems to have ended. An article that appeared in 1899 describes a system in which the scavengers sold their collected night soil to a contractor for 25¢ per load of 200 gallons. The contractor shipped the wastes by barge eight or ten miles below the city where he sold them to farmers for $1.67 per 1,000 gallons. At the turn of the century farmers purchased over 12 million gallons per year and used it to grow crops such as cabbage, kale, spinach, potatoes, and tomatoes. According to one reference, "little smell" arose from either the pits where the fertilizing material was stored or the lands to which it was applied.

To a large extent medical science in the nineteenth century posed no barrier to the use of human wastes directly on the soil. For most of the century, doctors believed that infectious diseases were caused either by the corrupted state of the atmosphere (miasmatic theory) or by specific contagia stemming from decayed animal or vegetable matter. Sanitary reformers about the time of the Civil War were insisting on the removal of filth from towns and cities because they believed that these wastes either generated epidemic disease or threw off "exhalations" that promoted disease. Actually, this belief went back for centuries. The twelfth century *Regimen Sanitatis Salernitanum* gave this advice:

> Though all ill savours do not breed infection,
> Yet sure infection commeth most by smelling.
> Who smelleth still perfumed, his complexion
> Is not perfum'd by Poet *Martials* telling.
> Yet for your lodging roomes give this direction.
> In houses where you mind to make your dwelling,
> That neere the same there be no evill sents
> Or puddle-waters, or of excrements,

A makeshift privy that brings back few pleasant memories for those who remember when.
(From Henry J. Kauffman's *The American Farmhouse.*)

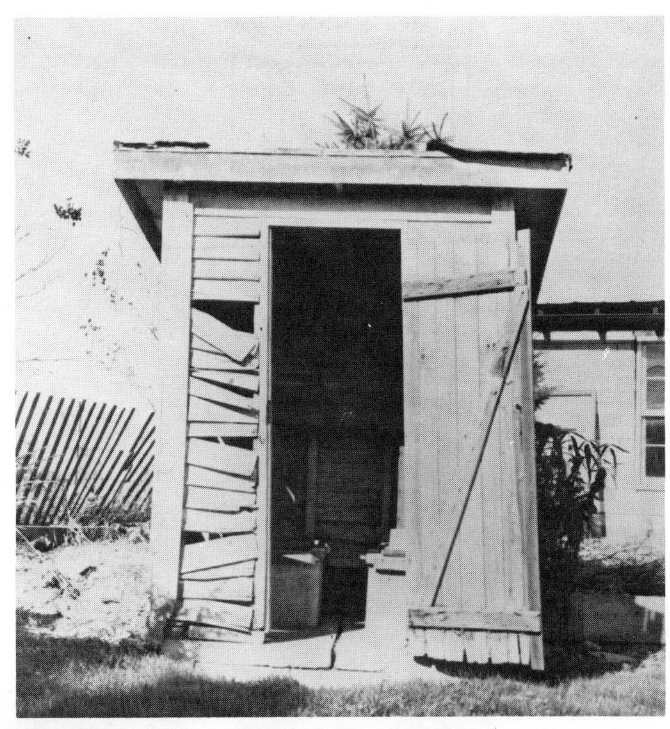

View of backyard in Pittsburgh before the city sewered up.

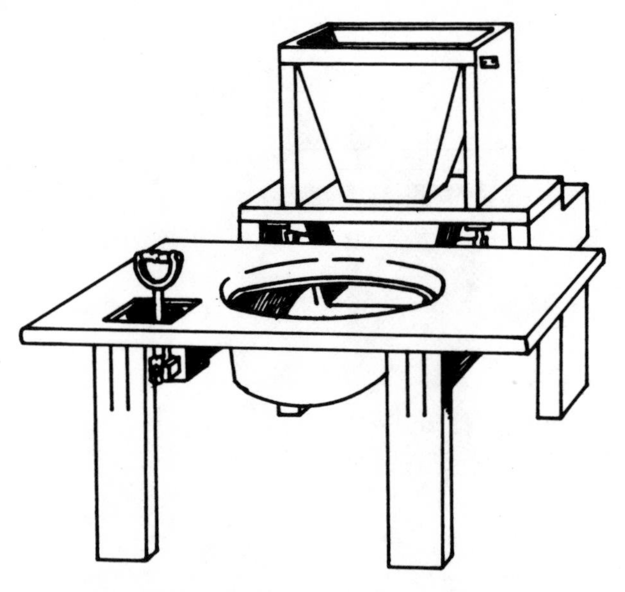

Moule's earth closet, 1860.

Portable water closet with pump and copper pail, 1882.

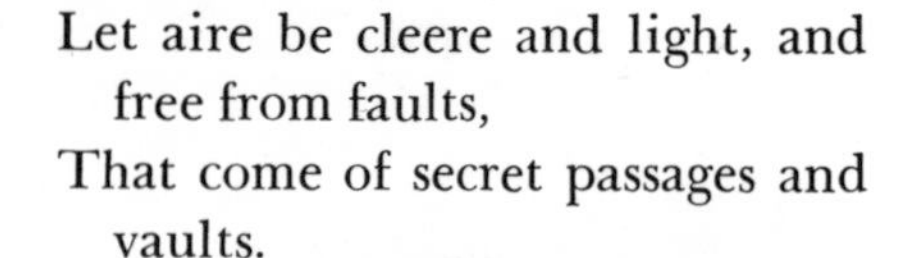

Let aire be cleere and light, and
free from faults,
That come of secret passages and
vaults.

When applied to the land "in the open country," however, as the Massachusetts Sanitary Commission of 1850 argued, the wastes were "diluted, scattered by winds, oxidized in the sun: vegetation incorporates its elements." Miasmas and contagions would be dealt with by nature.

In most cities, however, as population grew, the scavenger system and the existing sewers were inadequate to deal with sanitary requirements. Many privies were never cleaned or cleaned only imperfectly. The opening of vaults for emptying, even at night, caused the emission of nauseating odors. Although the law required that night soil carts have tight lids, the carts usually left a trail of wastes. City residents were "stiffled by the intolerable stench arising from the filth spilled from carts devoted to the *dirty goddess.*" The wastes, if not sold to farmers, were often dumped into a local waterway, causing serious pollution problems. In 1839, for instance, New York diarist George Templeton Strong wrote that Coenties Slip was filthy enough to infect the whole city with yellow fever: "The water was saturated with filth and where the sun fell on it, it was literally effervescing—actually sending up streams of large bubbles from the putrifying corruption at the bottom." "The whole city's one huge pigstye," added Strong, "only it would have to be cleaned before a prudent farmer would let his pigs into it for fear of their catching the plague."

One of the solutions to the inadequacies of the privy vault and the cesspool was the earth closet, a version of which was used in various English and European cities. In its basic form, the earth closet consisted of a wooden seat with a bucket beneath and a hopper containing dry earth, charcoal, or ashes at the back. Pulling a handle released a shower of earth into the bucket. The earth closet had the advantage of being simple to empty and thus less likely to overflow as did the privy vaults. In addition, the mixture of dirt and wastes made an excellent fertilizer. But while the earth closet was widely adopted in England and on the Continent, it never became popular in America. As one of its chief American advocates, Colonel George E. Waring, Jr., explained, the earth closet was "distasteful to a mass of persons in the United States whose necessities demanded immediate relief."

Much more popular than the earth closet in America was the water closet. The water closet had a long history. Types of closets had been used by early civilizations such as the Minoan and the Roman, and Sir John Harrington had proposed a water closet in his *Metamorphosis of Ajax,* published in 1596. The first English patent for the water closet, however, was not issued until 1775 and the first American patent not until 1833. In 1842, Andrew Jackson Downing, one of the most popular architects in America, was maintaining in his writings that "no dwelling can be considered complete which has not a water closet under its roof." By 1860 the 16 largest cities of the nation as well as a number of smaller ones were served by water works, and many affluent residents of these cities installed water closets in their homes. Some of these water closets were connected to sewers, but most flowed into cesspools or privy vaults. Vaults and cesspools had often overflowed before the introduction of the water closet, causing a nuisance and a sanitary hazard; now the closets caused constant danger of soil saturation and the leaking out of wastes. In addition, for many people there was an added financial cost, since cesspools now had to be emptied almost monthly rather than once or twice a year. But there was an obvious solution for the problems caused by the water closet and the increased amounts of household wastewater—build sewers connected with private homes to carry off the wastes by water carriage. These systems were known as sewerage systems.

Historian John Duffy has carefully detailed this process of the growing demand for sewers in his *History of Public Health in New York City.* The opening of the Croton Aqueduct in 1842 increased water consumption throughout the city and caused the number of water closets in use to increase sharply. The existing law forbade the connecting of closets to the sewers which were for surface water. In 1844, however, Alderman Gale of the Common Council proposed that householders be permitted to connect closets to the sewers. He argued that this would "greatly . . . promote the health and comfort of the inhabitants," and in 1845 the Common Council amended the sanitary law. The connection of house-

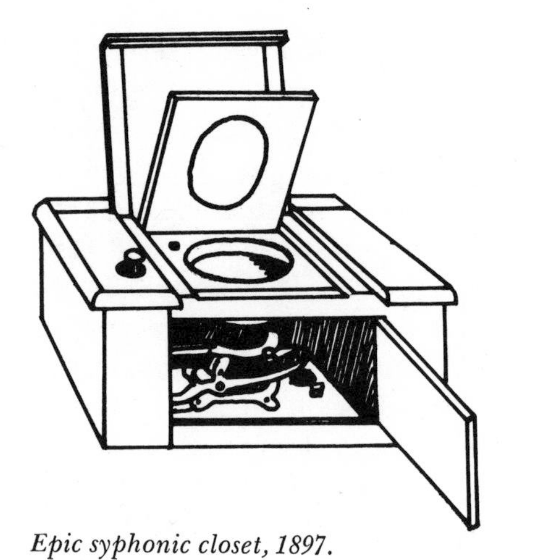

Epic syphonic closet, 1897.

Optimus valve closet, interior and exterior views.

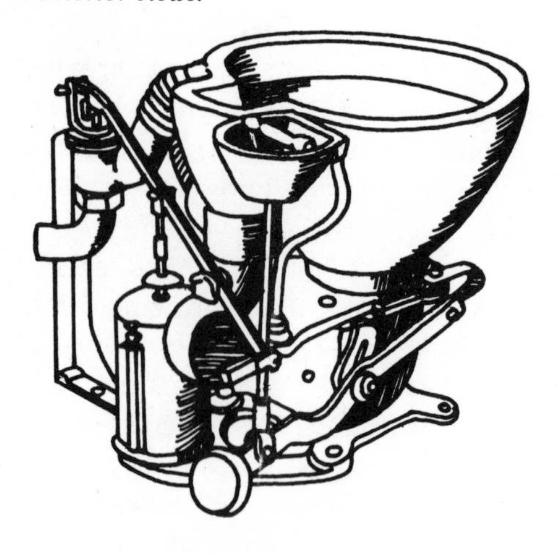

holds and water closets to the existing sewers, however, overburdened them and necessitated the acceleration of the city's sewer building activities. The process of sewering New York for both household wastes and surface water had begun, although the goal of a fully sewered city lay far in the future.

The process of change initiated in New York by the introduction of running water and the adoption of the water closet was duplicated in other cities. A city would introduce running water without constructing a sewerage system; householders would connect their water closets and sinks to cesspools which would then overflow; sanitary problems and general nuisance would result; and the costs for emptying cesspools would accelerate. It was this sequence of events in cities that produced a demand for the building of sewerage systems to remove human wastes in water carriage.

But while the introduction of running water and the adoption of water closets in cities was critical in creating a demand for the building of sewers to remove human wastes, considerations of public health were also important. By the middle of the nineteenth century the so-called "filth theory" of disease had strong support among sanitarians. Originally popularized by the great English sanitary reformer Edwin Chadwick, it attracted the backing in this country of sanitarians such as Lemuel Shattuck and Colonel George E. Waring, Jr. Some, like Waring, emphasized the necessity of removing human excrement from the vicinity of the house before it putrified and generated "sewer gas."

Regardless of the correctness of the filth theory, it had a beneficial impact on the public health by insisting that sanitation be improved. This demand for cleanliness was reinforced by altering beliefs about the causes of epidemics such as cholera and yellow fever. As historian Charles E. Rosenberg has noted, while many Americans accepted the cholera epidemics of 1832 and 1849 as a product of God's will and a punishment for sin, by the epidemic of 1866 society was much more prone to regard cholera as a social disease that could be cured by sanitary reform.

While sewerage systems that provided for the removal of a combination of storm water and household wastes ("combined systems") were constructed in cities like Chicago and Brooklyn in the 1850s, it was not until 1880 that the first system providing for household wastes alone was built. This was the Memphis, Tennessee, system designed by Colonel George E. Waring, Jr. The Memphis system was primarily the result of the yellow fever epidemics that devastated the city in 1878 and 1879. The virulence of the fever was blamed on the filthy condition of the city, and in 1880 the city authorities and the National Board of Health recommended that a sewerage plan proposed by Waring be adopted. The Waring so-called "separate system" provided for the removal of human wastes but not for storm water. It was based upon Waring's theories about the dangers of sewer gas and the need to speedily remove human wastes from the home.

An early sewer—little more than a stinking, open hole.

Waring's Memphis system and his many articles on sanitary questions, as well as the writings of other sanitarians, helped stimulate interest in sewer building in the late 1870s and the 1880s. Popular journals such as the *Atlantic Monthly, Harper's Weekly,* and *Scribners* carried articles on sanitation, as did more specialized periodicals such as *Popular Science Monthly* and *Scientific American.* In 1877 the first sanitary engineering journal published in the United States began appearing under the title *The Plumber and Sanitary Engineer.* By 1881 the journal had changed its name to *The Sanitary Engineer* and still later to *Engineering Record,* marking the progress

A not atypical early street scene, with children playing in an open sewer, next to a dead horse. No wonder infectious diseases were rampant.

of professionalization of this group of engineers. Many of the articles that appeared in the 1870s and 1880s discussed the English and European experience in regard to sewers and sewage disposal, hoping to provide insights into the American situation. Particularly useful were the reports of several royal commissions on river pollution, sewage, and sewage disposal, which reported the choices open to cities concerned with sewage disposal.

In an article and an editorial published in 1873, *Scientific American* summed up the various sewage disposal options suggested by the British reports. The journal took the position that the profitable disposal of sewage was subordinate to the "problem of disposing of the noxious substance in such a manner that it shall not breed disease or nurture pestilence in the narrow and confined limits of thickly populated districts." It recognized, however, that sewage had potential value and calculated that the worth of the "annual voiding of an average individual" ranged from $1.64 to $2.01. During the nineteenth century in England hundreds of patents were issued that described the production of a useable manure by combining sewage with various materials. The most popular method was known as the "ABC process" since it involved mixing sewage with alum, blood, and clay; other processes combined sewage with chemicals such as lime, sulphate of iron, or mineral phosphates. While these so-called chemical methods of sewage disposal had some popularity in England, the most widely used approach was that of sewage farming.

Sewage farming was actually an old practice. From about 1800, several towns in Devonshire, as well as Edinburgh, Scotland, had irrigated neighboring agricultural lands with their sewage. The leading advocate of sewage farming in England in the middle of the nineteenth century was Edwin Chadwick, who in 1842 advocated the use of untreated sewage as field manure. Chadwick believed that the sale of urban sewage to farmers would pay for the cost of maintaining urban sewage systems. In 1865 the British Sewage of Towns Commission, appointed in 1857 to inquire into the most beneficial and profitable

Prompted by snide references to Thomas Crapper, the inventor of the automatic flush toilet valve, Executive Director Paul W. Eastman of the Interstate Commission of the Potomac River Basin has recently found definitive proof that not only did Crapper live, but he did in fact operate a large foundry and sanitary engineering firm.

Eastman read a disparaging article about Crapper and his invention in the May 1975 Smithsonian. *In the article, the author said he was "convinced Crapper is a myth," created by a British author who wrote a biography of him called* Flushed with Pride *published in 1969.*

Believing Crapper had been handed a bum rap, Eastman took time during a London vacation to find Crapper's grave, in Elmer's End Cemetery (now known as the Beckingham Crematorium) on the outskirts of London. Eastman then proceeded to make a crayon rubbing of the gravestone and now gladly shows it to anyone doubting the existence of the great plumber. (Crapper lived from 1837 to 1910.)

Earlier, the ICPRB director had found a drain cover with Crapper's name on it, located just outside the apartments of the Dean of Westminster Abbey in London. . . . The "Marlboro Works" mentioned on the drain cover refer to Crapper's foundry in London, a photograph of which appears in Flushed with Pride.

Eastman has, since returning from England, had several of his rubbings framed and reproduced, and is more than willing to describe the fascinating aspects of Crapper's prolific life with anyone who comes by.

He has also attempted to present a rubbing to a prominent congressman long associated with the Potomac, at a meeting, but was ruled out as "in poor taste" by the organizers of the meeting—a prominent women's organization also long associated with the Potomac. Apparently, the stigma attached to Crapper's invention lives on.

Potomac Basin Reporter

method of sewage disposal, reported in favor of the land disposal method; similar recommendations were made by other English commissions throughout the remainder of the century. By 1880, 19 English cities with a total population of 738,191 disposed of their sewage on agricultural land. On the Continent, Antwerp, Berlin, Brussels, Paris, and Milan all had sewage farms.

In America during the 1870s, several New England institutions, starting with the Augusta (Maine) State Insane Asylum and the Concord (New Hampshire) Asylum, began using their sewage to grow crops. The first municipality to use a sewage farm as a means of disposing of its sewage was Lenox, Massachusetts, where Colonel Waring built a system in 1876. The next system was in Pullman, Illinois, in 1881, followed by Pasadena (1888), Colorado Springs (1889), and Salt Lake City (1895), as well as smaller towns. In 1899 George W. Rafter, in a work on *Sewage Irrigation,* listed 24 municipal sewage farms serving 280,000 people. Among the crops grown on sewage farms were potatoes, wheat, oats, barley, carrots, and other vegetables and fruits, while Italian rye grass was very common.

Sewage farming had appeal because it suggested that wastes could be disposed of in a profitable manner, and because alternative sewage disposal technologies were expensive and inadequate. However, during the 1890s a number of new technologies developed that appeared to do as efficient a job of sewage disposal as sewage farms without the land requirements and the disagreeable (and possibly unhealthful) aspects of the farms. One of the most important new technologies was intermittent filtration, developed by the Lawrence Experiment Station of the Massachusetts State Board of Health. In addition, sanitary engineers and scientists formulated other chemical, mechanical, and bacterial processes. By the turn of the century, more cities were using intermittent filtration and other methods, such as septic tanks, than sewage farming. During the first decades of the twentieth century, methods using bacterial forces, such as activated sludge, became the most popular forms of sewage disposal in the United States.

By the turn of the century, a flood of public and scientific attention focused upon the problems of sewers and sewage disposal. The affirmation of the germ theory of disease and the belief of sanitarians that "the excreta of man and other animals are the principal original vehicles of infection and contagion" stimulated efforts toward sewering towns and cities. Between 1890 and 1909 the miles of sewers in the United States increased from about 8,000 to over 25,000 and the population served by sewers from about 16 million to over 34 million. While the building of sewers was a health improvement over the use of privy vaults and cesspools, it also created immense problems of sewage disposal. The critical difficulty was that the great majority of cities, in line with the theory that "running water purifies itself," dumped their raw sewage into rivers and lakes without treatment. The most serious aspect of this practice, aside from the esthetic problems and the destruction of recreational facilities, was that many cities drew their drinking supplies from the very rivers into which other cities disposed of their raw sewage. In those cities downstream from the sewage outlets of upstream cities, such as Pittsburgh or Newark, typhoid death rates soared to incredibly high levels. Especially important in drawing attention to the dangers of river pollution was William T. Sedgwick of the Massachusetts Board of Health, who demonstrated the water-borne nature of typhoid fever germs.

In states such as New York, Ohio, and Pennsylvania, legislatures, under the prodding of public health authorities, passed legislation to compel cities to stop disposing of raw sewage in rivers. This legislation, however, was restricted in application to cities newly installing sewerage systems or extending existing ones, and even here it was severely limited. Attempts, for instance, by the Pennsylvania Department of Health in the years from 1909–1912 to compel Pittsburgh to treat its sewage in order to receive a permit to extend its sewer system failed. Most municipalities continued to dump raw sewage into waterways. In 1909 only 19 cities with a population over 30,000 attempted to treat their wastes. Sewage treatment technology was too expensive or uncertain to persuade municipalities to invest large sums of money for the benefit of downstream cities.

Simultaneously with these developments, large advancements were made in the field of water filtration. Much of the experimental work in this area was done at the Lawrence Experiment Station of the Massachusetts Board of Health, which had previously performed important explorations in the field of sewage treatment. By 1912, over 28 percent of the urban population (10,806,000 people) were drinking water filtered by either sand or mechanical filtration methods, while the water supply of many other urbanites was being chlorinated. In those cities with filtered water, typhoid fever deaths, as well as those from other water-borne diseases such as diphtheria or infant diarrhea, dropped precipitiously. Increasingly it became clear that even though cities did not "purify" their sewage, death rates from water-borne disease could be reduced.

The question as to whether municipalities should purify their sewage, filter their water, or do both became a crucial issue in the second decade of the twentieth century. Public Health authorities on state boards of health tried to compel cities to purify their sewage before disposal but sanitary engineers maintained that, given the primary goal of reducing typhoid death rates, water filtration was sufficient. Sewage purification, they held, was costly and reasonably ineffective in reducing pollution. It was "absurd to insist, as so many of the would-be conservators of public waters do," wrote the editors of *Engineering News* in 1910, "that sewage must always be purified before it is sent into the water." The elimination of "gross nuisances" was ample purification. In his book *Clean Water and How to Get It,* Allen Hazen, one of the nation's pioneers in the water quality movement, expressed the point of view of most of the country's sanitary engineers when he wrote the following:

> It is . . . both cheaper and more effective to purify the water, and to allow the sewage to be discharged, without treatment, so far as there are not other reasons for keeping it out of the rivers. It seems unlikely that a single case could be found where a given and reasonably sufficient expenditure of money wisely made could do as much to improve the quality of a given water supply when expended in purifying sewage. . . , as could be secured from the same amount of money in treating the water. Usually I believe that . . . one dollar spent in purifying the water would do as much as the dollars spent in sewage purification.

Sewage treatment, therefore, might pay dividends in "comfort and decency," but it would be costly and not necessarily effective in saving lives.

To a large extent, in this dispute, given the differential in costs, the advocates of water filtration rather than sewage purification were more convincing. Cities, therefore, made much more rapid progress in developing water filtration facilities than they did in building sewage treatment plants. In 1924, for example, nearly 24 million people were served by water filtration plants; six years later, in 1930, only 18,000,000 people had their sewage treated to any extent. As late as 1940, more than 66 percent of the people living in communities with sewers discharged raw sewage into water courses with little more than fine screening.

In the period after World War II the rate of both sewer building and sewage treatment expanded fairly rapidly, with the greatest increase occurring after the provision of federal construction subsidies in 1957. In 1968 the Federal Water Quality Administration reported that the sewage of about 131 million people received treatment, while that of approximately 10 million people was still discharged into waterways without treatment. Of this former group of 131 million, the sewage of over 91 million people received advanced rather than primary treatment. About another 50 million people still lived in unsewered places and used methods such as septic tanks to dispose of their wastes.

But while these developments in the construction of sewage treatment plants sound encouraging when gauged by the standards of 1900, the problem today is far more complicated. Many of the treated sewage effluents, from the more advanced treatment plants, are rich in inorganic materials such as carbon dioxide, nitrate, and phosphate which support the growth of algae. The algae bloom and die creating eutrophication difficulties for lakes and rivers that receive the sewage effluent. Another problem concerns an important by-product of the treatment processes, sewage sludge. Sludge is created in enormous volume and is either incinerated or disposed of on the land or at sea. All these methods of sludge disposal, however, have undesirable results: incineration creates air pollution, land disposal is expensive and creates nuisances, and ocean disposal has caused the fouling of many beaches. And finally there is the difficulty created by synthesized organic compounds that are not susceptible to ordinary methods of treatment and about whose environmental effects engineers and scientists have little or no information.

These problems of sewage disposal have caused some ecologists to recommend abandonment of the water carriage system of waste removal and adoption of alternatives such as waterless, composting toilets. For many other ecologists, however, given the expected continuance of the water carriage system, the solution to the sewage disposal problem is to return to the sewage farming methods of the nineteenth century, but with greatly improved technology. Barry Commoner argues, for instance, that "clearly the ecologically appropriate technological means of removing sewage from the city is to return it to the soil." Actually, while most sanitary engineers rejected sewage farming at the beginning of the twentieth century because it was land intensive and also suspect from a public health perspective, it has expanded slowly since World War I. It has developed especially in the Western states where sewage is valuable for irrigation as well as manurial purposes. But while many ecologists argue that sewage farming is the solution to the sewage disposal problem, some environmental engineers maintain that the land-disposal method would destroy the soil mantle, contaminate ground water supplies, and result in immense land acquisition programs. In spite of these objections, the entrance into the debate of the Environmental Protection Agency and the Army Corps of Engineers on the side of land disposal suggests that serious attempts will be made to make land disposal one of the foremost methods of urban waste disposal.

The search for an ecologically sound and economical method of sewage disposal goes on, and the National Commission on Water Quality has projected that between 100 and 400 billion dollars will be spent combating water pollution in the next ten years. Many of the issues concerning methods of sewering and sewage disposal that arose in the late nineteenth century have again become the focus of public and scientific attention. Ironically, however, it was the very manner in which these early problems were dealt with that has helped create our present predicament. Choices about the disposal of urban wastes and the pollution of the land and water made in the past have again made the question of sewage disposal—as Colonel George E. Waring, Jr., called it in the late nineteenth century—one of "the great unanswered questions of the day."

Chapter 2

How We're Handling Our Wastewater Now, and Alternatives for the Future

As Joel Tarr has so aptly explained, the introduction of a water-borne waste system and the convenient flush-and-forget-it water closet introduced one problem as it began to create, perhaps, even a bigger one. While flush toilets and sewers certainly made life in the mid-nineteenth century more sanitary and a lot more pleasant on the eyes and nose, few, if any, at that time could foresee the day when our cities would outgrow their wastewater systems. Even in the 1840s and 50s sewers were expensive, but at least they were thought to do the job they were designed for. There are a number of weaknesses in the sewer system that have already caused some health problems, and centralized treatment could conceivably cause more as our populations and the wastes they produce grow while our water and land dumping areas remain the same size. Let us explain.

• It's costly. Huge amounts of money are spent yearly to treat wastes that shouldn't be there in the first place. We have no firm figure for the total expenditures needed to provide traditional disposal systems for all urban areas in our country, but we have seen estimates ranging from tens of billions to the hundreds of billions. Costs for the Blue Plains Sewage Plant in Washington, D.C., totaled $20 million in 1976, according to the plant manager. Installation of new sewers alone to replace failing septic tanks for 1,200 suburban homes in the Maryland area around Washington, D.C., cost about $4,400 per home, and the yearly maintenance rates for each household are about an additional $250.

The 1976 EPA survey, released in early 1977, estimated that it will cost $96 billion to install the most advanced wastewater treatment techniques throughout the country, and another $54 billion for construction of separate systems for storm water runoff. The federal government is mandated to fund 75 percent of sewer and treatment facilities construction, but it is not willing to pay this huge amount.

• It's energy-intensive. The energy requirements to run a plant are very high. For instance, the Blue Plains plant each day uses anywhere from 17,500 to 4,000,000 khw of electricity, 232 tons of chemicals

---

Carol Hupping Stoner is Executive Editor of Rodale Press Book Division. She has written articles for Organic Gardening and Farming® *and has edited a number of books, including* Producing Your Own Power *and* Stocking Up.

(which in themselves are energy-intensive to use), and 35,600 gallons of fuel oil a day to treat 296 million gallons of sewage water.

• It uses lots of water. Large quantities of water are necessary to keep the system going,* and naturally, as our population increases, more clean water will be needed to flush away the larger amounts of wastes we create. While our population potential really has no limits, our clean water supply is finite. The 1985 sewage flow estimates for Prince Georges County in Maryland exceed the available water supplies by 321 million gallons of water a day. When too much water is withdrawn from subsurface acquifiers to flush away wastes, wells and other surface waters such as lakes and streams begin to dry up, and the ground water table is lowered. In coastal regions, this pulls the seawater into underground acquifiers causing saltwater contamination of drinking water supplies.

This whole sewer-caused water shortage has taken an interesting turn on Long Island, New York. Fertilizer runoff and cesspool seepage have caused the accumulation of detergents and nitrates in ground water supplies on the 121-mile-long island. To alleviate this problem, officials have stepped up construction of sewer networks and centralized treatment plants that treat wastewater and discharge it to sea.† While this move has lessened the pollution of ground water, it has created other problems that are just as or perhaps more serious.

By discharging treated wastewater into the sea instead of returning it to ground water supplies, both Nassau and Suffolk's fresh water supplies have been seriously threatened. Lakes, ponds, and streams on Long Island are drying up and the water table gets lower each year. Depletion of these ground waters also affects offshore areas into which these waste waters discharge. The island's south shore bays, which are not as salty as the waters on the ocean side of the barrier beaches, are ideal environments for a variety of fish and bird life. Most notable is the hard-shell clam which supports a $100 million industry and provides more than 60 percent of the country's hard-shell clams. However, wildlife is threatened because the salinity in the bays has been increasing as the fresh water supplies on shore decrease.

Something must be done to stop this problem, and several alternatives have been proposed. Fresh water could be transported to the island from above New York City, in the Hudson River area. But there are even more costly and complicated suggestions. One is to dam up the entire Long Island Sound and convert it into a fresh-water lake. Others include towing in ice-

* Marin County, California, had a real problem on its hands in spring 1977 when it cut back its water consumption by 57 percent because of the severe drought there. The sewer lines became clogged because not enough water was being flushed down toilets and poured down tub and sink drains. Officials there had to truck in low-grade, undrinkable water and flush it down the sewers to help the drier sewage flow through the flat, low-gravity areas.

† As of 1974, 55 percent of Nassau County has sewers, and by 1983 it is estimated that 98 percent of the county will have them. Suffolk County is only 7 percent sewered, and there are no present estimates as to if or when the rest of that county (which is one of the fastest growing in the whole United States) will have sewers.

During the 50's we lost land to the interstate highway system. In the 60's we lost land to suburban sprawl. In the 70's we're losing land to the sewage treatment facilities, all of which require flat farmland. You can't build a sewage lagoon on a hillside.

Bob Bergland,
U.S. Secretary of Agriculture

bergs and artificially inducing precipitation. Rather than bringing in more water, Long Island might be better off looking at ways they could conserve and better utilize the water that they do have, including alternative wastewater management methods that use less water than sewers and centralized treatment plants.

Despite the large investments in money, energy, and water, today's sewer system does a poor job of doing exactly what it's designed to do: manage our wastes so that they will not present health and environmental problems.

• For instance, it pollutes our ground water supplies. About one percent of the raw sewage entering sewage lines leaks through cracks and gaps in the sewer lines into the surrounding soil and eventually works its way into our surface water.* Such leaks are difficult to locate and expensive to repair. During heavy rains when the quantities of wastewater reaching the plant would exceed its capacity, regulator valves and plant bypasses allow sewage overflows to spill directly into surface waters. It's not uncommon for these regulator valves to be left open even in dry weather.

• It can intertere with the ecosystem of our lakes. The effluent from treatment plants introduces unwanted nutrients into our waterways that cause excess algal growth and can lead to the premature death of lakes. Although advanced wastewater treatment was designed to eliminate this trouble, it has not been shown to be reliable except for land-based treatment technologies.

• It can cause landfill problems which can lead to other environmental problems. These occur with the buildup of large amounts of sewage sludge that in some situations can be contaminated with heavy metals or toxic chemicals by industrial plants which discharge without proper supervision into municipal sewers.

• It pollutes our off-shore areas and beaches, as it did Jones Beach, New York, during the summer of 1976. Although ocean dumping will be banned as of 1981, it is still being done now.

• It pollutes our air. One way to get rid of sewage sludge is to incinerate it, despite the fact that the incineration process is air polluting. It's also energy-intensive (natural gas is most often used to burn the sludge), still leaves us with an ash to dis-

* Hopefully this percentage will get smaller as the Water Quality Amendments passed in 1972 (PL 92–500) make funds available for the repair of sewer lines and the adoption of large holding tanks to eliminate or reduce the amount of raw wastes that bypass the treatment plant.

pose of, and makes the use of sludge as a resource impossible.

There are steps that can be taken—and have been taken by some cities—to both provide a valuable fertilizer and prevent or minimize these kinds of landfill and ocean dumping problems at the same time. For many years Milwaukee has treated and dried it and sold it in 50-pound bags as a fertilizer under the name of "Milorganite." Now other cities are following suit: Philadelphia has Philorganic and the U.S. Department of Agriculture has an encouraging experiment underway at its Beltsville, Maryland, station to compost sludge from the Blue Plains treatment plant in Washington, D.C. The sludge is mixed with wood chips and composted for about three weeks. The temperature in the piles rises to 150°F (78°C) and the resultant produce is odorless and potentially pathogen-free.

A number of experiments are being conducted to pipe liquid sludge to farmlands and forests where it is sprayed as a fertilizer. This has a double advantage: the waste is recycled to the land, and water, filtered by the soil, replenishes the water table. But the long-range effect is not yet clear; there may be undesirable buildup of heavy metals and toxic chemicals in the soil. Probably the only way such sludge can be truly safe is for household sewage to be separated from street runoffs and industrial wastes. Sewage systems built before the 70s have industrial wastes, the oil from urban streets, street trash, and the excess nitrogen that rains and sprinkling wash off lawns and golf courses all travelling through the same sewers and winding up together at the same treatment plants.

Newer systems, under order from the Environmental Protection Agency, must have their sanitary (residential) lines separated from their storm sewer lines, and industries serviced by these new systems must treat their own wastewaters before discharging them through municipal sewer lines. While such a separated system does make the sludge from residential sewers safer to use for land treatment, it does just as much or more potential damage to our discharge waters and land areas, since the storm sewer lines are not hooked up to any treatment plant, but are discharged raw—oil, nitrites, nitrates, and all.

- Most serious of all environmental problems is the risk of introducing disease and other organisms into the drinking water of nearby communities. Bacteria can be controlled by standard chlorination procedures, but there is real doubt about eliminating viruses that can transmit infectious hepatitis, polio, intestinal flu, and other related diseases. Such water-borne diseases get into our water supply primarily from our body wastes.*

Chlorine, which is used to treat water, has proven to create its own health hazards. In 79 of the 79 cities whose waters were tested by the Environmental Protection

* Must reading for all interested in water quality is an important article in the December 1975 issue of *Environmental Science and Technology* by three virologists from the Baylor College of Medicine, Charles F. Gerba, Craig Wallis, and Joseph L. Melnick. The article points out that more than 100 different enteric viruses are known to be excreted by human beings, and states: "Studies have shown that enteric viruses easily survive present sewage treatment methods and that many can persist for several months in natural waters."

Agency, the chlorine in the water created cancer-causing substances as it reacted with organic elements in the water.*

Also, chlorine reacts with our wastewater to produce volatile chloroform, which enters our atmosphere where it can react to destroy the ozone that protects us from dangerous ultraviolet radiation. There are indications that chlorine poses greater threats to this ozone layer than all the aerosols that we use.

- It can encourage land development and housing sprawl. The present-day interceptor sewers are designed to handle a larger capacity of sewage so that they will be big enough to accommodate projected community growth. These large systems are naturally more expensive than a system that is just large enough to handle the current waste capacity. Communities cannot afford adequate-sized systems, no less oversized ones, and in order to pay for them, local officials are forced to encourage rapid development in the sewered area.†

Even well-designed and well-operated sewage treatment plants are not infallible. Such large, highly centralized operations are subject to power failures, equipment breakdowns, and employee strikes, any one of which can result in direct discharge of millions of gallons of untreated sewage into our waterways. In 1970 the Council on Environmental Quality estimated that municipal sewage loads would nearly quadruple within 50 years. But what are the alternatives?

Well, the most obvious is on-site treatment of wastes through septic tanks systems. Unfortunately, the septic tank is not *the* perfect solution. For one thing, not all soils are suitable for septic tanks. The land in densely populated areas could not possibly filter properly the wastes from apartment buildings and row houses. And septic tanks are not safe in areas where the water table rises dangerously high in rainy times. According to North Carolina state officials, only 30 percent of that state's soils is suitable for septic systems. State soil scientists estimate that septic tank failures are a major problem in 85 of North Carolina's counties. (Fifteen percent of North Carolina's rural population is using privies or dumping wastes directly into drainage ditches or nearby waterways.)

Septic systems have other drawbacks. Thirty to 80 percent of the suspended solids are retained in the tank, and because septic tanks are anaerobic (oxygen-free) very little decomposition takes place there. These solids can clog up the drainage field or remain in the tank and reduce its volume. Annual inspection is recommended, and opening up the septic tank every two to three years on the average for sludge removal is necessary. The material removed (septage) may be taken to municipal treatment plants, contributing to overloading problems and later disposal problems of regular sewage. Or this septage may be disposed of on land, sometimes causing ground water and surface water pollution. Land disposal problems of this sort can be alleviated if the septage is first composted,

(continued on page 28)

* EPA report to Congress, December 1975.

† The Council on Environmental Quality, July 1974 report, *Interceptor Sewers and Suburban Sprawl*.

Using Sewage Sludge in the Garden

Using sewage sludge in the garden has been, and still is, a controversial subject. Based on the findings up to now and conversations with many of the people doing research in this area, we, the editors of *Organic Gardening and Farming,* can safely recommend the usage of most sewage sludges in the garden.

Many people we've talked to who are considering using sewage sludge in the garden are concerned about the possible transfer of pathogens from the sludge to garden vegetables. We'd like to point out that most of the health problems created by such pathogens have been the result of eating food fertilized with raw human manure, and there's considerable difference between raw human manure and sewage sludge. Sludge is the end product of a waste treatment process which is designed to greatly reduce the pathogen content of the incoming human wastes. In almost all cases the sludge made available as garden fertilizer has gone through a multi-step treatment. It is first activated and subjected to a period of biological breakdown, after which it is settled, and then it is placed on sand drying beds for a period of time. This treatment renders almost all viruses inactive, and subsequent storage and composting of the sludge kills off any surviving pathogens.

At this time we have not been able to find any research that supports the claim of pathogen persistence in composted sludge. Any sludge that has been composted, or has been activated, dried, and stored for a period of time may be considered safe from pathogen contamination. If you have access to raw sludge, you should compost it and store it for 6 months to a year before using it.

A bigger concern than pathogen content should be *heavy metal content.* Some sewage sludges, mainly those receiving industrial effluent, may have heavy metal contents so high that the sludge is unsafe for garden use. The four metals that gardeners should be most concerned with are zinc, copper, nickel, and cadmium. The first three are of concern because they are phytotoxic to plants at relatively low levels. Cadmium is of concern because plants tend to accumulate this metal in their leaves, and ingesting the leaves can, over time, lead to an unsafe buildup in the body of large supplies of cadmium. For these reasons, sludge with relatively high levels of cadmium should never be used on food crops, and sludge containing high levels of zinc, copper, and nickel should be carefully monitored to avoid a toxic buildup, both in the soil and in the plants.

Since all these metals are so toxic, unacceptable amounts are incredibly small. For instance, a sludge considered to have a high cadmium content may have 75 parts of cadmium for every million other parts of the sludge. Because of the small numbers involved no layperson could possibly do an accurate heavy metal content analysis at home; laboratory equipment is absolutely necessary.

When discussing limits for these metals, we can only give you general figures because the actual uptake of a metal by a plant will depend on many factors, like the small pH of a soil, type of plant grown, and the cation exchange capacity of the soil. And there are many things that no one knows about the way heavy metals behave in the soil. Currently, no federal agency is settled on guidelines for heavy metal content of sludges. The USDA has done most of the research in this field, but this agency does not want to be responsible for establishing and possibly enforcing the eventual final guidelines. The figures given below, which are the result of USDA work and establish safe and acceptable total yearly accumulations, should be used only as general guidelines:

Metal	Kilograms per hectare
Cadmium	5 to 20
Zinc	250 to 1000
Copper	125 to 500
Nickel	50 to 200

In most cases the results of sludge analyses will be in a parts per million basis. In that form, the upper limit for cadmium would be 25 ppm for sludges to be used for food crops, and 50 ppm for sludges to be used on ornamentals or tree crops. At those limits the sludge could be applied on a regular basis for many years without harmful buildups in the soil. If you cannot get a copy of an analysis of the sludge, you should not use it, unless you are absolutely sure that there are no industries dumping heavy metals into the system.

Another way of telling whether the sludge is safe is the cadmium/zinc ratio. If the ratio is 1/100, or in other words, there is one part cadmium for every one hundred parts zinc, you may use the sludge. Such a ratio acts as a safety guard because if there were enough cadmium in the sludge to harm you or your animals, the greater amount of zinc would kill off the plants first. If this should happen the soil would have to be limed to bring up the pH, and soil and plant samples should be analyzed.

Another concern of those thinking about using sewage sludge is the possible presence of industrial chemicals. Such items as PCBs and PBBs are noted problem-causers in sludge because they can be taken up by plants and then passed along in the food chain. Ask the director of the sanitation plant from which you will be getting your sludge if these chemicals are present. If the sludge contains more than 10 parts per million of PCB or PBB we do not recommend you use it. Luckily, this should not be much of a problem for most people, since for the last five years the discharge of both these chemicals has been carefully controlled.

Concerning the actual use of sludge, we recommend that it be applied just like any manure. It is best spread in the fall and allowed to overwinter, or applied in the early spring, 4 to 6 weeks before planting time. For safety, you should keep the soil well limed to maintain a small pH of 6.5 which will further inhibit metals uptake. You would not wish to apply fresh sludge to root crops that you'll be harvesting that year and eating raw, nor to vegetables whose leafy greens you'll be picking shortly after fertilization.

and there are some towns that have realized the value of such pretreatment. In Massachusetts, three towns (Swansea, Rehoboth, and Seekonk) are planning an eight-acre municipal plot for composting the septage of about 30,000 residents. The compost will then be sold at about five dollars a ton as fertilizer.*

The Alternatives

Unfortunately, most people assume that we're stuck with sewers and septic tanks, even if they're not perfect. Few realize that there are other ways we can

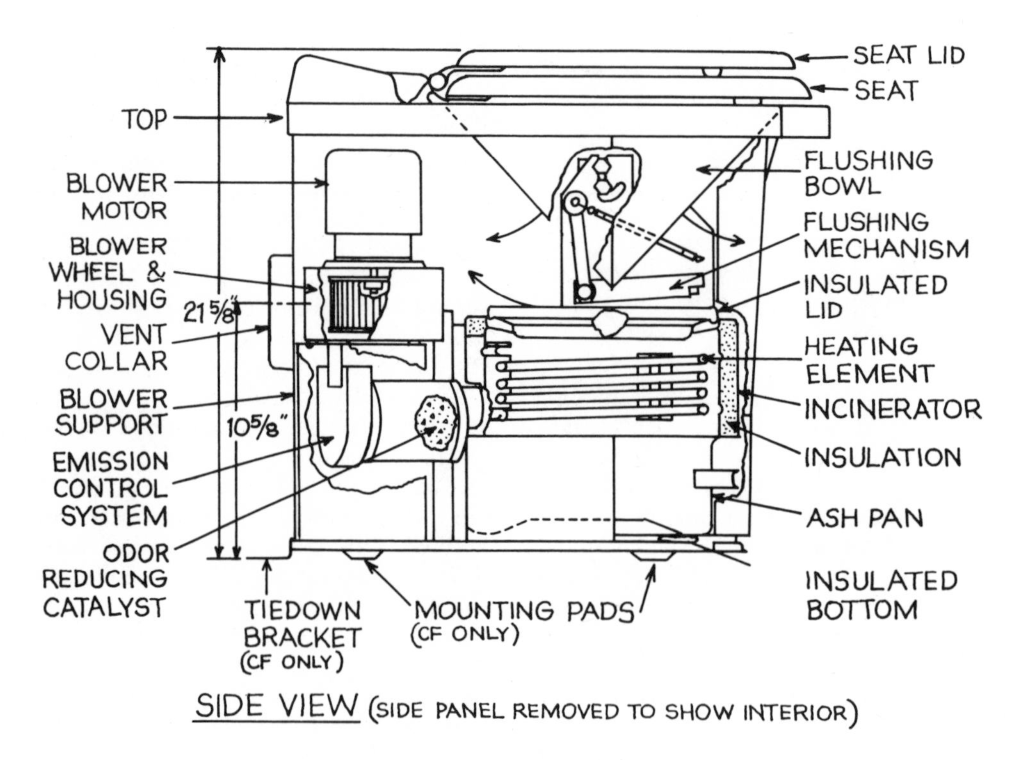

Cross section of an incinerating toilet.

* "Something for Nothing," *Massachusetts Farm Bulletin* 19 (1977).

handle our bathroom wastes. Harold Leich, in his talk before the Sixth Annual Composting and Waste Recycling Conference,* listed some of the options open to us now. Let me expand upon them and list a few others that deserve attention.

Incinerating Toilets

Fired by oil, electricity, or piped or bottled gas, these toilets reduce body wastes to a sterile ash. All require periodic removal of the ash and some require that a paper liner be put in the bowl before every use. Prices range from $400 to $600 a unit. Tens of thousands have already been installed in vacation homes and other places where a sewer connection is not possible. Now an increasing number are being used in year-round houses, especially in Australia, New Zealand, and Japan.

These toilets consume considerable amounts of energy and those that require bowl liners have this extra expense, but they completely solve the water-pollution problem. Inevitably there is some release of gases and odor into the air, but a few new incinerating systems boast of discharging nothing but carbon dioxide and water vapor.

Biological Toilets

A U.S. company makes a recycling toilet based on biological principles of waste digestion; body wastes and toilet tissue are turned into water by the continuous action of enzymes and bacteria, both aerobic and anaerobic, in the base of the commode. Flushing is done by a hand pump. The effluent is said to be clear, odorless water containing no pathogenic organisms, and can be disposed of in a dry well beside the house. There is no residue, so no sludge removal is needed. Once a week a package of freeze-dried bacteria and enzymes is added to the toilet. Every two years the charcoal filters must be replaced. Prices are in the $400 to $500 range; no electricity is needed. The company states the units are adaptable to high-rise buildings.

Montgomery County, Maryland, and Fairfax County, Virginia, tried out a number of these units in rural homes having no other sanitary facilities. The tests were reported as failures because of complaints about odors, but the company explains the problem as one of mis-use—homeowners used chlorine-based cleaners in the commodes and the chlorine reportedly upset the biological breakdown process.

Another company is developing a biological toilet which requires addition of a bacteria package every two weeks. It makes use of both aerobic and anaerobic digestion, and uses extensive filtration to clean the water for recirculation to the toilet, which uses only a fraction of a gallon a flush. An electrical connection is needed. Once every two years the unit must be pumped out because of residues.

Oil-Flush Toilets

These are closed-loop toilet systems that use a white, low-viscosity mineral oil

* This conference, cosponsored by Rodale Press, *Rain,* and the Oregon Museum of Science and Industry, was held in Portland, Oregon, May 11–14, 1976. Proceedings can be found in *Compost Science,* vol. 17, nos. 3 and 4 (Summer 1976).

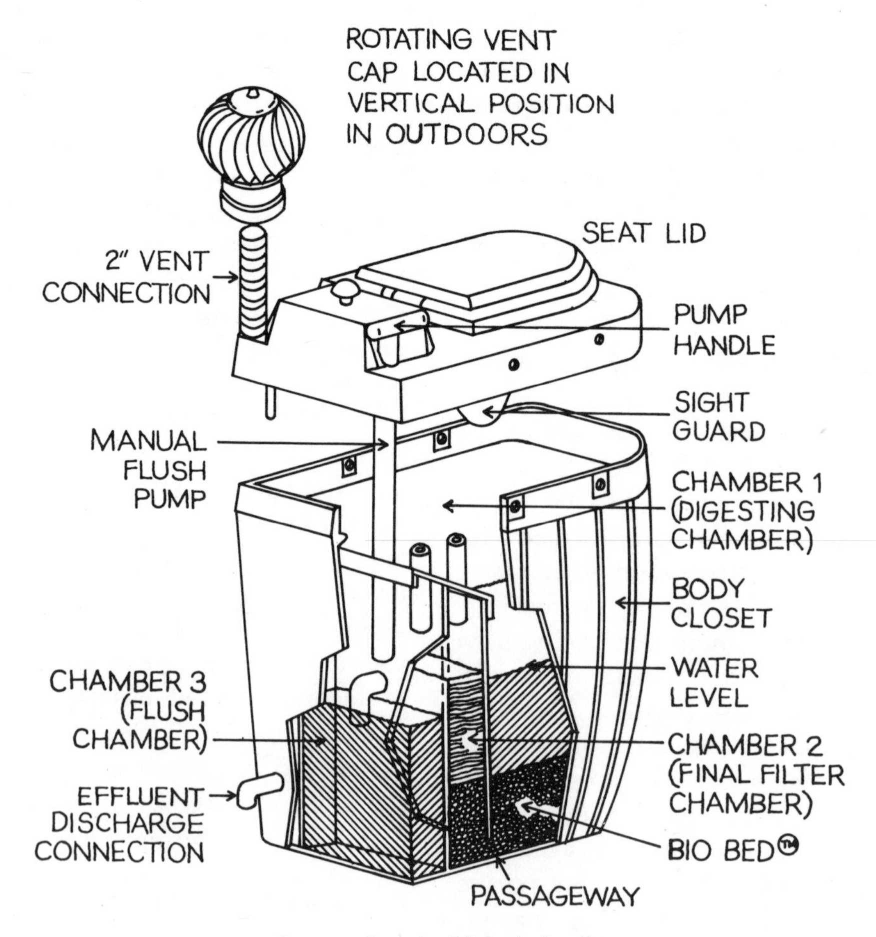

Cross section of a biological toilet.

as the flushing medium. The toilet bowl looks much like the conventional commode, the only difference being that the inside is coated with Teflon. Wastes are carried to a gravity separation tank where the oil floats to the top and the water-saturated wastes sink to the bottom. The oil is drawn off the top through filters containing chlorine crystals and activated charcoal and recirculated to the commodes. It remains clear and odorless, and bacteria are controlled to nearly zero.

Many marine installations have been made, and the system has been adapted for high-rise buildings ashore. It is now also being used in public rest rooms in national and state forests and parks, especially in locations where water is scarce or soil conditions discourage septic tanks and drain fields. One of the most unusual of these installations is in Custer National Forest, where solar panels are used to provide electricity for pumps that move the oil from the holding tanks back to the eight toilet fixtures. The U.S. Forest Service is also testing near Yellowstone Park a solar-assisted oil-flush toilet it designed.

An oil-flush system is now being marketed that can be used in single-family homes; it has a 400-gallon holding tank with a capacity to store wastes from four people for up to a year.

Ultimate disposal of wastes from an oil-flush system can still be a problem. Today they are usually trucked to a municipal treatment plant or taken to farmland

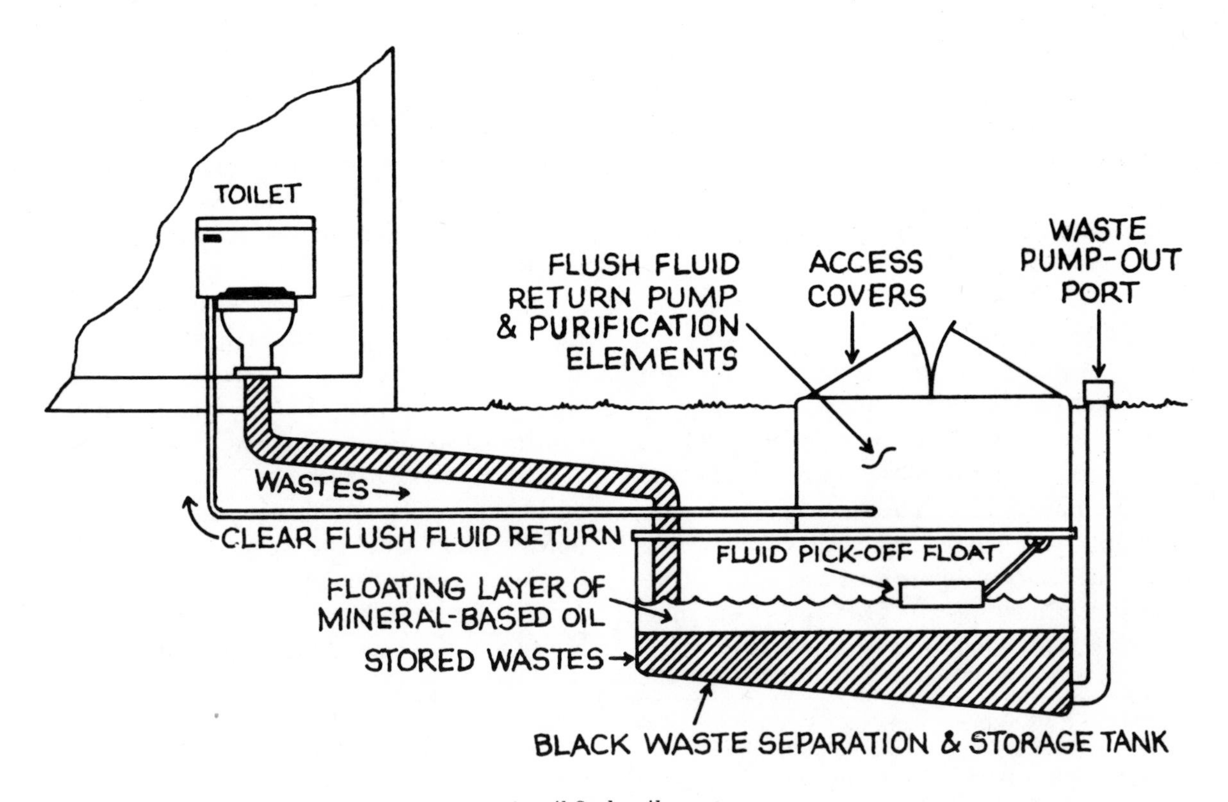

An oil-flush toilet system.

for agricultural use, but one company is working out plans for generating methane gas from the wastes.

Vacuum Systems

Swedish and American companies make toilet systems that use the vacuum principle to transport wastes from the commode to the place of treatment and/or discharge. The system can be hooked up to a holding tank, sewerline, incinerator, or septic tank. An electric pump maintains a vacuum in the line. About a liter of water is needed per flush, since differential air pressure moves the wastes through the line. Although the system does save water, the electricity needed to power these systems can cost as much as it saves in water. The system can be adapted to a variety of needs, from single-family houses to apartments and whole new towns. More than 20,000 such toilets are now in use.

Aerobic Tanks

A number of U.S. companies are now marketing disposal systems that rely on aerobic rather than anaerobic (septic) action to digest the wastes in the tanks and tile drain fields. These units can be as simple as a septic tank with air bubbling through, or as sophisticated as a small, activated sludge plant. They may contain as many as five chambers or as few as just one. In order to speed up decomposition, solids are broken up in any one of a number of ways: rotating blades or disks chop them up, high pressure jets forcibly push them against the tank wall, or air is blown through the liquid vigorously enough to churn them up. Any solids that remain are either filtered, settled down, or skimmed off so that they don't get carried with the effluent to the discharge field.

In contrast to a septic tank, an aerobic tank does not generate foul odors and the effluent carries a charge of dissolved oxygen that helps to prevent clogging of the drain field. The effluent is semiclean, odorless, and acceptable in some states for surface or stream discharges. Although aerobic tanks are more expensive than septic ones, the cost can be balanced out to some degree in those states that allow the installation of one-third less drain field if an aerobic system is used.

Aerobic units are relatively new and still being tested. Although they do a better job of breaking down sewage than septic tanks, they need more maintenance. Routine inspection, oiling and replacement of vanes, seals, or filters is necessary, and problems can occur should the air compressor break and the system go septic.* Certain aerobic systems have been approved for installation in Pennsylvania, Maryland, Kentucky, Maine, Ohio, Alaska,

* The Appalachian Regional Commission has tested these aerated systems in individual homes where there has been chronic failure in the existing septic systems. These tests in Boyd County, Kentucky, and similar tests in Oakland, Maryland, have been successful in preventing clogging in the soil. However, in one West Virginia county "where approximately 80 home aeration units had been installed, within five years 75 percent of the units were no longer operable. The poor operation record was based on (*a*) equipment breakdown with no realization by the property owner of this occurrence; (*b*) cost of power, causing the property owner to turn equipment off; and (*c*) noise of the equipment, causing the property owner to also turn the equipment off." (Letter from William W. Bradford, PE, Div. of Sanitary Engineering, State of West Virginia, November 28, 1975.)

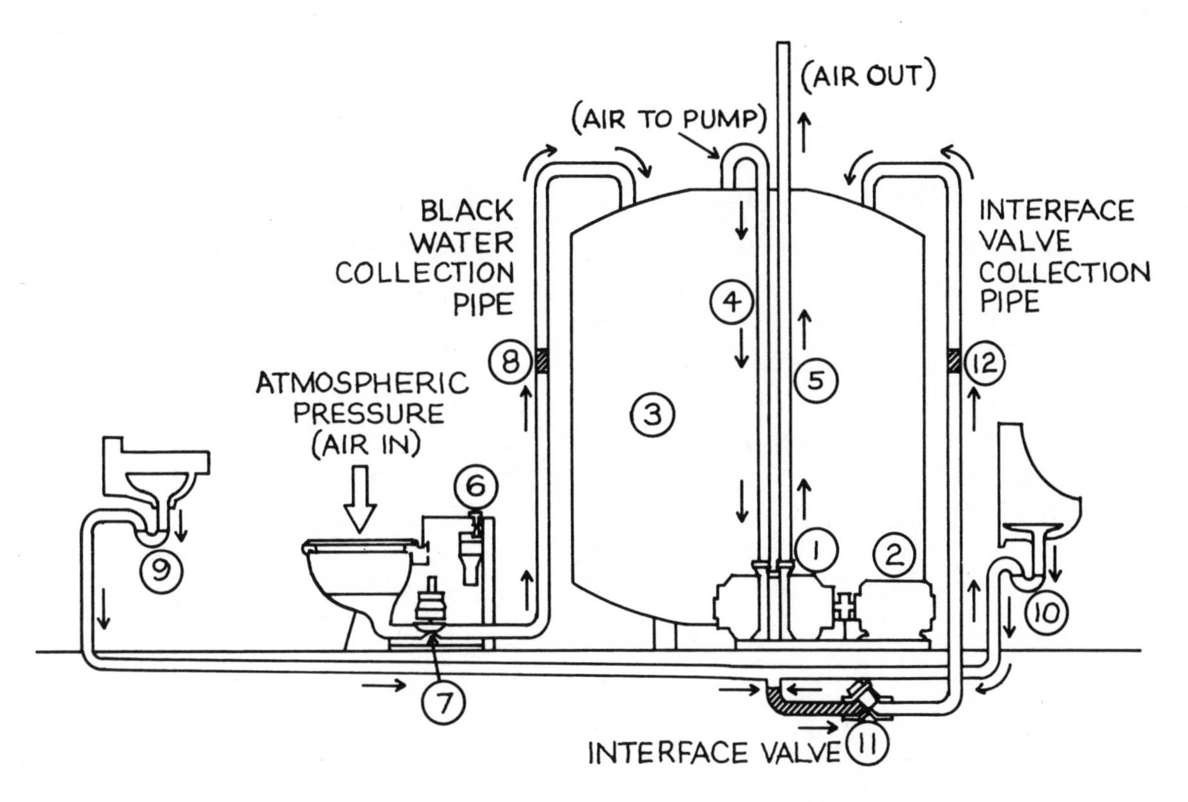

TOILET...

ELECTRIC MOTOR (1) DRIVEN VACUUM PUMP (2) MAINTAINS VACUUM ON STEEL VACUUM TANK (3) BY REMOVING AIR FROM VACUUM TANK THROUGH LINE (4) AND THEN DISCHARGING TO ATMOSPHERE THROUGH LINE (5).

USER PRESSES BUTTON (6) ON TOILET TO FLUSH.

DISCHARGE VALVE (7) OPENS ALLOWING ATMOSPHERIC PRESSURE TO FORCE SEWAGE THROUGH PIPING (8) TO THE VACUUM TANK (3).

FRESH WATER CLEANS BOWL AND ALSO ENTERS THE VACUUM LINE (8) PASSING ON TO THE VACUUM TANK. DISCHARGE VALVE (7) CLOSES AND FRESH WATER REFILLS TOILET BOWL RESERVOIR.

SINK AND URINALS...

SINK WASTE (9) AND URINAL WASTE (10) GRAVITY FLOW TO INTERFACE VALVE (11).

WHEN INTERFACE VALVE HAS SUFFICIENT LIQUID WASTE VOLUME TO ACTIVATE, INTERFACE VALVE OPENS AND SEWAGE PASSES TO VACUUM TANK IN LINE (12).

INTERFACE VALVE CLOSES AUTOMATICALLY AFTER 2–4 SECONDS AND IS READY FOR REUSE.

TOTAL ELAPSED TIME IS 7 SECONDS BEFORE TOILET IS READY FOR REUSE.

A vacuum system that handles wastes from toilet, urinal, and bathroom sink.

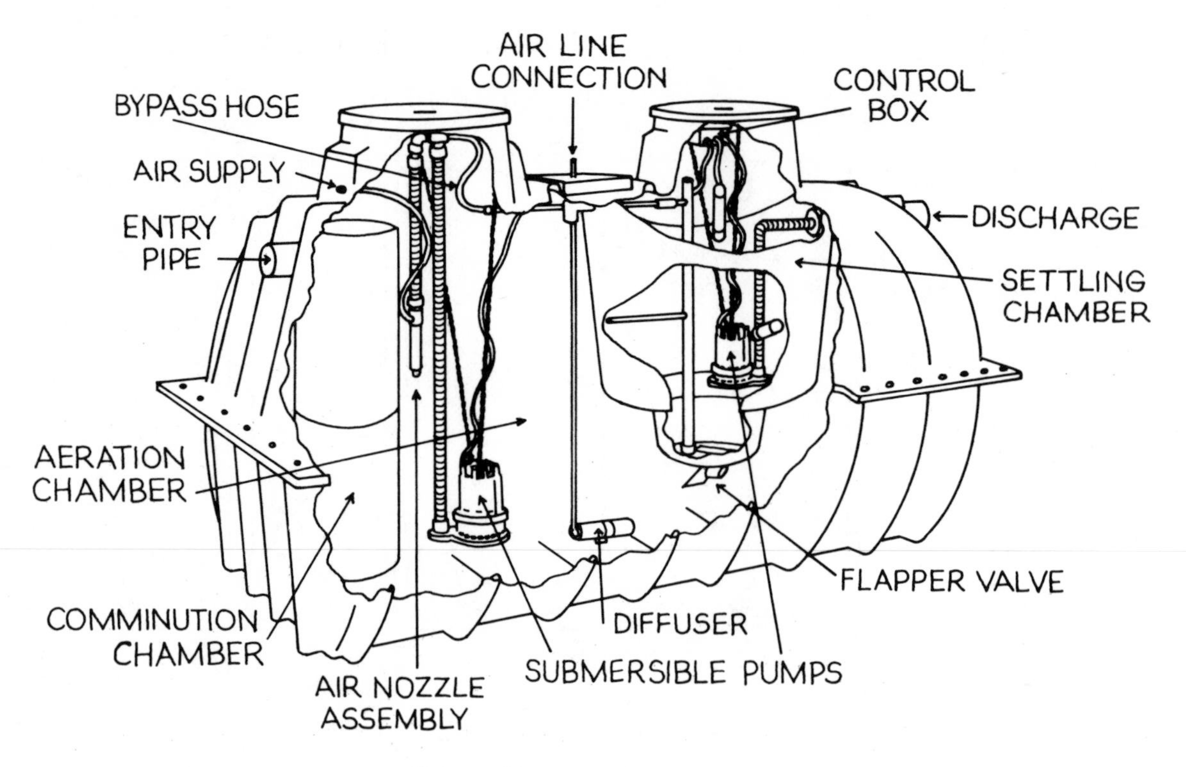

A complete aerobic wastewater treatment system.

New York, and in many other states on a county basis.

True enough, such aerobic systems use as much water to flush the toilets as do traditional systems, but one manufacturer is making an aerobic tank which filters the effluent to the point where it can be safely recycled back to the toilet tank.

Pressure Sewers

To many sanitary engineers, the most promising alternative for small communities faced with expensive sewer hookup and maintenance fees is the centralized collection of septic tank effluents by means of pressure sewers.

Unlike conventional sewer systems, pressure sewers do not take the place of septic tanks, but rather work with them. Household sewage enters a septic tank (sometimes termed "interceptor tank"), which provides excellent pretreatment by acting as a grease trap, extended sedimentation basin, and storage chamber with some solids decomposition. Effluents spill into a sump, and after they reach a prescribed level, a pump activates. Effluent discharges through a 1¼-inch feeder line to a pressure main. Check valves and gate valves are located strategically throughout the system to assure ease of maintenance and repair. Air release valves are also strategically placed to vent pockets of entrained air. Effluent transport requires small and few lift stations to move it to final treatment and disposal by conventional means—lagoons, extended aeration, sand filtration, or whatever.

Proponents state several advantages to pressure sewers. They claim that they are less costly than regular systems because the existing on-site septic tanks take care of the pretreatment. They work along with water conservation measures because they are not as dependent on water for transportation of wastes as are conventional sewers, which rely only upon water and gravity to move

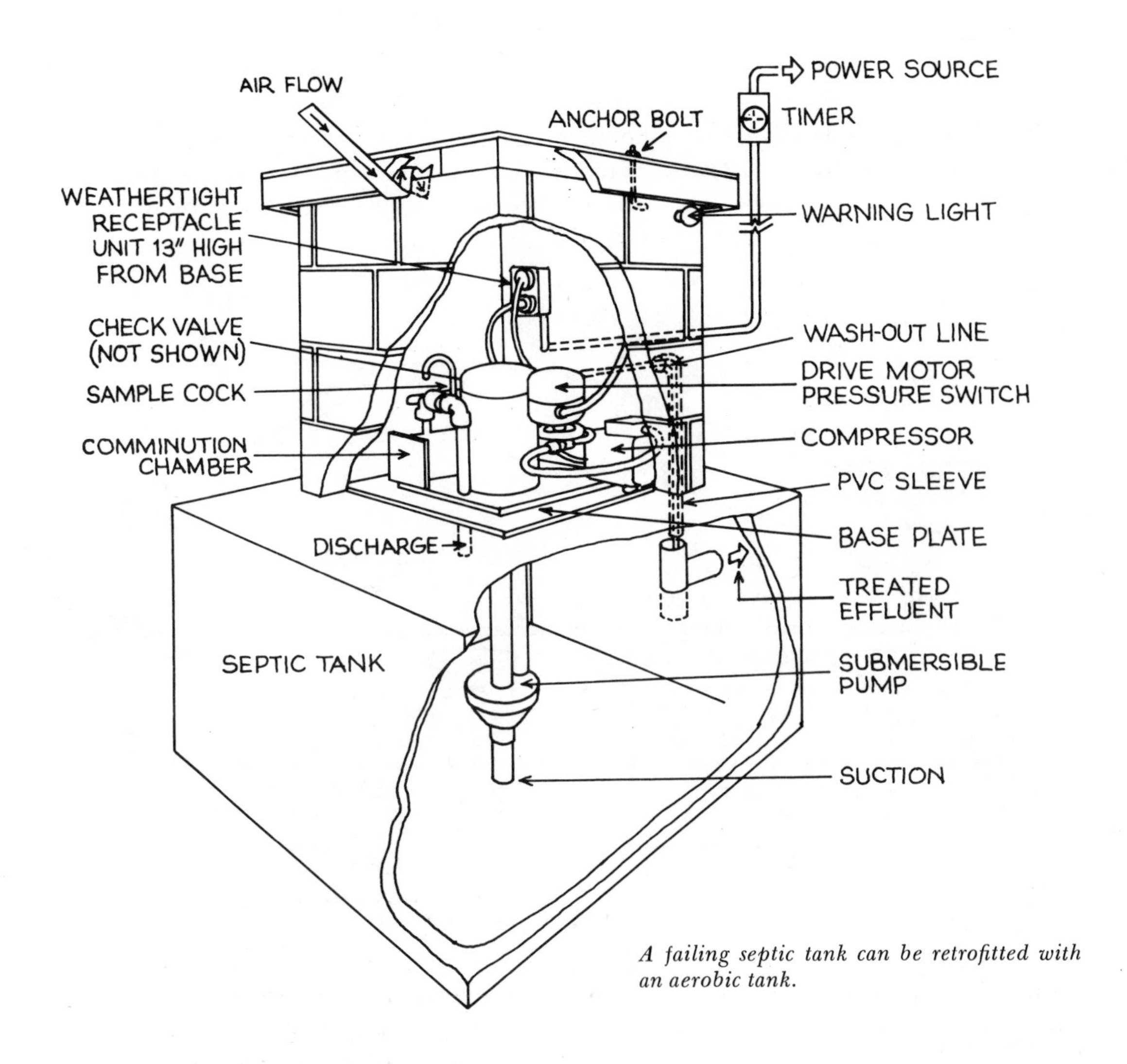

A failing septic tank can be retrofitted with an aerobic tank.

wastes. (As was mentioned earlier in this chapter, severe water conservation practices backfired a bit in Marin County in early 1977 when gravity sewers in low-lying areas clogged.) Since each individual home's wastewater outflow can be metered separately and charged accordingly, those homeowners who practice water conservation will save on sewer maintenance bills.

The CANWEL System *

A household waste disposal system called CANWEL (Canadian Water Energy Loop) that incorporates a few different methods of treating wastewaters with as little energy as possible, is being developed by the Ontario Research Foundation at the request of the Central Mortgage and Housing Corporation, a Canadian agency that corresponds to the U.S. Department of Housing and Urban Development.

In the CANWEL system wastewater from the household is cleaned (or "polished") to one of two levels: (*a*) the first level of water quality is good enough to recirculate to the toilet tanks and to use for such chores as washing cars and sprinkling lawns—in total about 60 percent of all domestic water use, with any surplus being safely discharged into storm drains or surface waters; and (*b*) the higher level of water quality can be safely used for all cooking and drinking purposes.

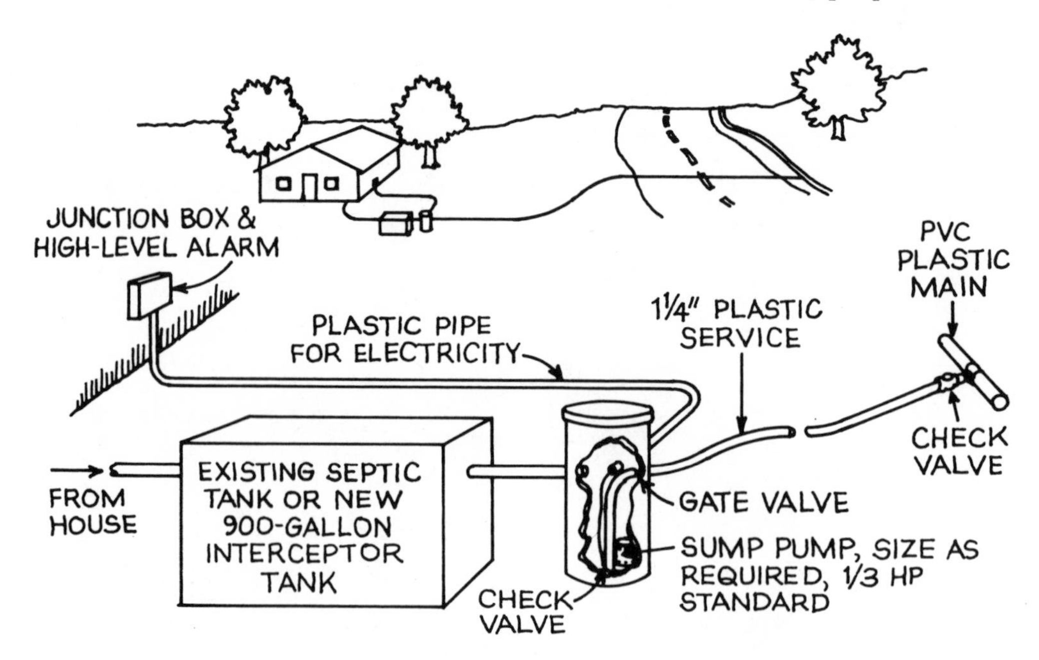

The main components of a pressure sewer system.

* Much of what follows originally appeared in Harold Leich, "Canwel Can Do?" *Compost Science* 17 (1976): 21.

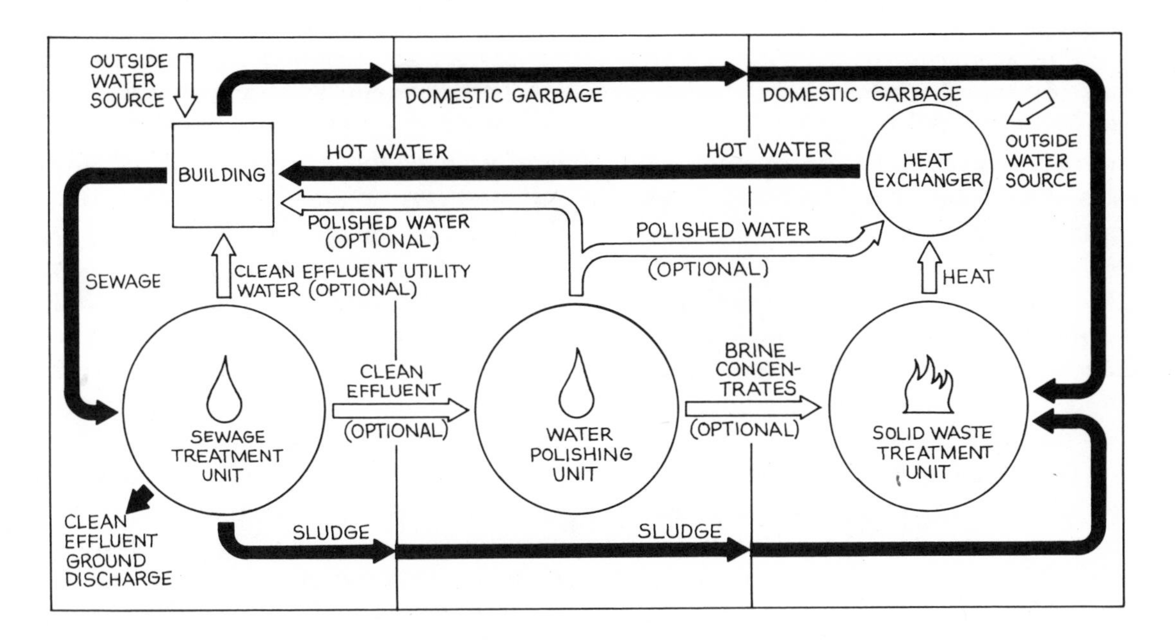

THE SEWAGE TREATMENT UNIT

TO PRODUCE A SUPERLATIVE EFFLUENT SUITABLE FOR UNDILUTED SURFACE DISCHARGE

THE SOLID WASTE UNIT

TO RECOVER AND UTILIZE THE MAXIMUM ENERGY IN GARBAGE AND TO DISPOSE OF SEWAGE SLUDGE, WITH NO ENVIRONMENTAL DAMAGE

THE WATER POLISHING UNIT

TO REDUCE, TO LEVELS SAFE FOR DRINKING, HAZARDOUS CONTAMINANTS NOT REMOVABLE BY CONVENTIONAL MEANS

A schematic of the CANWEL waste management concept.

The sewage is treated by a combination of physical, chemical, and biological processes, and the last of these relies on controlled aeration in the presence of a "mixed microbial population" to convert organic wastes to carbon dioxide, nitrogen gas, and cell mass. The water is polished by filtration and reverse osmosis techniques to reduce contamination levels in the effluent. Brine concentrates from the polishing process may be incinerated along with the sludge and garbage.

Ozone rather than chlorine is used as the antiseptic agent in purifying wastewater. (Recent research has raised the point that chlorinating water supplies may create new carcinogenic substances in the water.)

Sewage sludge and kitchen garbage are incinerated together, with the resultant heat used to supply about 80 percent of the hot water needs of the household.

If proved out in actual use, the new system has great advantages over what we do today: no sewer lines would be needed to carry away liquid wastes; no central sewage treatment plants would be required; total water use in the home would be greatly reduced because of the recirculation features; any effluents released to ground or surface waters would be markedly cleaner than present effluents from treatment plants; most hot-water needs of the family would be met without additional energy input; and cost of refuse collection would be lower since no garbage would be included. And all this at a lower cost to the public than what we do today.

There are several flexibilities in the CANWEL system; for example, use of the two levels of water quality in the household is an optional feature, and the solid waste system can be used separately from the water-purification system. CANWEL may be designed to serve an apartment house or office building or an entire small town, and work is going forward to develop a system for a single house. Some features of the design could be used to improve operation of existing municipal treatment plants so that we can still take advantage of the present system in which we have already invested so much. CANWEL's features are especially appropriate for new suburbs and new communities.

The system today represents 15 years of research, and a prototype sewage treatment unit has operated successfully for some time at the Ontario Research Foundation. The next testing stage will be to install the whole integrated system in a Toronto apartment building.

The Domestic Sewage-Methane Cycle

Robert Mueller, a former senior scientist with the National Aeronautics and Space Administration, has proposed what he calls the Domestic Sewage-Methane Cycle (DSMC)—a system which integrates a sanitary waste collection service and a community bio-plant in which wastes are transformed into methane gas and fertilizer. Human and kitchen wastes are collected in cannisters which are connected to waterless toilets in home bathrooms. When full, the containers are sealed and picked up and

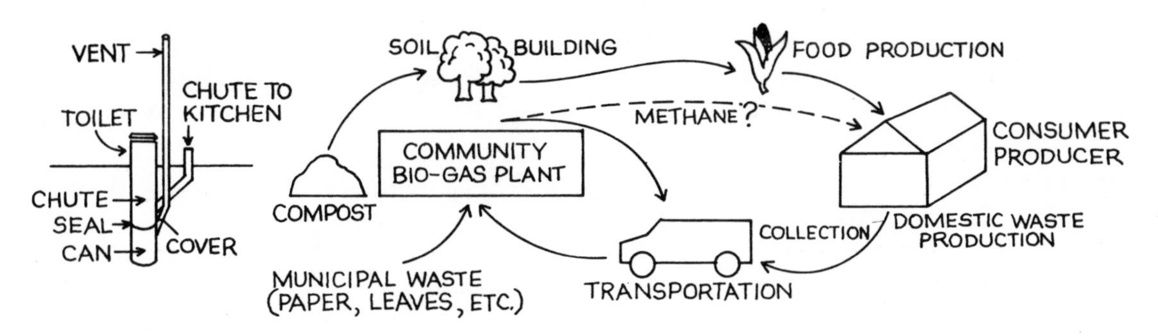

Mueller's Domestic Sewage-Methane Cycle.

homeowners are given replacement cans. Pickup might be once or twice a week and operate much like garbage collections, the wastes are deposited in community bio-gas plants, and the cannisters are steam cleaned and reused.

Such a system would not only save water, it would also produce a fuel—methane gas—which is produced during the anaerobic decomposition of the slurried wastes, as well as a valuable fertilizer, in the form of the slurry which is removed after methane production. For the Washington, D.C., area alone, 2.5 million cubic feet a day would be generated*—enough to heat 10,000 single family homes.

An Algal Regenerative System for Single-Family Farms and Villages †

Even more futuristic than the DSMC is a self-contained living system that according to its designers, Drs. Golueke and Oswald, provides waste disposal and nutrient recycle for four people, one cow, and 50 chickens. Aside from the people and the animals, the principal components of the system are an anaerobic digester, a series of algal growth chambers, a sedimentation chamber, sand beds, a solar still, and a gas exchanger. The operation of the system involves the charging of all manure, urine, wasted food, night soil, and cleanup water into the digester shortly after they are produced, or at least once daily. In the digester, fermentation once established continues on a steady basis, as does gas production. Particular care would have to be exercised to avoid unnecessary loss of useful components. Therefore, all solids, liquids, and gases must be recycled or consumed. Complex substances are decomposed in the digester. Products of this decomposition are organic acids, ammonia, CO_2, and methane. The methane is stored for use as needed. The addition of the nutrients to the digester displaces soluble substances into the algal culture, where the latter serve as a fertilizer for algal growth. The methane, under slight pressure, is used as fuel in cooking. Carbon dioxide formed

* Robert Mueller, "Getting Rid of Sewage—Isn't There a Better Way?" *Environmental Action* (March 29, 1975): 3.

† For a more complete description of this system, see C. G. Golueke and W. J. Oswald, "An Algal Regenerative System for Single-Family Farms and Villages," *Compost Science* 14 (May-June 1973): 12.

by the combustion of the methane is vented by convection to the algal culture, where a part of it is used as a carbon source by the growing algae. Algal slurry is fed to the cow and constitutes its sole source of drinking water, thereby forcing it to consume algal protein in the wet form. Algal slurry not consumed by the cow is removed from the trough and is spread over sand beds. The dewatered and dried algae can be used on the site for chicken feed or to augment the algal slurry feed for the cow, or it can be sold.

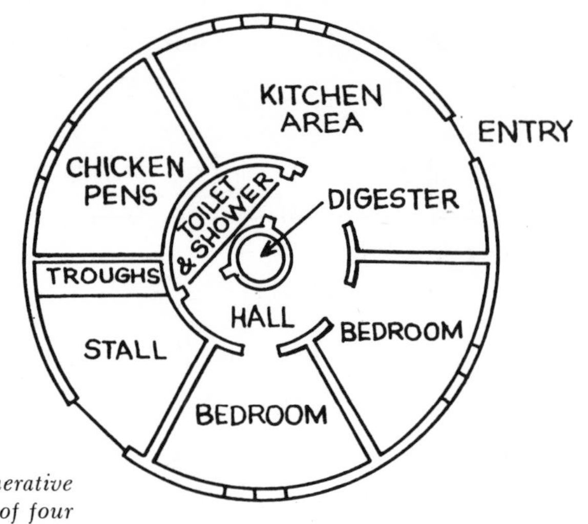

Golueke and Oswald's algal regenerative system. A dwelling unit for a family of four and their livestock that incorporates a microbiological recycle system for water, nutrients, and energy in a convenient and hygienic environment.

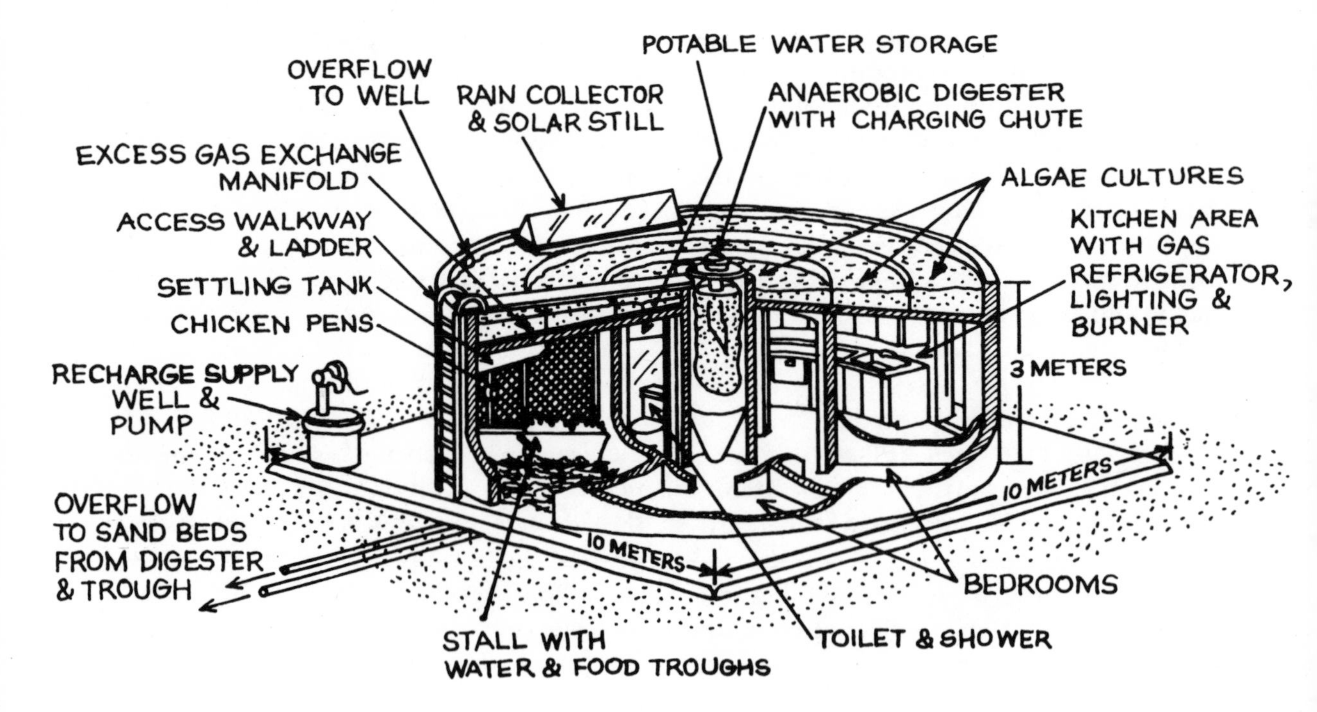

Using the space below the culture as living quarters serves to shelter both the humans and the animals from the elements. The algal culture and digester provide a buffer against rapid change in temperature for the occupants; and the metabolic heat given off by the occupants would, in turn, supply some warmth to the algal culture and digester during cool periods.

While the system has not been adopted and used in a real-life situation, it is beyond the conceptual stage because its components have been shown to work effectively together in laboratory and pilot scale studies.

Composting Privies and Toilets

Since a good portion of the rest of the book examines these units in depth, I won't dwell upon them here. Suffice it to say that they are designed to decompose toilet (and, in most cases, kitchen and some garden) wastes into a soil amendment. There are a number of different types and they vary in complexity, but they all are designed to encourage good ventilation and evaporation, and make storage and decomposition of human wastes safe and convenient. Although commercial models were introduced into Canada and the United States only a few years ago, they have been installed in several thousand vacation homes, a few hundred year-round houses, and in some public facilities on an experimental basis.

The little unit monitoring that has and is being done is hardly adequate to really evaluate these toilets. Some very early poor test results were clearly due to faulty installation and a misunderstanding of conditions necessary for good decomposition. Even some distributors of commercial units don't know enough about the product they're selling to give proper advice to their customers. Ecos, a large Boston-based distributor of several models, has started composting toilet workshops for its local representatives to make sure that they're aware of the most up-to-date findings and have all the proper information to install and help maintain the units they sell.

Government officials, with the exception of those in California, Maine, and a handful of other progressive states, have given little attention to commercial composting toilets and even less to owner-built units, because of lack of time, scant knowledge of the process behind these toilets, or a reluctancy to consider seriously a wastewater treatment system that doesn't handle laundry, kitchen, and bath waters as well. Only a few states have given blanket approval for their use. Elsewhere the toilets are being installed without official approval or under experimental permits issued either by state or county governments.

Despite the growing pains, composting toilets have generated more interest than all the other conventional toilet alternatives combined. Indeed, there is a small, but growing composting toilet movement afoot. The reasons are many, as you'll discover in later chapters.

Chapter 3

A Short Lesson on the Principles of Composting

Biological decomposition is a natural process that is going on all the time—on forest floors, sanitary landfills, refuse piles, lawns, and even in our stomachs. Composting is a form of biological decomposition that takes place under controlled conditions so that the decomposition process is as complete as possible (reaching a state of stabilization) in a relatively short time.

If you're an organic gardener you know something about composting. But it's probably not enough if you want to have a well-running composting toilet or privy. If an outdoor pile isn't made or maintained correctly there's little harm done. But throw the moisture level off too much in your toilet or privy, neglect to maintain a reasonable C/N ratio, or ignore the need for ventilation and you could be in for trouble, or at least some unpleasantness. You need to know something about the process so that you can make sure the necessary "controlled conditions" are met. A good composting toilet or privy design will take care of these conditions to some extent, as you'll read later, but equally or perhaps more important is how you help to maintain the composting process itself.

There are two types of composting: that which takes place in the presence of oxygen (*aerobic*) and that which proceeds in its absence (*anaerobic*). Aerobic composting is characterized by its rapid decomposition rate, only slight foul odors, and high temperature. Anaerobic composting on the other hand, is known for the bad odors that come from the hydrogen sulfide and other sulfur compounds that result. Temperatures don't rise nearly as much as in aerobic composting; they are not high enough to have an effect on pathogens or parasites in animal manures or human wastes. The pathogenic organisms do eventually get destroyed by other unfavorable environmental conditions and biological antagonisms, but the destruction is slow and can take as long as six months to a year. Infectious hepatitis virus may last as long as two years.

Comparing the two, it should be obvious that both gardeners and composting toilet and privy owners want to have an aerobic process taking place in their gardens or homes. Which process they are really going to have depends upon the "controlled conditions" mentioned earlier. These are temperature, moisture, oxygen, the carbon to nitrogen ratio, and, to a lesser extent, the pH (or acidity/alkalinity) of the decomposing matter.

Paraphrasing Ray Poincelot, in his article "A Scientific Examination of the Principles and Practice of Composting," * composting itself starts with different kinds of organic matter that contain their own natural bacteria and fungi. When conditions are favorable, these microorganisms will grow and start the process of aerobic decomposition. During their growth, the microbes use some of the available carbon, nitrogen, and other nutrient elements. As this life cycle proceeds, the temperature begins to increase from the heat generated by these biological oxidations. Since the organic matter acts as an insulator, much of the heat is retained in the compost pile. As decomposition slows down, the pile cools off. The chemical constituents in the wastes are altered as a result of this microbiological activity; this changed form of organic matter is know as humus and is a soil conditioner and fertilizer. (In the case of composted feces and urine, the finished compost will be much lower in nitrogen than the fresh excreta. However, it will be in a form much more useable by plants. Plants can use nitrogen only as ammonia, or as nitrates or nitrites, which are only produced during decomposition. Much of the nitrogen in raw excreta spread over the land will escape as a gas and never be available to plants.)

Temperature is one of the important conditions in this decomposition process. Initially, the material is at the same temperature as the surrounding air. As the microorganisms grow, the temperature rises, until the inside of the pile reaches temperatures as high as 160°F (71°C). Temperatures of 122°F (50°C) and higher are necessary for killing pathogenic organisms. Temperatures above 131°F (55°C) will tend to inactivate the organisms responsible for the continuation of the composting, and activity will decline until optimum temperatures are reached again. In any event, the temperature gradually declines and returns to lower temperatures of about 104°F (40°C).

Several factors can affect the temperature. If the compost pile itself is too small, the ratio of surface area to volume will be great and the pile will "cool off" because it will not be able to keep the heat generated by the process inside the pile. The minimum size pile should be two feet high, but a more ideal size is about four feet high. In cold temperatures the pile should be higher than this in order to hold on to its own heat.

Up to a certain point, the smaller the particle size of the materials added to the pile, the faster the decomposition and the higher the temperature.† This is because the increased surface area of smaller but

---

* In *Compost Science,* September 1974: 24–31.

† The ideal particle size really depends upon the material. If the particle size of wet garbage is too small it will turn into a soupy mess. However, with materials like straw and wood, the smaller the better. The ideal size for straw to be mixed with feces and urine is about one inch.

Temperature and Time of Exposure Required for Destruction of Some Common Pathogens and Parasites

Organism	Observations
Salmonella typhosa	No growth beyond 46°C; death within 30 minutes at 55°-60°C and within 20 minutes at 60°C; destroyed in a short time in compost environment.
Salmonella sp.	Death within 1 hour at 55°C and within 15-20 minutes at 60°C.
Shigella sp.	Death within 1 hour at 55°C.
Escherichia coli	Most die within 1 hour at 55°C and within 15-20 minutes at 60°C.
Entamoeba histolytica cysts	Death within a few minutes at 45°C and within a few seconds at 55°C.
Taenia saginata	Death within a few minutes at 55°C.
Trichinella spiralis larvae	Quickly killed at 55°C; instantly killed at 60°C.
Brucella abortus or *B. suis*	Death within 3 minutes at 62°-63°C and within 1 hour at 55°C.
Micrococcus pyogenes var. *aureus*	Death within 10 minutes at 50°C.
Streptococcus pyogenes	Death within 10 minutes at 54°C.
Mycobacterium tuberculosis var. *hominis*	Death within 15-20 minutes at 66°C or after momentary heating at 67°C.
Corynebacterium diphtheriae	Death within 45 minutes at 55°C.
Necator americanus	Death within 50 minutes at 45°C.
Ascaris lumbricoides eggs	Death in less than 1 hour at temperatures over 50° C.

Source: C.G. Golueke and P.H. McGauhey, *Reclamation of Municipal Refuse*, Sanitary Engineering Research Laboratory Bulletin 9 (Berkeley: University of California, June 1953), p. 73.

Although valuable as a guide to the relative effect of temperatures on pathogens and parasites, this chart should not be taken as gospel. According to Dr. Golueke, the temperatures represent wet heat, and the chart assumes that the bacteria are dispersed uniformly throughout the pile so that each is exposed to the high heat; this is rarely the case. Actual kills seldom reach 100 percent.

more materials means there is greater susceptibility to microbial invasion and increased availability of oxygen. It also makes for better heat distribution and less heat loss to the surrounding environment.

As explained above, the temperature of the pile is regulated by the activity inside it. If any one of the other conditions (moisture or oxygen level, or C/N ratio) is not favorable, the pile won't reach ideal temperatures.

Moisture affects the composting process, too. In theory, the ideal moisture content of the pile would be 100 percent,* but this would leave no spaces for air, and the compost would go anaerobic. The only way to achieve an aerobic pile with a 100 percent moisture content would be to make a slurry out of the compost and bubble oxygen through it—a very impractical idea for most setups.

Most experts agree that an optimal moisture content is between 50 and 70 percent. As a guideline, organic wastes with good moisture contents should feel damp, but not soggy. If a pile goes much below 40 percent the organic matter will not decompose rapidly, and if it's above 60 percent, the compost tends to become anaerobic, unless it contains good amounts of strong fibrous material, like straw and peat moss, that adds bulk to the pile, thereby creating air spaces.

Turning or rotating the pile (as in a drum composter or privy) can also help to control an unfavorably high moisture content by aerating the pile and distributing the moisture evenly throughout so that pockets of anaerobic activity can be eliminated.

Sufficient *oxygen* is also very important. If the oxygen level becomes too low, the aerobic microorganisms die or become inactive and are replaced by anaerobic microbes. And we've already told you why that would not be desirable. Exactly what the minimum oxygen level necessary to maintain aerobic conditions is is not known, although apparently the oxygen level can drop to as low as 0.5 percent in some materials without producing an anaerobic situation.†

Proper aeration of the pile through turning or rotating, favorable moisture content, and enough bulky materials will assure a good oxygen level.

Another important aspect of successful aerobic composting is the relationship between the *carbon/nitrogen ratio* and the rate of organic matter decomposition. Microorganisms require carbon for energy and growth and nitrogen for protein synthesis and growth. Since the best diet for microbes is about 30 parts carbon per 1 part nitrogen, a C/N ratio of 30:1 is ideal for composting. This ratio can be stretched a bit on either side so that the best ratio lies between 25:1 and 35:1. If the ratio is above 35, the process becomes inefficient and the compost requires more time for completion. In addition, a finished compost with a high C/N ratio is not good for

(continued on page 49)

* C. G. Golueke, *Composting* (Emmaus, PA: Rodale Press, Inc., 1972), p. 26.

† R. P. Poincelot and P. R. Day, "Rates of Cellulose Decomposition during Composting Leaves Combined with Several Municipal and Industrial Wastes," *Compost Science* 14 (1973): 23–25.

Composition of Human Excrement

Human feces without urine

Approximate quantity

0.3-0.6 pound (135-270 g) per capita per day moist weight
0.08-0.16 pound (35-70 g) per capita per day dry weight

Approximate composition

Moisture content			66-80%
Organic-matter content (dry basis)			88-97%
Nitrogen	"	"	5.0-7.0%
Phosphorus (as P_2O_5)	"	"	3.0-5.4%
Potassium (as K_2O)	"	"	1.0-2.5%
Carbon	"	"	40-55%
Calcium (as CaO)	"	"	4-5%
C/N ratio	"	"	5-10

Human urine

Approximate quantity

Volume: 1¾-2¼ pints (1.0-1.3 liters) per capita per day
Dry solids: 0.12-0.16 pound (50-70 g) per capita per day

Approximate composition

Moisture content			93-96%
Organic-matter content (dry basis)			65-85%
Nitrogen	"	"	15-19%
Phosphorus (as P_2O_5)	"	"	2.5-5%
Potassium (as K_2O)	"	"	3.0-4.5%
Carbon	"	"	11-17%
Calcium (as CaO)	"	"	4.5-6%

Source: Gotaas, *Composting,* p. 35.

Approximate Nitrogen Content and C/N Ratios of Some Compostable Materials (Dry Basis)

Material	N (%)	C/N
Urine	15-18	0.8
Blood	10-14	3
Fish scrap	6.5-10	—
Poultry manure	6.3	—
Mixed slaughterhouse wastes	7-10	2
Night soil	5.5-6.5	6-10
Activated sludge	5.0-6.0	6
Meat scraps	5.1	—
Purslane	4.5	8
Young grass clippings	4.0	12
Sheep manure	3.75	—
Pig manure	3.75	—
Amaranthus	3.6	11
Lettuce	3.7	—
Cabbage	3.6	12
Tomato	3.3	12
Tobacco	3.0	13
Onion	2.65	15
Pepper	2.6	15
Cocksfoot	2.55	19
Lucerne	2.4-3.0	16-20
Kentucky blue grass	2.4	19
Grass clippings (average mixed)	2.4	19
Horse manure	2.3	—
Turnip tops	2.3	19
Buttercup	2.2	23
Raw garbage	2.15	25
Ragwort	2.15	21
Farmyard manure (average)	2.15	14
Bread	2.10	—
Seaweed	1.9	19
Red clover	1.8	27
Cow manure	1.7	—
Wheat flour	1.7	—
Whole carrot	1.6	27
Mustard	1.5	26
Potato tops	1.5	25
Fern	1.15	43
Combined refuse, Berkeley, Calif. (average)	1.05	34
Oat straw	1.05	48
Whole turnip	1.0	44
Flax waste (phormium)	0.95	58
Timothy	0.85	58
Brown top	0.85	55
Wheat straw	0.3	128
Rotted sawdust	0.25	208
Raw sawdust	0.11	511
Bread wrapper	nil	—
Newspaper	nil	—
Kraft paper	nil	—

Source: Gotaas, *Composting,* p. 44.

the soil because the bacteria in the compost will rob the soil of nitrogen to make use of this excess carbon. If it is below 25, the excess nitrogen is converted to ammonia and wasted into the atmosphere. (The smell of this ammonia, by the way, should be familiar to everyone who has ever used a poorly functioning outhouse like the kind you might find at busy camps and outdoor festivals. If material abundant in carbon, like leaves, sawdust, or even shredded paper, were available for throwing down the hole after each use, this odor could probably be lessened. But since it's not, the wastes that pile up do not contain enough carbon to balance their nitrogen, and therefore ammonia is constantly being produced.)

The best way to assure yourself of a good C/N ratio is to add a great variety of wastes to your pile: manures or human wastes, grass clippings, straw, kitchen scraps, leaves, garden wastes, etc., and mix them up as much as possible by layering or mechanical mixing.

A less important composting condition is the *pH or acidity/alkalinity of the pile.* It will affect the fungi and bacteria population, but it makes little sense to fool around with the pH. The pile will probably be a bit on the acid side which is lower than the ideal of neutral, because nitrogenous material tends to be slightly acidic. But "sweetening" the pile (usually with lime) to create a more neutral pH will cause some of the valuable nitrogen in the pile to escape in the form of ammonia. Just as with the C/N ratio, a neutral pH can be achieved most naturally and safely by making sure a variety of materials make up the pile and that aeration is sufficient.

Chapter 4

Composting Privies

Composting human wastes has long been a part of Asian tradition. Indeed, it probably got its start in that part of the world. For centuries people in Japan, China, Korea, and other Far Eastern countries have been collecting their own wastes in holes or buckets and then combining them with animal and plant wastes in piles and trenches where they decompose until they're needed to fertilize the land. Many communities and individual homes now have more sophisticated setups whereby human excreta is composted for months right in the holes in which it was originally deposited. These composting pit privies, as they're called, are particularly interesting to us because they are the precursors to the composting privies being designed and built in North America and about which we'll talk later.*

The composting pit privy is nothing more than a large hole in the ground, covered with a slab or squat plate and perhaps enclosed in a small shelter. Excreta and other organic matter are added in approximately a 1:5 ratio, and when the pit is almost full, it's leveled with organic matter and topped with earth. The slab or squat plate and shelter are moved to a new location and a second pit is then used. When the second pit is full, it's covered in the same manner as the first. The first pit is then uncovered, the decomposed matter removed, and the pit reused. Two pits might also be dug side by side and share the same shelter so that it is not necessary to move the privy housing back and forth over the separate pits.

Safety Precautions

When built and maintained properly, such privies serve their purpose well enough. They get raw human wastes out of the way and enable them to decompose, at least partially, so that they're safer and easier to handle. However, odors and flies

(continued on page 54)

* If you want to read more about composting pit privies in Asia and Third World countries, see the following good sources: Edmond G. Wagner and J. N. Lanoix, *Excreta Disposal for Rural Areas and Small Communities* (Geneva: World Health Organization, 1958); Uno Winblad, *Compost Latrines, A Review of Existing Systems,* Alternative Waste Disposal Methods, P.O. Box 1588 Dar Es Salaam, Tanzania, 1975; Minimum Cost Housing Group, *Stop the Five Gallon Flush* (Montreal: McGill University, 1976).

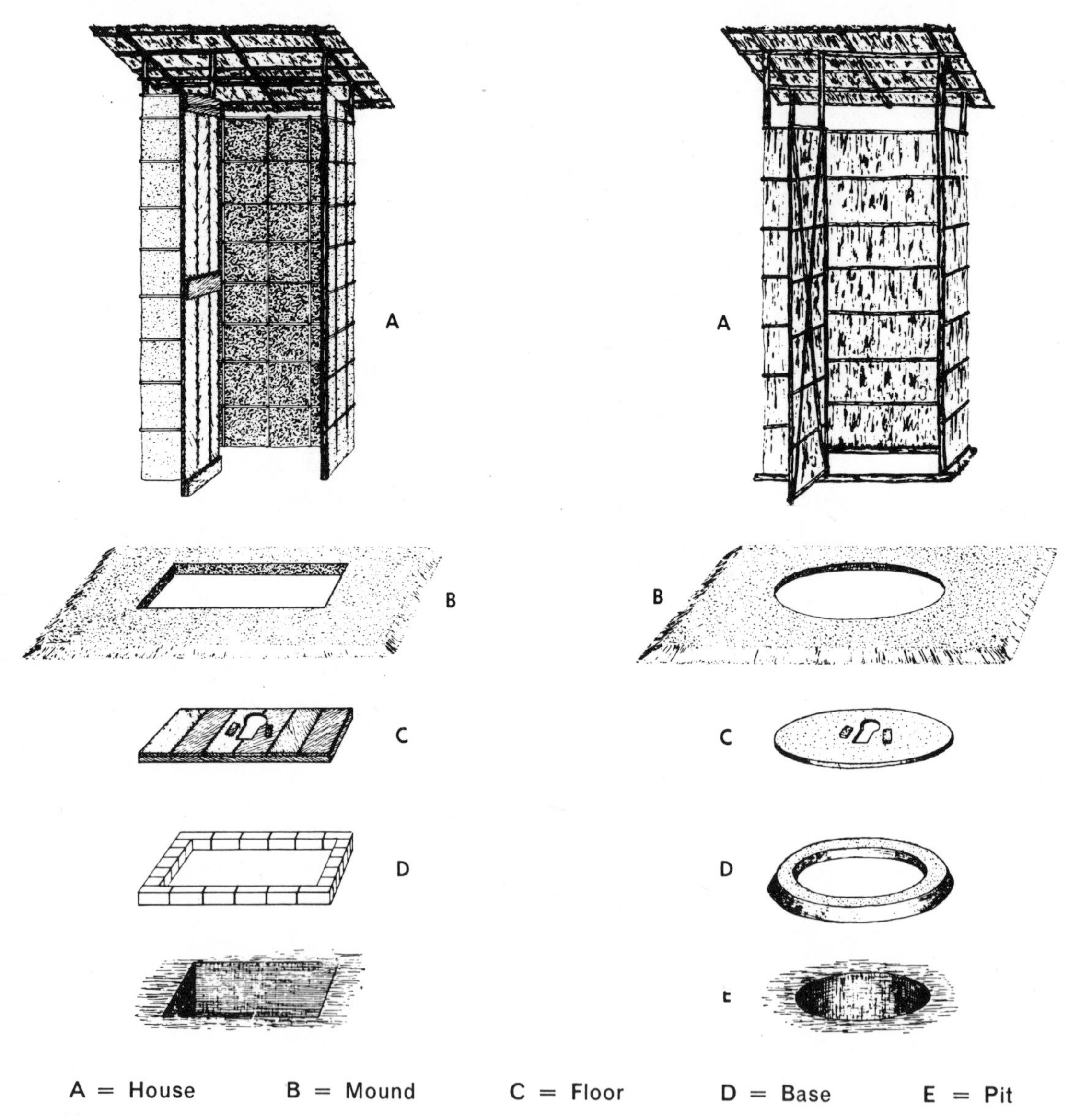

The principal parts of two types of sanitary privies.
(From Wagner and Lanoix, *Excreta Disposal for Rural Areas and Small Communities,* p. 44.)

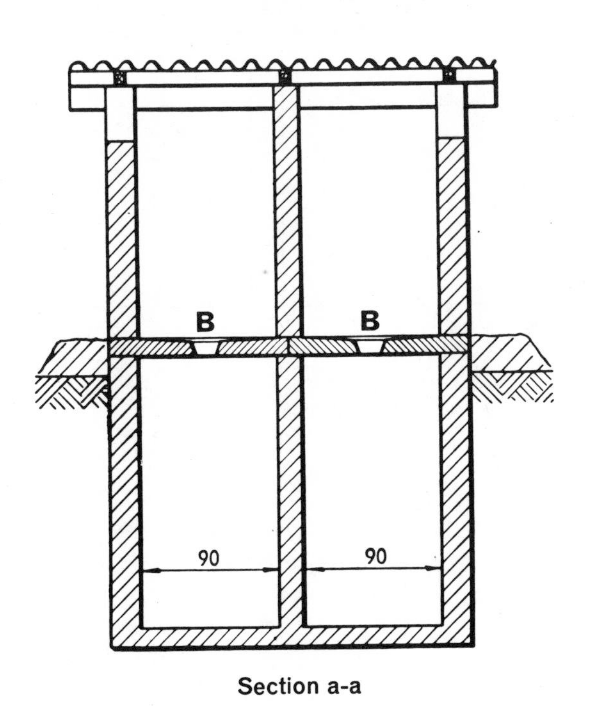

Section a-a

Measurements shown are in centimetres

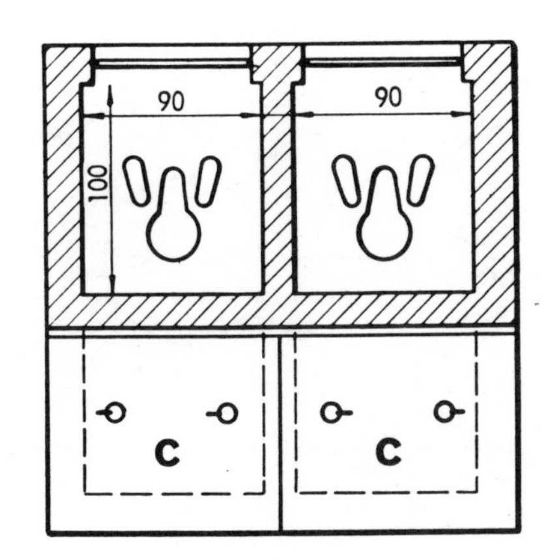

Section b-b

A = Two vaults

B = Squatting slabs

C = Removable covers

D = Step and earth mound

Double-vault latrine.
(From Wagner and Lanoix, *Excreta Disposal for Rural Areas and Small Communities,* p. 118.)

are not uncommon, and health problems associated with these wastes, like cholera, typhoid, dysentery, diarrhea, and hookworm diseases, are still quite possible. Organizations like the World Health Organization spend a great deal of their time, energy, and money on improving sanitary setups and teaching people using such privies necessary safety precautions.

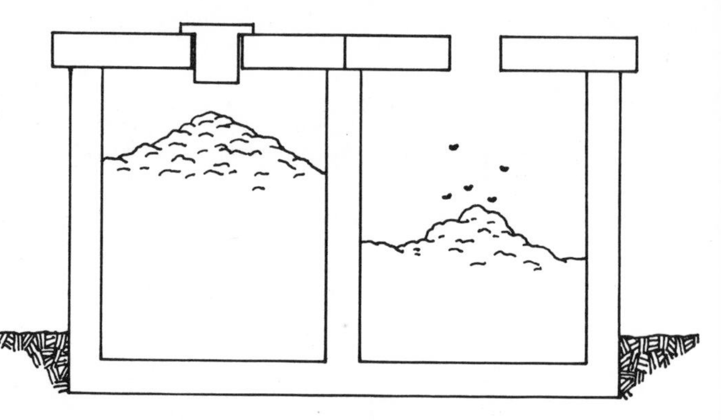

North Vietnamese composting privy.

Any acceptable method of excreta disposal must provide a barrier between raw excreta and possible means for the transmission of disease. Disease from the feces of infected persons can be carried to new hosts through contact with soils, water, animals, or hands.

Fecal matter may directly pollute drinking water. Cholera epidemics in the Orient have been traced to the use of raw human manure as fertilizer, washing into drinking supplies. Leaf and root vege-

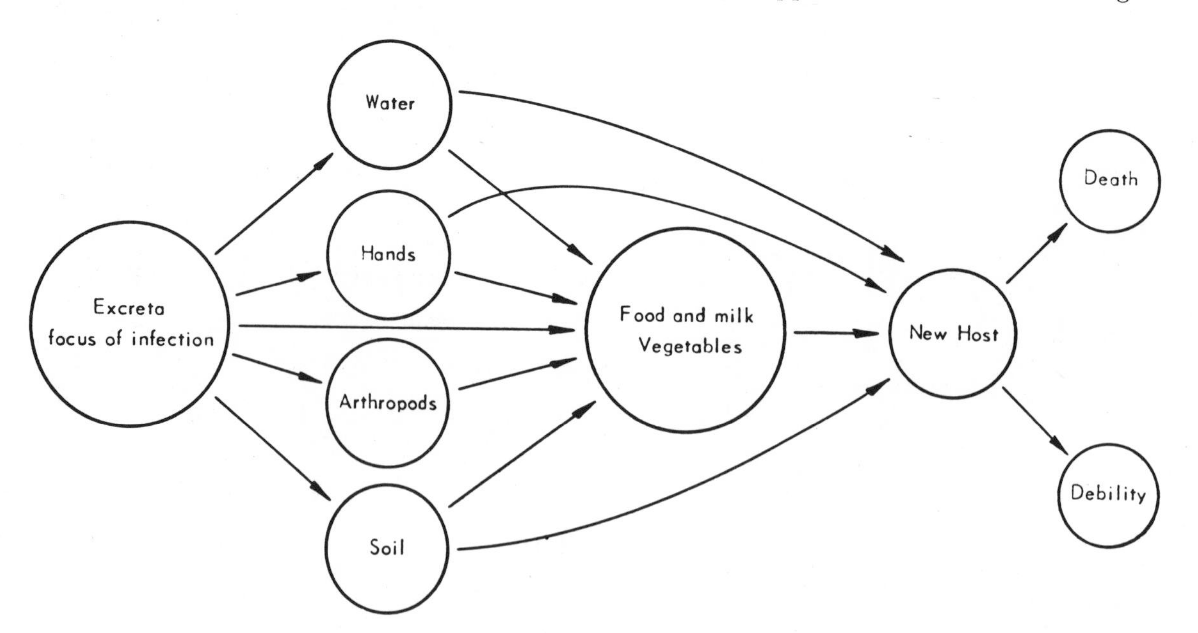

Transmission of disease from excreta.
(From Wagner and Lanoix, *Excreta Disposal for Rural Areas and Small Communities,* p. 12.)

tables, grown in infected soil, can transmit disease. Insects and rodents that come in contact with infected material transmit disease by contaminating food stuffs. Unwashed hands that have been in contact with infected soil, water, or feces are a common carrier.

It's generally agreed that the composting privy designs that follow are superior to the composting pit privies we've just mentioned. Since they are watertight, pollution of ground water is not a problem. There's also little chance that water will seep up from the ground into the wastes and interfere with the composting process. Ventilation is supplied, and turning or rotating the wastes is possible, which means that aerobic decomposition should be taking place in several locations in the pile, and this is superior to the slower and less complete anaerobic decomposition that goes on inside the pit privies. Also, the privy designs that follow make removal of decomposed wastes easier because they have access doors.

But we don't want to mislead you: no matter how much of an improvement they are over composting pit privies, they're

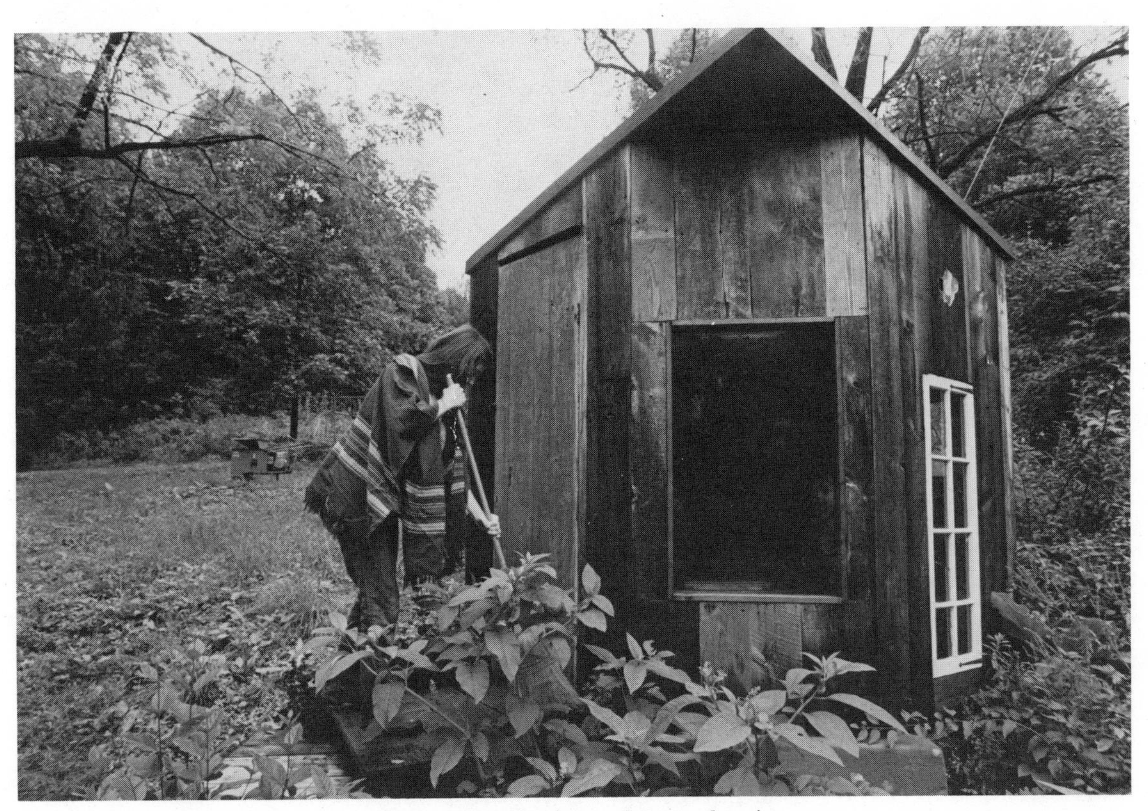

Homemade privy in rural Pennsylvania.
(Photo by Bob Griffith.)

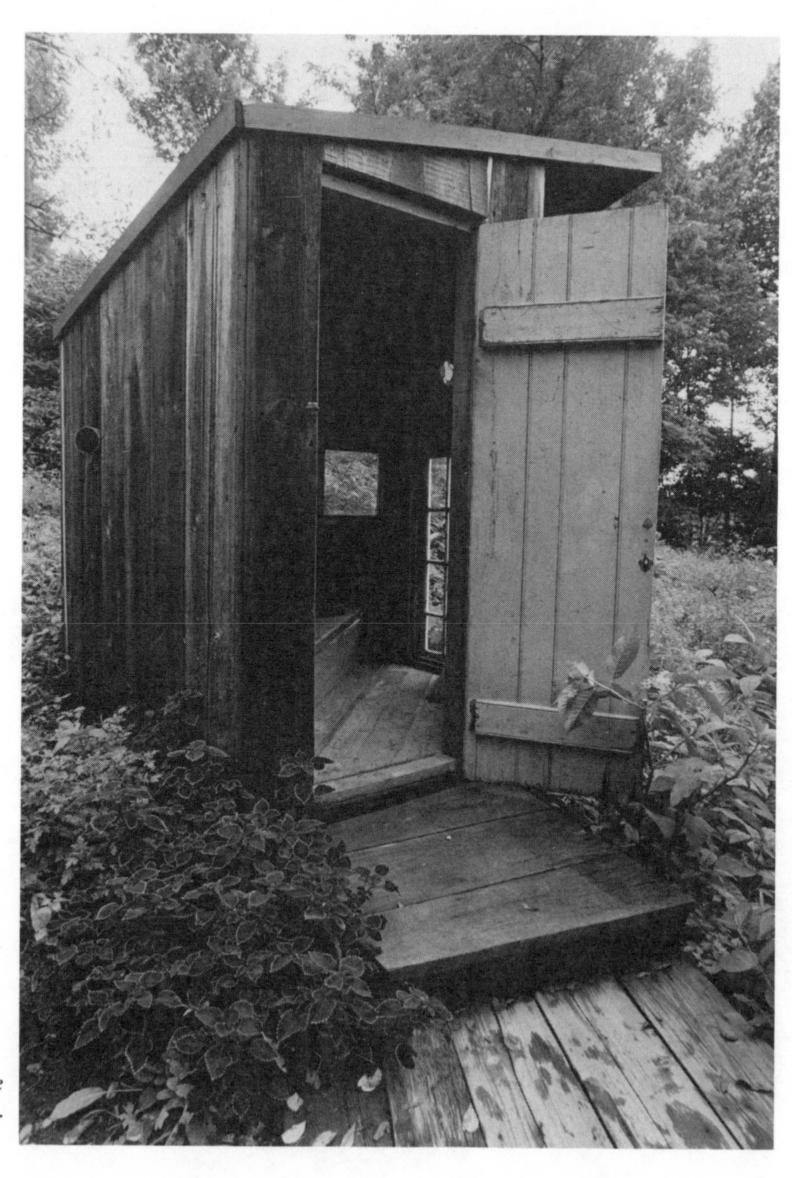

More than just an outhouse—a place to be alone and take in a bit of nature. (Photo by Bob Griffith.)

still a potentially great health hazard. While Asia has a long tradition of recycling human excreta, we do not. As Joel Tarr explained in chapter 1, outhouses and night soil collectors had but a brief glory in most parts of America. Handling human wastes is nothing to take lightly, and we have a lot to learn.

Composting Privy Basics

If you aren't afraid of work and aren't put off by coming into pretty close contact with your own feces and urine, a composting privy may be your answer to an alternative to a flush toilet and septic tank. That is, if you can convince yourself and your local health department that you can build a unit that is fail-safe. Sanitation and public health experts have developed criteria and designs for a number of types of acceptable low-cost composting privies that are in use throughout the world. Keep the following basics in mind should you decide to design and build your own system:

1. *Excreta cannot come into contact with surface soil, surface water, or ground water.* In the privies that follow, feces, urine, and organic matter go directly into impervious containers such as 55-gallon drums and concrete boxes that are sealed from contact with the ground. No water enters the systems (greywater must be handled separately) and the liquids in the pile escape as vapor through the vent or are oxidized by microorganisms.

2. *Excreta cannot be accessible to insects, animals, or children.* Insect screening at the vents prevents flies from entering. If the privy is freestanding from the house, it should be provided with a screen door. The main insect problem is flies, and they can pass through a ⅛-inch crack. Flies are attracted by smell and seek light. Sprinkling sawdust on fresh material, and of course, keeping the lid or cover down when the privy is not in use will help prevent a fly nuisance.

3. *There should be no noticeable odor or unsightly conditions.* There will be no odor if the design and operating instructions for each privy are followed carefully. Make sure the cover is tight and the vent is unobstructed. If odor becomes noticeable it is due to one or more of the following reasons:

- Pile is too small or of wrong proportions, unable to maintain hot temperatures.

- Too wet; add more dry sawdust or peat moss and turn and mix the pile.

- Too high nitrogen; add more sawdust or peat moss or other carbonaceous material—too much nitrogen smells like ammonia.

- Not enough oxygen; turn the pile—anaerobic process will smell like rotten eggs.

4. *Compost technique should be simple. Material should not be handled directly. Maintenance should be minimal.*

In the drum composting privy the wastes are turned as the whole drum is mixed and aerated as it is rolled along the ground.

In the other privies, the wastes should be turned with a pitchfork often, about every other week. They should be turned so that the outer materials are moved inside the pile, and the materials in the center are moved to the outside. In these privies, the entire pile is shifted to an adjacent storage compartment and removed six months later. The pitchfork used to turn the pile should be stored inside the compartment and reserved solely for this job.

5. *Construction must be durable.* The privy container, be it a steel drum or a concrete or cement box, should be impervious to weather, bacterial action, and other conditions.

6. *Finished material must be free from pathogens and safe to use as a soil amendment.* Laboratory and field experiments confirm that pathogens cannot survive the normally high temperatures of aerobic composting, nor do they survive very long in material that is allowed to age.* Proper composting and lengthy exposure to the elements are the cornerstones to purification. Beyond this, only sterilizing all finished material with heat to kill all microorganisms, good and bad, can guarantee complete safety.

Here now are some of the best composting privies that have been built and are being used in North America. Rather than try to paraphrase them, I'll let the designer-builders themselves describe their privies.

Ken Kern's Composting Privy

Some years ago I first read of the use of human excrement as soil fertilizer in F. H. King's 1909 Asian travelogue, *Farmers of Forty Centuries* (Rodale Press edition, 1972). King reported that excreta used as fertilizer made it possible for the

---

Ken Kern is a designer, mason, surveyor, and author of The Owner-Built Home *and* The Owner-Built Homestead *(both available from Charles Scribner's Sons, New York, or from the author at Box 550, Oakhurst, CA 93644).*

* Westerberg and Wiley (*Applied Microbiology*, December 1969) inoculated sewage sludge in an aerobic composter with polio virus, salmonella, ascaris eggs, and *Candida albicans.* The temperature of 116° to 130°F maintained for three days killed all indicator pathogens. Gotaas (*Composting*) confirms similar experimental results, indicating that few organisms are able to survive tem-

Handling and Composting Human Wastes

Composted human manure can be safely used if the following precautions are taken:

1. Its use should be confined to fertilizing the soil in orchards and to the culture of upright berry bushes (e.g., raspberries), of corn and other plants, the edible portion of which is *well above* the surface of the soil. (Remember rainfall and other watering may splash soil particles on the base of plants as high as a foot above ground level.)

2. Any fruit, berry, or vegetable which may accidentally touch the ground should be cooked. The reason for the strict precaution is that the minute size of these organisms permits them to become lodged in the numerous pores, the waxy coating, or the corky outer layers of vegetables and fruits in such a manner that even repeated washing fails to dislodge the organisms.

3. Tilling the soil should be separated from harvesting the fruits by a thorough washing of the hands.

4. Human manure should be used only after it has been composted and thoroughly aged. As Bob Rodale wrote in February 1972 *Organic Gardening and Farming:* "We recommend turning the pile at least three times in the first few months, and then once every three months thereafter for a year. Then wait until fall to apply the compost to land you'll plant in the spring." It is well to avoid the use of this compost on ground from which root crops will be eaten raw. If you farm, you can use privy compost on pasture, allowing some time for the material to settle into the soil before grazing.

5. Do not rely upon an isolated bacteriological examination. It takes but a single contribution by a carrier to contaminate the feces being collected.

As far as the actual conduct of an operation involving composting human manure is concerned, these precautions should be observed: Use an "approved" compost process, one that heats the pile sufficiently. Be extremely meticulous about personal hygiene before, during, and after setting up or working the compost pile. If possible, a set of clothing should be reserved for use when working the pile. The shovel (or fork) used to manipulate the manure should be reserved solely for that job. It is *extremely* important that flies be excluded from the operation. Remember that flies are important vectors (carriers) of disease organisms.

Until the compost process has been completed—at least six to eight weeks—the pile should be sealed from the environment. It should be protected from rainfall, and no seepage should be allowed to escape. This precaution is especially important if the operation takes place on a hillside. It has been

(*continued*)

peratures of 120°F for more than an hour. He suggests that natural "biological antagonisms" in the pile negatively affect the survival of pathogens.

Other evidence indicates that simple aging kills pathogens. Rodale reports (*Organic Gardening and Farming,* February 1972, p. 45) experiments by Bernard Kenner of the Environmental Protection Agency. Raw sewage inoculated with salmonella was applied directly to the soil. Indicator pathogens survived a maximum of 21 weeks.

demonstrated that runoff water can carry salmonella and other pathogens for distances as much as 1,500 to 1,800 feet. Coming under the category of nuisance rather than health hazard is the matter of odor.

If a privy is vented the vent can be a channel not only for venting air, but also of odors. The important point here is that while a well-motivated individual may become inured to objectionable odors arising from his operation, his neighbors may not. Odors may be kept down, if they arise, by covering the excrement with dirt, peat moss, or wood ashes after each use.

As a closing note, I would like to refer to Bob Rodale's recent account of his observations on agriculture in China. He noted that human manure is used extensively in Chinese agriculture. However, he also observed a quite widespread incidence of intestinal diseases, of which the various forms of dysentery were the most common. Proper composting and lengthy exposure to the elements are the two cornerstones of purification.

Clarence Golueke, Research Biologist,
Sanitary Engineering Research Laboratory,
University of California

Reprinted from *Organic Gardening and Farming,* December 1973, p. 96.

Chinese to maintain a large population of animals and humans on small acreages. In one Chinese province visited by King, he found a family of 12, a donkey, a cow, and two pigs subsisting on a two-acre parcel of excreta-fertilized, cultivated land. With this degree of husbandry we might therefore expect to support as many as 3,072 people, 256 cows, 256 donkeys, and 512 pigs on a square mile of similarly tilled land. King considered this Chinese practice of land use nothing less than an art.

But the art of reusing human waste largely escaped me until the summer of '61 when a house design project for a client provided me with the challenge. At this time I was unfamiliar with alternative methods for excreta disposal—except that the word "privy" held for me, as it did for others, the stereotyped connotation of being inconvenient, smelly, and cramped. My house-building client, however, insisted that I draw into his plans a composting privy to be attached to his dwelling.

As might be expected, some structural and functional mistakes and miscalculations were made in the design of my first composting privy. Yet these errors have been corrected in ensuing years, while a few of the basics employed in this initial attempt have since proven satisfactory and remain unchanged. For instance, where my first privy was merely attached to a house, my current design is integral to the structures I build. In fact, as the accompanying drawing illustrates, my latest privy design is very much an architectural factor as well as feature of the whole building complex.

When I first began building this type of privy the structural material I selected was reinforced concrete. This has proven entirely satisfactory. A center pipe is used

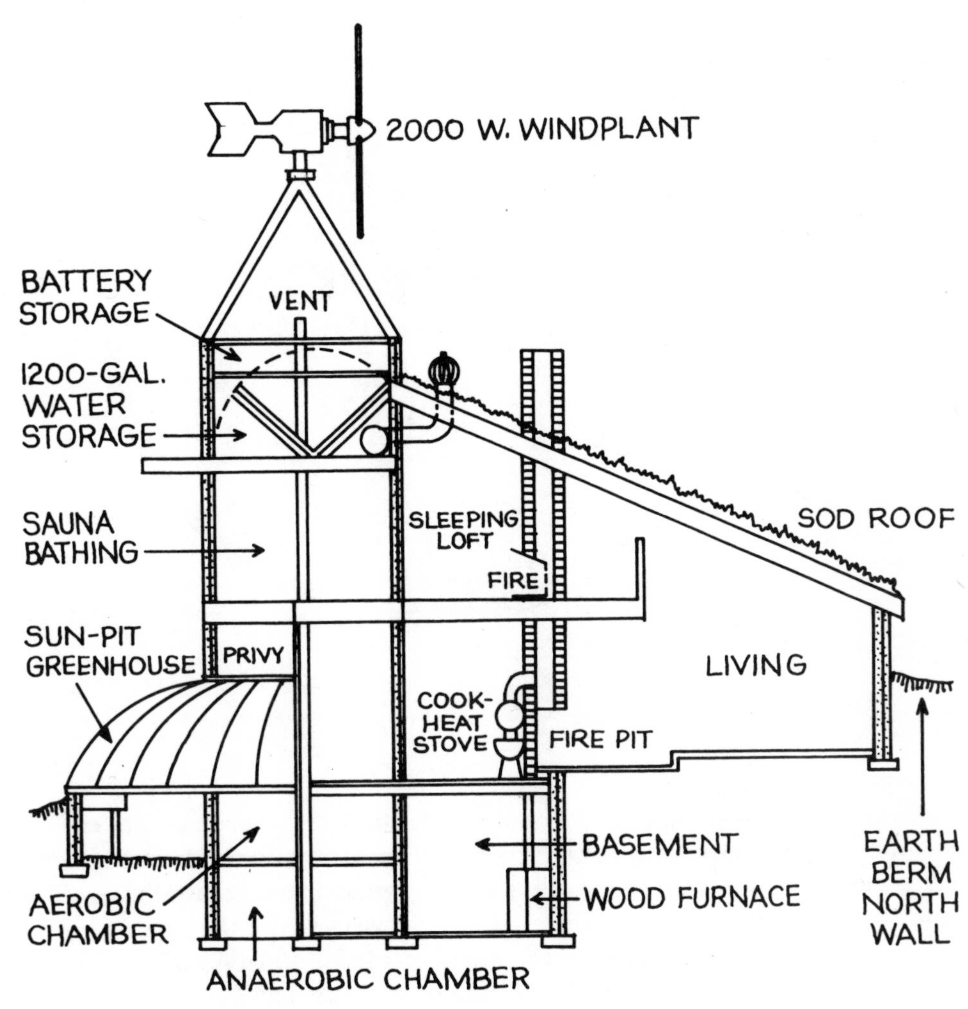

Side elevation of Kern's latest house design.

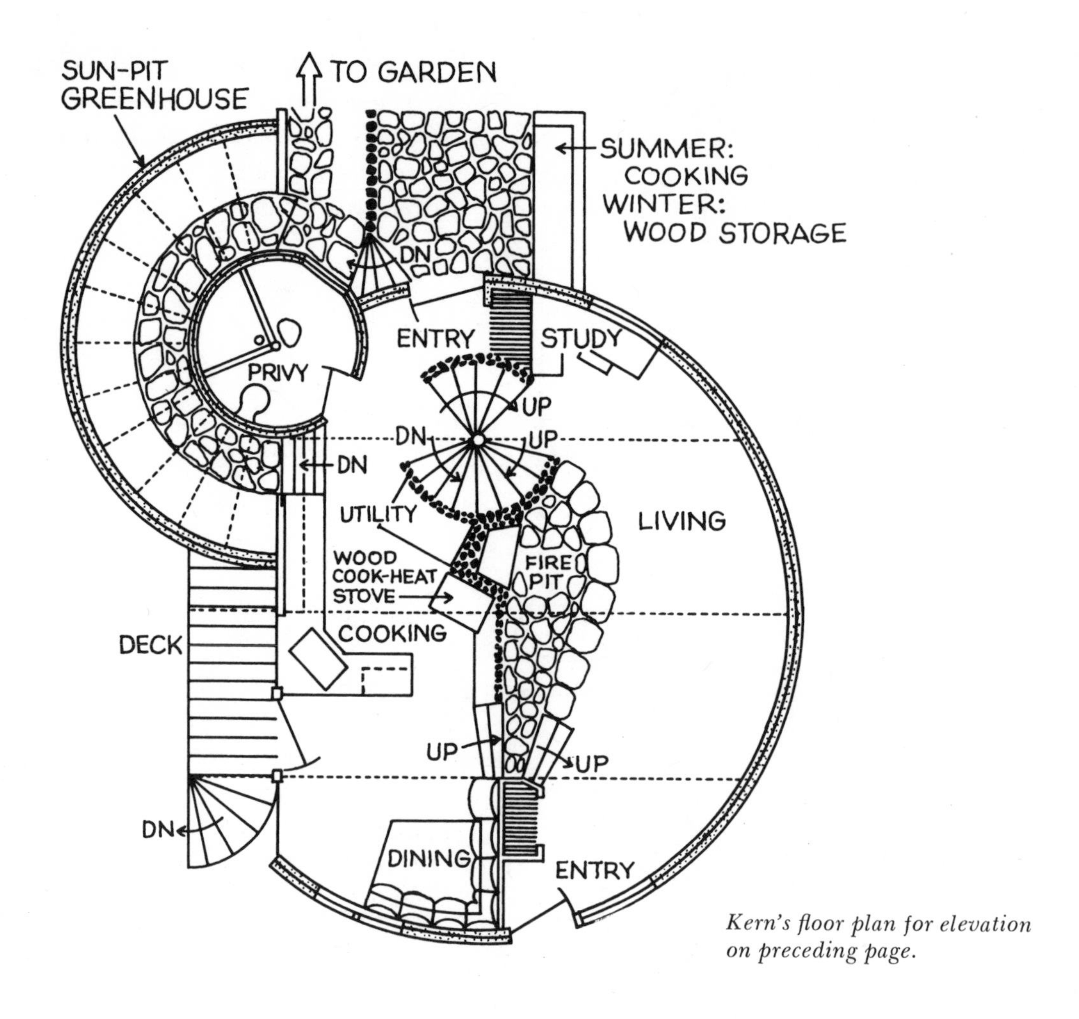

Kern's floor plan for elevation on preceding page.

as the forming guide and later becomes a member of primary structural support when the walls are completed. This pipe functions thereafter as the central drain and as air vent for the composting chamber. All fixtures such as water lines, the wash basin, the shower head, and the device for flushing the squat toilet hang from this central pipe. Privy builders thus find it practical to centrally locate all pipes and wires where they are accessible, apart from exterior load-bearing walls.

A baffle-controlled, two-sectioned receiving chamber is another pertinent de-

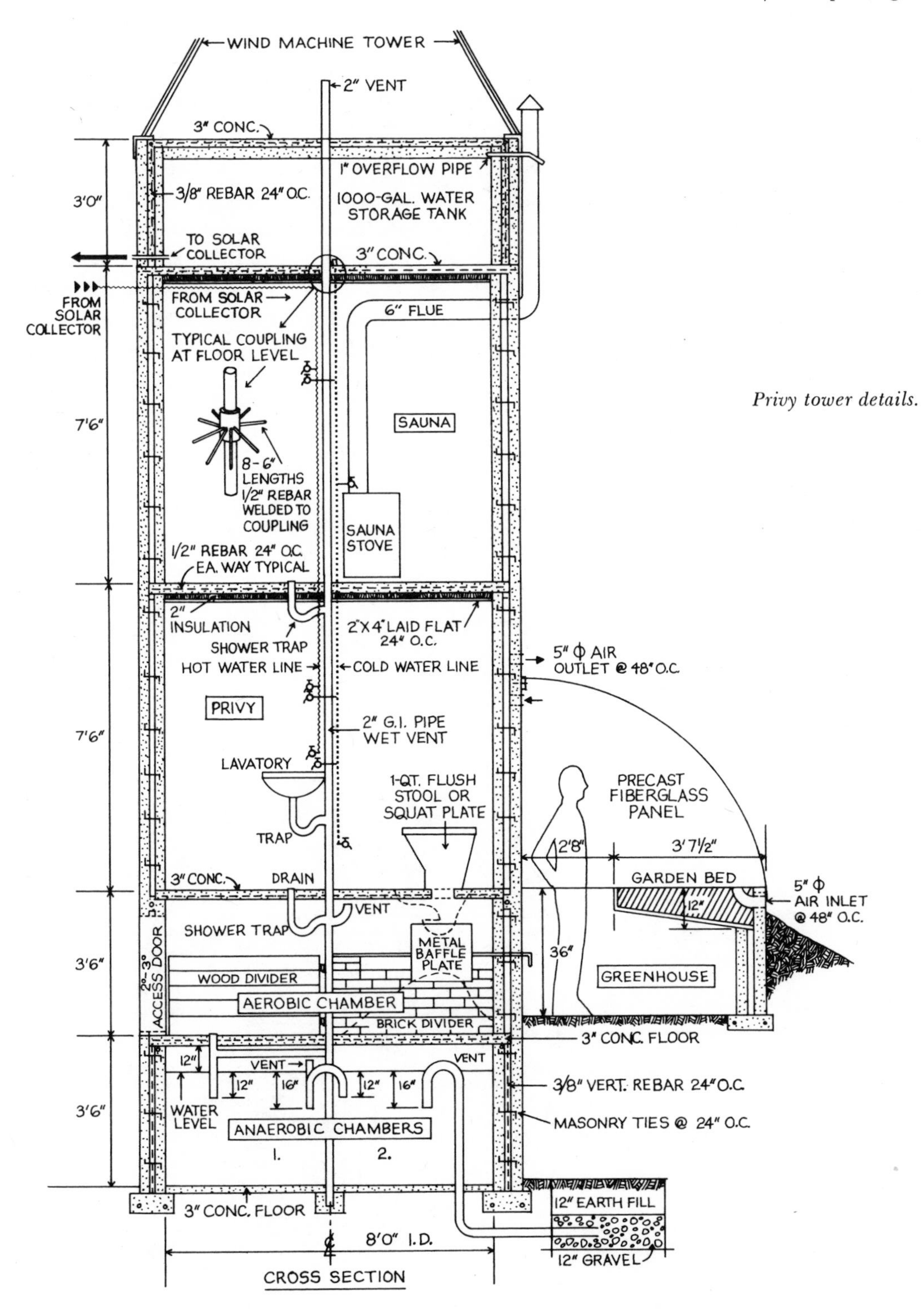

Privy tower details.

sign feature worked out for my more recent privy constructions. Every six months this baffle plate is turned, diverting feces into an adjacent chamber. During the six months that the second chamber receives fresh donations, material in the first chamber ripens.

The fixtures I have chosen for installation in my composting privy require minimal water for their use. It is possible, for instance, to save many gallons of water by discreet showering rather than tub bathing. For that matter, how can one expect to cleanse the skin in water that becomes progressively soiled? Soaking cleansed bodies in a hot tub is, however, therapeutic, esthetic, and indispensable, and a wooden barrel is provided for this type of occasional communal bathing in the sauna included in this design.

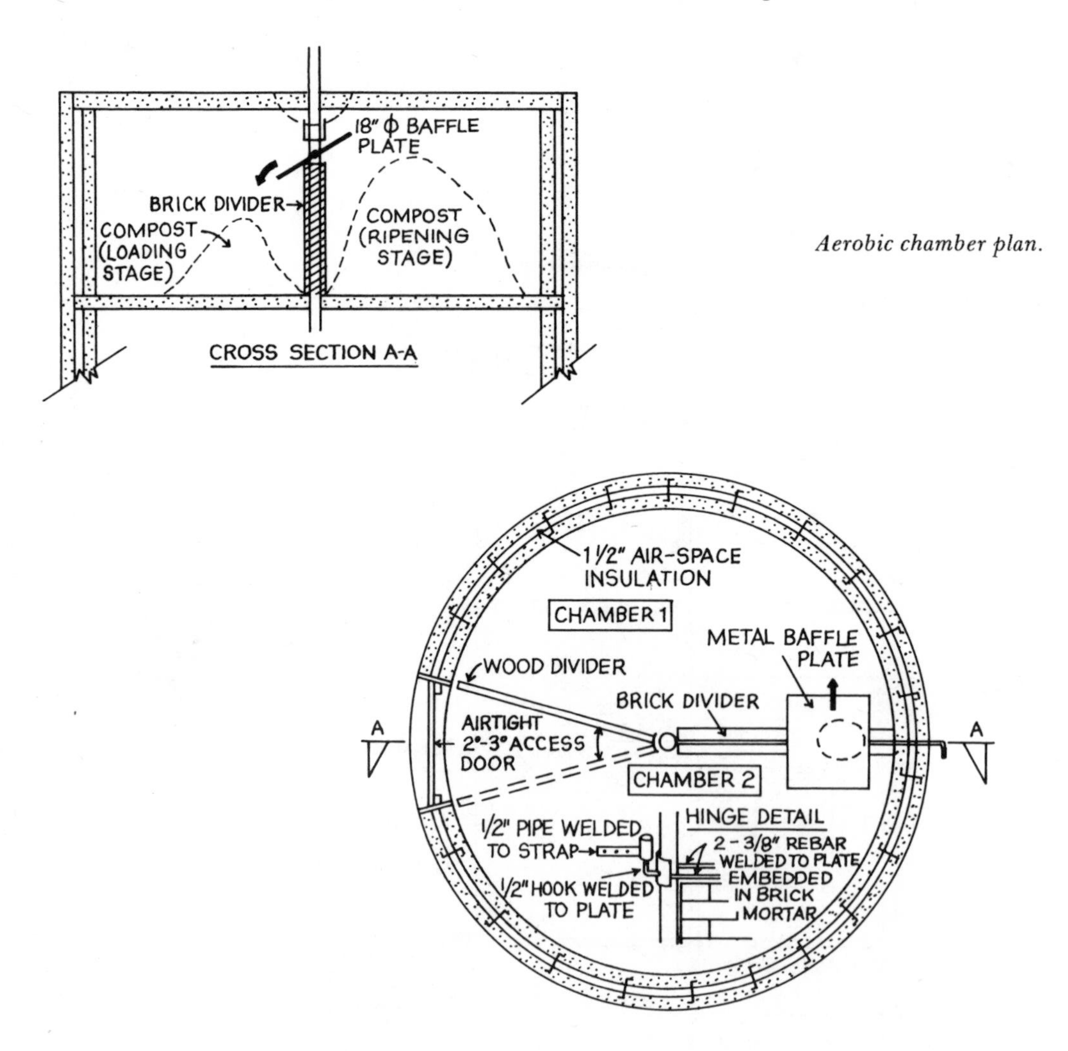

Aerobic chamber plan.

Much water can also be conserved by replacing the standard toilet with a fixture using a minimal quantity for flushing excreta into the composting chamber. Flushing customary large quantities of water over decomposing feces in this chamber would be destructive of composting action in the pile.

A number of one-quart flush toilets are currently marketed, any one of which is satisfactory for use in my privy design. I prefer, however, to use a simple squat plate fixture—although it took a while to locate one suitable for use in my earlier privies. None were manufactured in the United States at that time, and those plates produced in France and Italy were expensive and weighty to ship. At one time I tried to fabricate a privy plate of my own construction, following plans provided by WHO and using cement as the material for construction. The resulting heavy product tended to be unmanageable during installation and defied finishing smooth enough for cleansing in use.

Next I secured an aluminum mold from the Ministry of Health in Thailand. Using this mold I contracted with a local plastics manufacturer to produce a gel-finished, lightweight fiberglass squat plate. The finish on this item is entirely cleanable and these plates are readily portable. They are distributed at cost for $20, including postage and handling.

Those who use these plates must, of course, assume the squat position to evacuate. When this position is taken it favors immediate, uninhibited release of the contents of the body's excretory organs, thereby completely emptying the bladder and the lower bowel tract. Another bodily response to this positioning is that the quantity of each evacuation decreases while periodicity increases, relieving the tendency for chronic constipation and bladder infection so common to North Americans. Anthropologists studying primitive societies have found that their people invariably assumed the squat position to eliminate. Few doctors today will refute the fact that this habit of squatting to evacuate is preferable to the modern habit of trying to eliminate in the constrained position imposed by the ordinary, chair-height toilet stool.

For those who cannot, however, easily attain and hold this position, as is often the case for many of the elderly, for the infirm, or for pregnant women, a grab bar is mounted to my privy's central utility pipe, within easy reach of the squat plate. By using this bar most people can lower themselves onto (and raise themselves from) a portable toilet seat—of median height on a sturdy aluminum frame—which can be placed directly over the squat plate.

By trial and error I learned some lessons for composting excrement in my years of experimentation with the use of this privy design. The most urgent problem encountered was the buildup of liquid waste in and about the composting mass in the receiving chamber. Installing a toilet fixture which flushes with minimal water only partly solved this matter. When one quart of water is used to flush waste material into the compost chamber each time it is used, an excess collects below and must be drained beyond the confines of this compartment. I first tried a stone-filled sump, but this device ceases to function and becomes a health hazard in the rainy season, particularly in areas of the country which have a high water table. Also, where inspection of building construction is requisite, sanitarians will not accept any handling of even the greywater from sink drains without its first undergoing anaerobic decomposition in the below-grade, sealed compartments of the so-called septic tank—which eventually drains into the soil from leach lines or hooks up to an ecologically disastrous public sewer system for emptying. Therefore, to comply with health department requirements for treatment and disposal of all wastewaters, my current privy design includes—in addition to the aerobic receiving chamber for the decomposition of solids—an anaerobic chamber and drain line. This additional construction takes little more effort to build and adds little to the overall cost.

Walls of my earlier privy designs were not adequately insulated, for I have since learned the value of a whole range of tem-

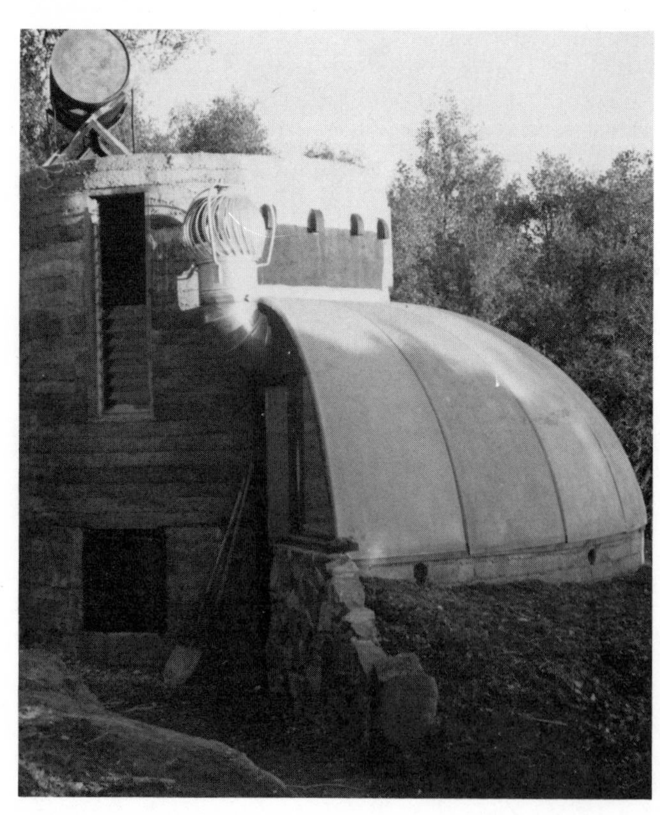

Kern's partially completed privy tower. The dome-shaped structure on the right is the sun-pit greenhouse.
(Photo by Ken Kern.)

peratures which must be activated within a working organic pile. Aerobic bacteria, with an affinity for oxygen for their life processes, live and work in the composting mass, generating degrees of temperature ranging from 90° to 140°F (32° to 60°C) while they consume carbohydrates and proteins. Even as much as 160° to 180°F (72° to 82.3°C) may sometimes result, destroying all resident pathogens in excreta. However, in privies installed in northern climate zones it has been learned that, during winter months when ambient (surrounding) air temperatures are below 50°F (10°C), virtually all ongoing aerobic bacterial activity ceases. No initial decomposition may begin nor may ripening of the mass be completed at or below this temperature.

To conduce to a relatively warm environment in which this activity may evolve, I now build my privies with a double concrete wall containing a dead air space. In cold, northern climates it is advisable to fill this inner cavity with Styrofoam insulation, especially in the area surrounding the compost chamber. In a current design which incorporates a pit greenhouse around a composting privy, the absorbent walls of the compost chamber collect heat produced in the greenhouse during the day, storing it for nighttime reradiation to plants while enhancing the warm environment of the compost chamber.

A particularly wet pile will need aerating or turning to provide aerobic bacteria with sufficient oxygen for their functioning. Urine and feces which are low in carbon require considerably more oxygen for their bacterial digestion than materials such as sawdust or straw, which are high in carbon.

It is therefore necessary to add organic matter, such as kitchen wastes, leaves, and fresh grass, to maintain this working ratio of carbon-to-nitrogen. For this purpose an outside door is provided for access to the receiving chamber where materials high in carbon may be tossed onto the pile. After 40 hours of composting activity, disease-bearing organisms are rendered lifeless and the finished material may be taken—in as short a time as a month—through this exit for use as soil fertilizer.

An important bonus received from the decomposition of organic wastes is found in the emission of that great fertilizer of the air, carbon dioxide. When fermentation is most active, about two-thirds of the carbon content of the mass is first consumed and then given off as carbon dioxide by bacterial organisms. Ordinarily these emissions from the working pile are vented into the atmosphere. At present, however, I am venting this resource directly into my greenhouse—with no detectable undesirable odor—for the benefit of a thriving crop of winter greens. I feel the use of this resource far outweighs the use of any methane gas generated from anaerobic decomposition for fairly limited use as a source of domestic lighting or refrigeration.

My relatively brief experience in building and managing composting privies has been drawn from both the science and the art of excreta composting in an attempt to make a safe nutrient with which to grow edible plants. Today, as in King's time, this art is largely limited to practice in a few Asian countries, China in particular. The World Health Organization estimates that in China perhaps 90 percent or 300 million tons of excreta—representing about one-third of all the plant fertilizer used on crops in that country—is processed each year with methods similar to those mentioned here. Such husbanding of this vital resource demands practice around the entire globe. Effluent-affluent Western nations must especially learn the art and science of using human wastes before it is too late to maintain a balance between the income and the out-go of organic materials within their respective borders.

The Farallones Two-Hole Composting Privy

The Farallones two-hole composting privy consists of a two-chamber concrete box with four feet by four feet by eight feet outside dimensions. Each chamber has a capacity of one cubic yard, which is ample to hold human wastes and additional organic materials produced by a family of four over a six-month period.

The design includes positive ventilation—air movement through screened vents in the privy access panel and up a flue that runs to the roof. A plywood access panel secured with wing nuts allows easy access to the bin for periodic turning of the pile. Weather stripping is applied to the inside edge of the access panel to make an insect-proof seal. The plywood top is fitted with an opening squat plate over one chamber to receive excrement, household wastes, and additional high-carbon content organic matter. The privy should receive no wastewater other than urine.

Once or twice a month the pile is turned and mixed with a pitchfork and/or flat shovel stored within the compost chamber and used exclusively for this purpose. After six months, the pile is turned to the storage compartment (the bin that doesn't have the squat plate over it) for at least six months of final composting and aging before it is removed for use in the flower garden or orchard. (As an added safety factor it is not recommended that the compost be used directly in the vegetable garden but on fruit trees and shrubs or ornamental plantings.)

The Farallones Institute is a small, independent association of scientists, designers, and technicians who are carrying out research and education programs in appropriate technology both at their Integral Urban House in Berkeley, and at their Rural Center in Occidental, California.

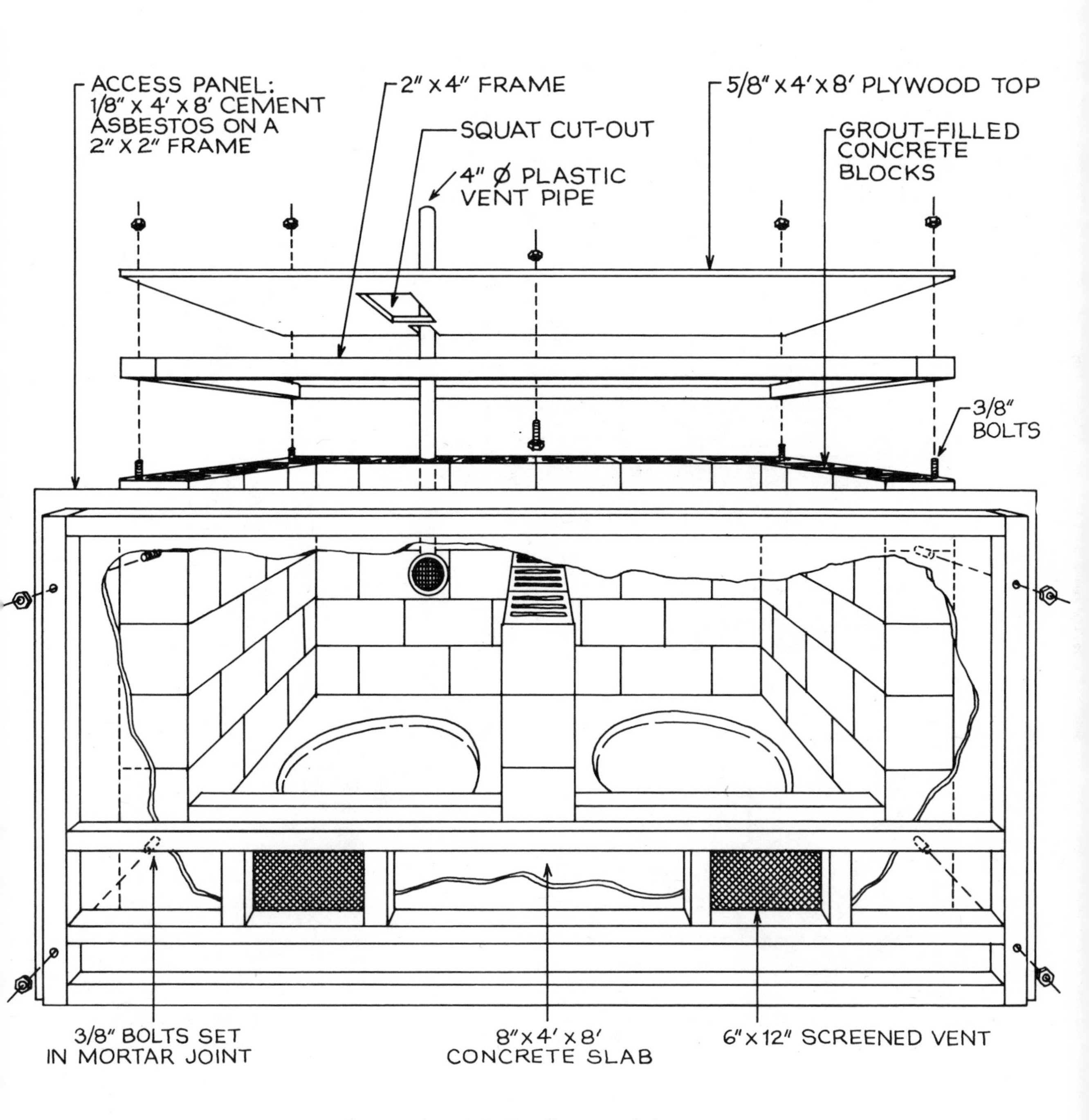

Cross section of the Farallones two-holer.

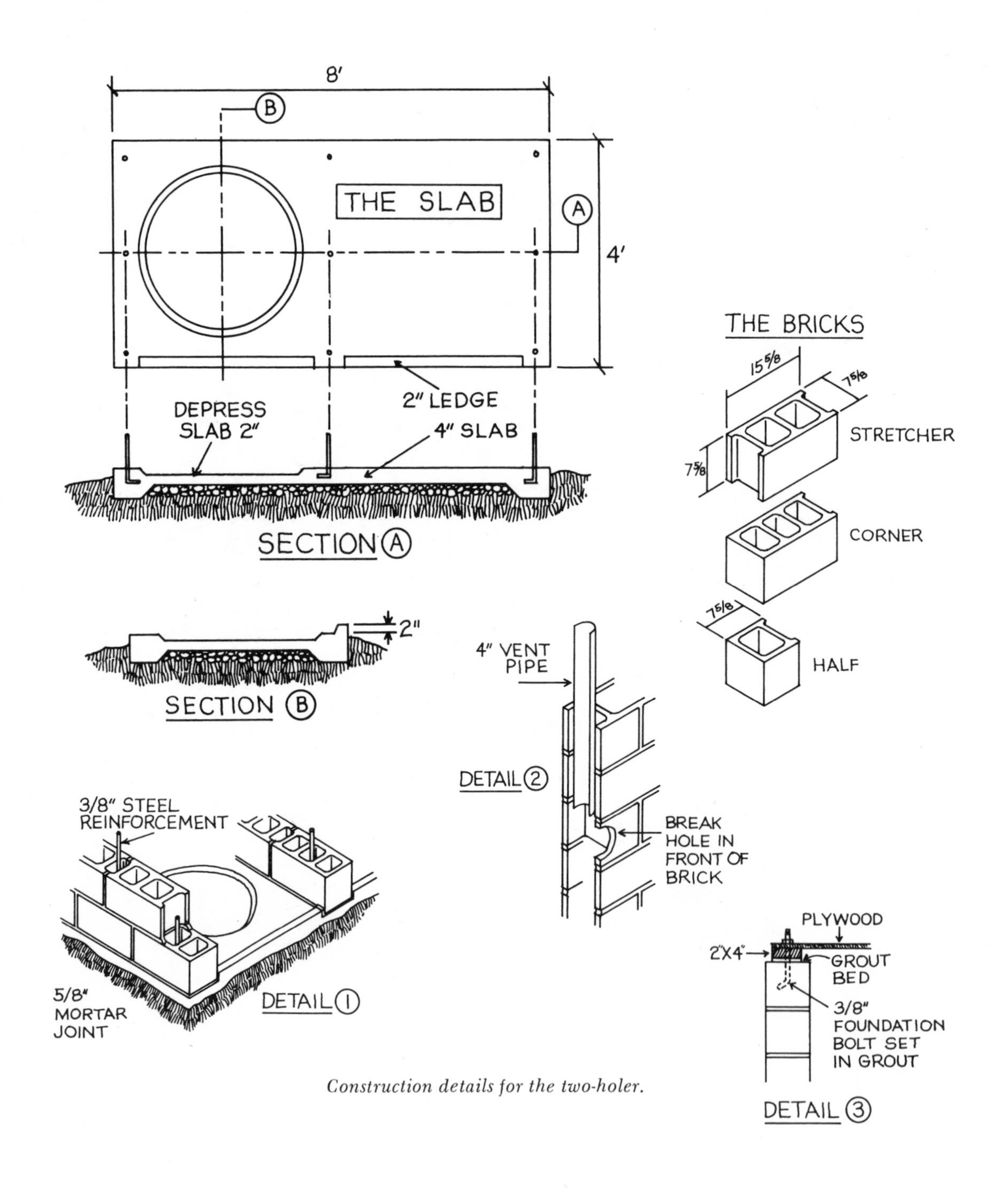

Construction details for the two-holer.

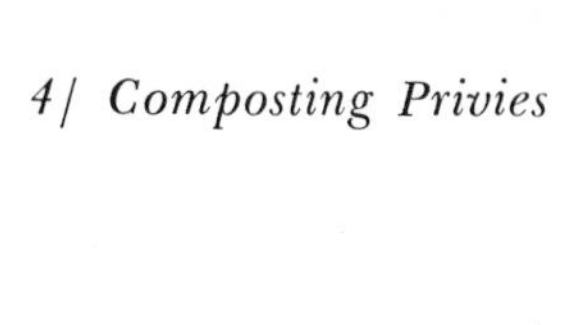

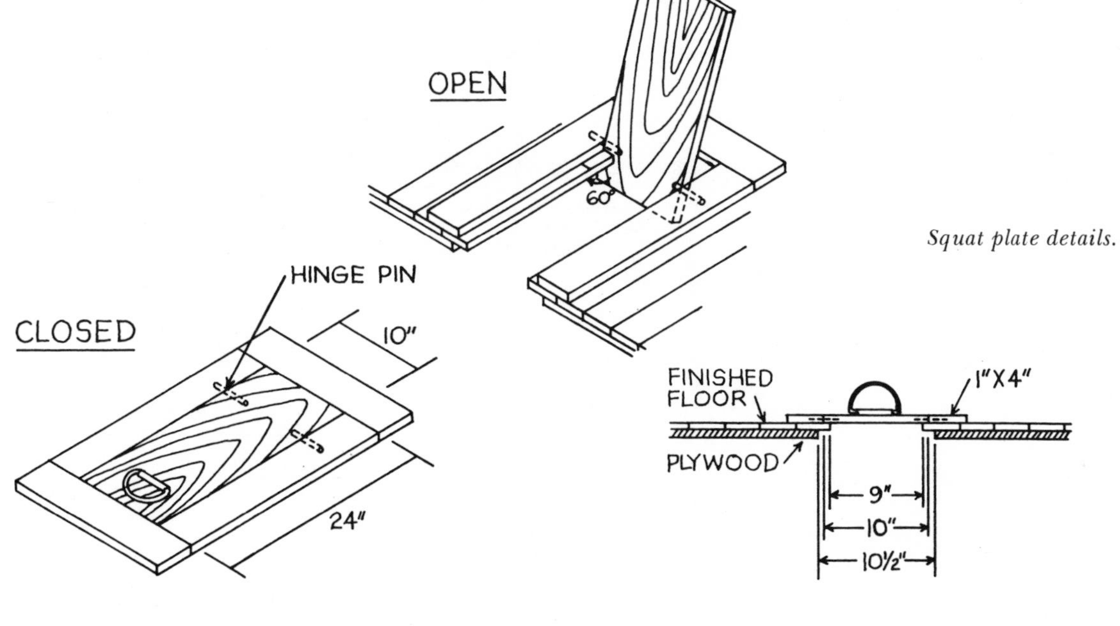

Squat plate details.

ROTATION PROCESS

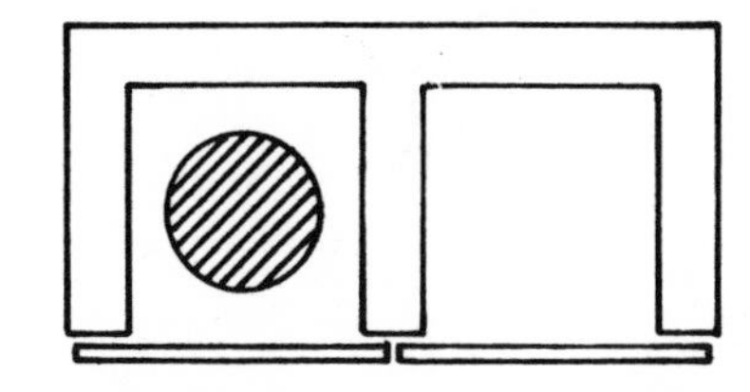

SIX MONTHS OF USE TURNED EVERY OTHER WEEK OR AS NEEDED.

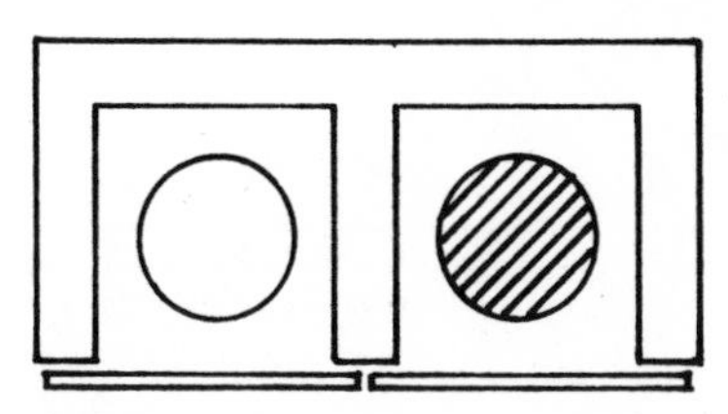

AFTER SIX MONTHS THE PILE IS MOVED TO THE AGING COMPARTMENT.

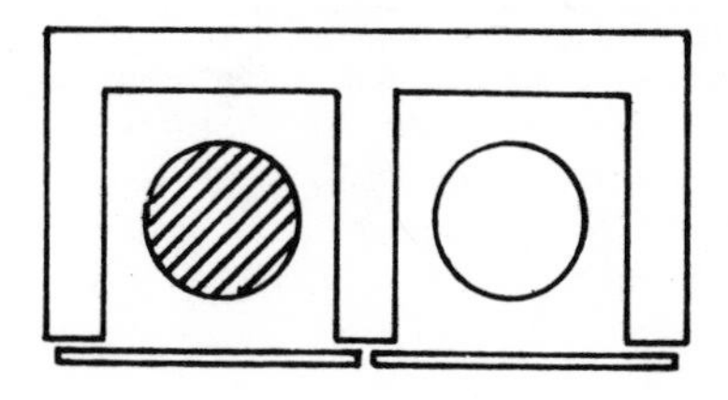

AFTER ONE YEAR THE FIRST PILE IS READY FOR USE IN THE ORCHARD. REPEAT PROCESS.

Recommended rotation process for use with the two-holer.

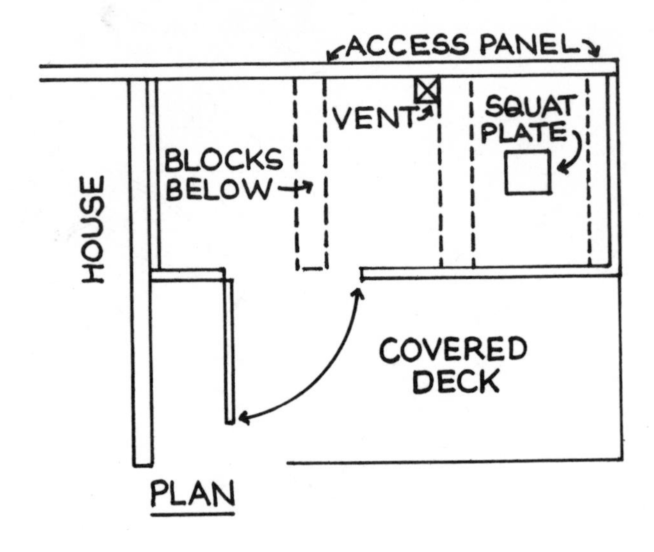

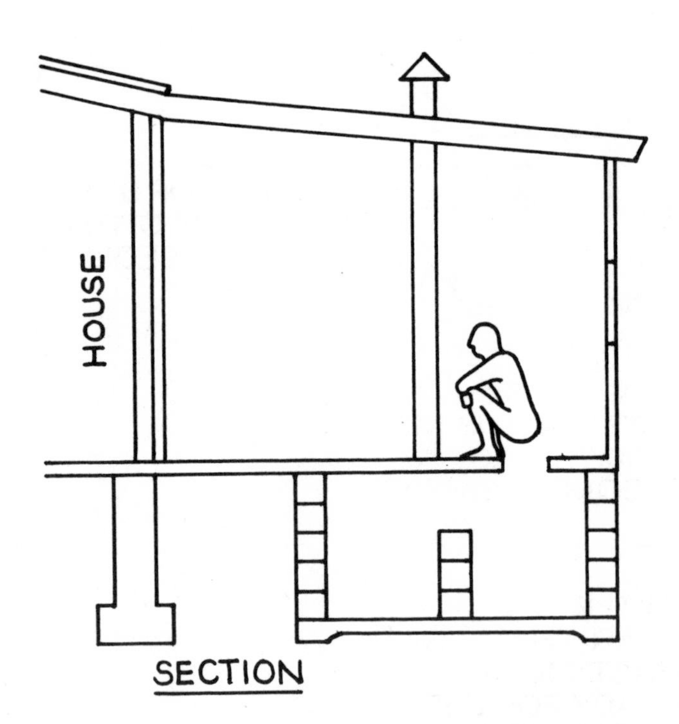

A floor plan and side elevation of the two-holer added onto an existing house. The access panel should be on an outside wall. For a house built on grade, adjust the floor level or dig out to provide access. In a house with a basement, the privy chamber could be built into that existing underground space.

Calculating the Size of Your Privy Compartments

Feces should constitute no more than 20–25 percent of the composting material. Human waste per person per day averages ½ pound feces (moist weight) plus one quart urine. A yearly average equals about 180 pounds feces, 80 gallons urine. At 11 pounds/gallon and 7 gallons/cubic foot, this equals 3 cubic feet feces, 10 cubic feet urine. Decomposition reduces this raw wet volume to 1/20 its original volume, or about *one cubic foot per person per year.* (We use 7,000–10,000 gallons of water per person per year to flush away what naturally reduces to something you could lug around in a five-gallon can!) Government sources say to size a privy at 2 cubic feet/person/year. Figuring a volume of other organic waste five times that of human waste, two 3-foot x 3-foot x 3-foot compartments would serve a family of four for a year.

Calculations based on a six-month use by four persons:

Daily Input

Feces: ½ pound/person/day (0.15 pound dry weight) 6% Nitrogen C/N ratio=7

Urine: 1 quart/person/day (0.15 pound dry solids) 16% Nitrogen C/N ratio=1

Kitchen garbage: 1 pound (dry basis) 2% Nitrogen C/N ratio=25

Sawdust and peat moss: 100 pounds/month 0.15% Nitrogen C/N ratio=400

Quantities for 150 days (six-month period) for four people

Feces:	.15×4 ×150= 90 pounds× 6%=	5.4% *N*× 7=	37.8 pounds *C*
Urine:	.15×4 ×150= 90 pounds× 16%=	14.4 pounds *N*× 1=	14.4 pounds *C*
Garbage:	1.0×150=150 pounds× 2%=	3.0 pounds *N*× 25=	75.0 pounds *C*
Sawdust/ peat moss:	600 pounds×.15%=	0.9 pound *N*×400=	360.0 pounds *C*
Total		23.7 pounds *N*	487.2 pounds *C*

C/N=21 (toilet paper raises this closer to 25)
Total mass @ 50 pounds/cubic foot=1 cubic yard

Farallones Institute

Variations on the Farallones Two-Holer

The composting privy for the Farallones Institute's Rural Center was originally designed to have just one hole and almost as an afterthought a second hole was added and the upper room divided and provided with two doors to serve more users. The vent was formed by the two interior walls. This interior space was lined with building paper for a sealed vent to prevent the entrance of vectors and the

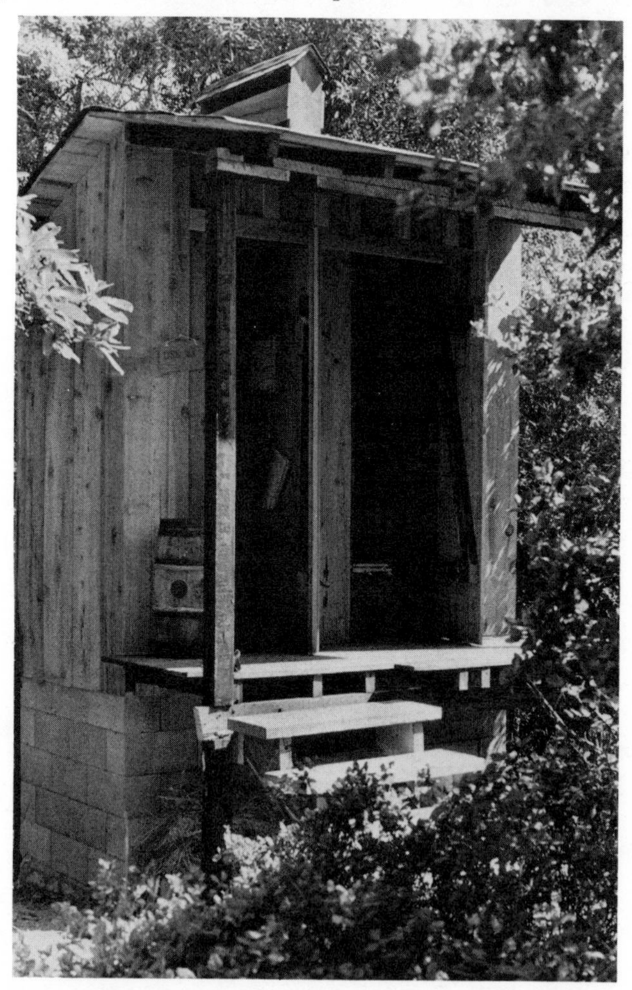

Front view of the modified two-holer at the Farallones Institute's Rural Center. (Photo by Peter Webster.)

Max Kroschel is in charge of the waste management systems at the Farallones Institute's Rural Center in Occidental, California.

escape of odors. The top of the vent protruded above the roof and was screened and capped, forming a cupola.

Little thought was given at that time to composting space for the large quantity of compost that would accumulate in such a frequently used privy. (Its initial use was by 35 to 40 people for three months.) After an initial development period involving several undersized bins with no ventilation, a large composting and storage bin was designed and constructed. Cement stabilized earth blocks for the walls were made in several workshops using a Cinva Ram block press. Plywood and redwood 2-by-4s were used for the lid and doors.

The composting bin is essentially another privy substructure. The lid is hinged to tilt up while loading and mixing. It has a large vent on top to exhaust the gases released in composting. Each compartment is fitted with a door individually, rather than one four-foot by eight-foot door as on the original privy plans. They are lighter

Rear view of the modified two-holer, showing the redwood storage bins off to the left. (Photo by Art Rogers.)

and easier for one person to handle and manage. The center partition between the bins is of wood, reducing the amount of heat-robbing mass and increasing the internal volume.

Because the storage bin was added after initial construction of the privy vault, no provision was made for turning the compost into the storage bin from within the enclosure. The material must be moved from one bin to another with a shovel. To facilitate cleaning up after turning and transferring compost, a smooth concrete slab apron was poured in front of the access doors.

The composting bin retains the compost material for about two months to 10 weeks. All the high temperature compost-

But What Do You Do with the Urine?

The Rodale Research and Development Group recently turned up some interesting facts about fresh human urine that suggest that it can be a valuable garden fertilizer.

How potent is it?

Human urine ranks very high in nitrogen content when compared to all compostable materials.

How much strength will be lost before use?

The real strength of human urine and livestock manures is in their chemical makeup when they are *used,* not when they are *excreted.* When used, human urine may in fact be higher in nitrogen than all livestock manures because urine is:

—more likely to be used soon after excretion,
—would be held in a sealed container, and
—is a liquid.

Thus human urine may decompose less, will not suffer from runoff and leaching by rain, and will not lose nitrogen during the time between spreading and plowing.

How much is produced?

Three to four people excrete just enough urine in six months for a garden fifty feet by fifty feet or a moderate-sized compost heap.

It may actually be better to keep urine and feces separate.

Unlike using animal manures which consist of both feces and urine, human urine should be used alone even though urine combined with feces has more nitrogen, phosphorus, and potassium and is collected more easily since only one toilet is needed.

ing takes place during this time period, and after 10 weeks it is no longer attractive to flies or other pathogen-carrying pests and has no noticeable odor. It is transferred into storage boxes for the final four months of aging.

The storage boxes are made of redwood one-foot by six-foot tongue-and-groove boards with a two-foot by four-foot frame. A tight-fitting plywood lid with a weather-stripping seal completes the pest-proof enclosure. We have six bins to store the aged compost until needed in the winter for planting trees.

Using the Privy

The bottoms of the two collection bins are primed with about three to four inches

---

Why separate them then? For one thing, human urine is a "minimal health risk entity" according to Dr. Clarence Golueke of the Sanitary Engineering Department of the University of California at Berkeley. "That is, the chances of catching anything from it are small," said Dr. Golueke. This is not the case with human feces which can carry bacteria, viruses, and other disease-causing parasites. For another, human urine is easier to apply daily so there is less chance for its nitrogen to decompose.

When and how should urine be applied?

Daily, for best results in the garden and compost heap. This can be done easily with human urine since it has already been collected in a container. If you wish to compost it, pour it directly on the pile. If you want to apply it directly to the garden, mix in water and gypsum and then sprinkle it on. The water dilutes the salt in urine so it won't burn the vegetable crops, and the gypsum neutralizes the urine's high pH. Work at the Farallones Institute has indicated that five parts water to one part urine works satisfactorily according to Sim Van der Ryn of the Office of Appropriate Technology in California.

Can urine from sick people be used safely?

Yes, if it is treated in some way to destroy the potentially harmful microorganisms which may be present.

One way to do this is to boil the urine for at least 15 minutes per 15 pounds at 250°F (121°C) to sterilize it, according to John Mandel of the Lehigh Valley Laboratory. But by doing this, some of the urine's nitrogen is lost into the air.

Dr. Wally Burge, USDA soil scientist, suggests that if a gardener wants to be sure he is always using disease-free urine he must even treat the urine from people who appear to be healthy because their urine may carry disease-causing microorganisms too.

Marc Podems,
Rodale R and D Group

---

of peat moss and four to six inches of straw. Peat moss absorbs the urine, and straw lightens the pile and helps the air to travel through it easily. After each use a one-pound coffee can (about two or three handfuls) of peat moss mixed with leaves is added to cover the fresh feces. To prevent excess moisture buildup in the privies from urine during high-use periods, a separate urinal is provided for men in the form of a simple five-gallon can.

Once a week, early in the morning before any fly activity, the access doors are opened for maintenance. A long-handled pitchfork and flat-bladed shovel stored inside the privy vault are used to turn, mix, and aerate the piles. With two piles accumulating, the most complete mixing oc-

The inside of the modified two-holer is pleasant and clean—a far cry from most people's conception of the inside of an outhouse. Note that the seat is low so that users must squat, which, in the owners' opinions, is a much more healthful position than sitting. (Photo by Susie Benson.)

curs when all the material is shoveled into one bin. The following week it is all turned back to the opposite side. In this manner all the new material is incorporated into the center of the pile each week.

One side of the privy has a toilet seat on a raised box and is used often by women as a urinal. It tends to be much wetter than the other side which has a squat plate. Additional material is added to this wetter side as needed to help absorb the moisture and prevent an anaerobic condition that could cause odors.

At a use rate of 10 to 12 persons daily, the two-holer fills up in about two months, and the material needs to be transferred to the storage bin for composting. Experience has shown us that the leaf-sawdust-urine-and-feces mixture straight from the privy vault doesn't compost well on its own. A new pile must be constructed in the storage bin using the privy mixture as a nitrogen source; straw, grass clippings, garbage, garden wastes, and other readily decomposable carbonaceous materials are added in layers, with the privy mixture of feces which should comprise only one-half to two-thirds of the bulk of the pile. A three-inch-diameter cylinder of wire mesh placed in the center as the pile is constructed helps the compost to breathe.

The composting action generally doesn't start in the collection bins. The bulk is too small and often too wet, so we don't add any of the carbonaceous materials until storage. Garbage attracts fruit flies and can cause odors during collection—another reason for adding it only during the storage composting phase.

A long probe thermometer enables us to read the internal compost temperatures from outside the bin. After two to three weeks when the pile has dropped in temperature to around 104°F (40°C), it is turned again to aerate it and to incorporate the outside masses of the pile to the inside for thorough composting and the resultant heating. Ideally, a third turning after the same amount of time would insure that all material had gone through a hot composting for adequate treatment. Complete mixing while turning is accomplished by having two bins in the composting compartment. The pile is made in one bin. After two or three weeks it is turned into the other bin with care taken to incorporate the outside material to the inside of the pile. The third turning returns it to the first bin. This total process takes about two months to 10 weeks. Then the pile is transferred to a redwood storage box for at least six months before it is used in the orchard.

At this time our county health department will not approve the use of privy compost on the soil surface. Until more extensive testing concludes the absence of human pathogens in the final compost they would prefer that we dispose of our sludge at the dump. But since we can satisfy the requirements for a sanitary landfill (i.e., six to 12 inches of soil cover) here at the Rural Center, they have agreed to our using it in a prescribed manner. We use the compost in planting our new orchard trees, and we can utilize all the privy compost generated in this manner for the next several years. By then we should have positive test results to allow its more liberal use.

Retrofitting a One-Hole Privy with Three Storage Compartments

After building and using the above privy we built another onto an existing structure. The space available was limited, but the design demonstrates the flexibility of the composting privy in application to existing buildings.

The basic construction of the substructure is the same as the one described above: cinder block walls on a dished-out concrete slab. Additional storage and composting compartments are incorporated, and ma-

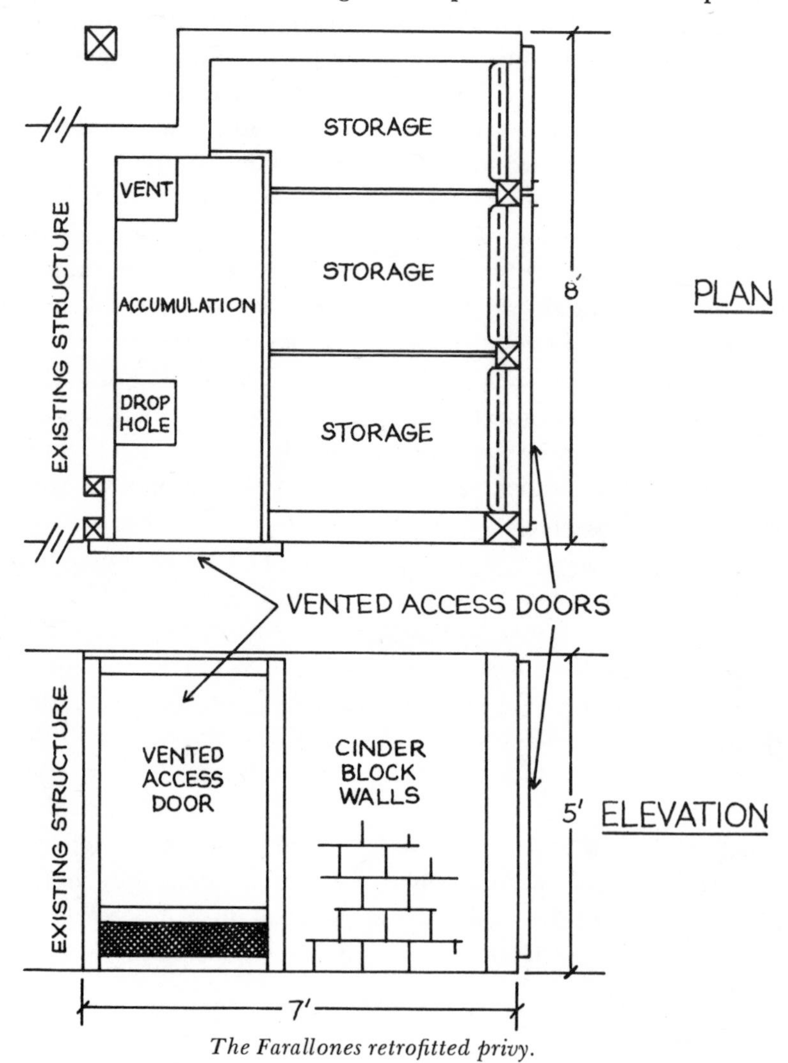

The Farallones retrofitted privy.

terial can be turned from one bin to another internally so the compost remains within the structure for a full six months after leaving the collection bin. This is a great improvement over the last privy. The partition between the collection bin and storage bins is wood to minimize heat loss and maximize space. The three composting and storage bins are separated by a heavy wire mesh (the type used for chicken and rabbit cages). This same mesh is used for the baffles in front of each bin and is much stronger and more durable than chicken wire. A much larger vent shaft provides more adequate breathing and venting of the composting piles.

The rest room itself has been nicely finished in sheetrock and wood paneling. The walls and ceiling are insulated and screened windows provide adequate light

This photo of the unfinished retrofitted privy shows the heavy wire mesh separating the composting bin from the two storage bins. (Photo by Susie Benson.)

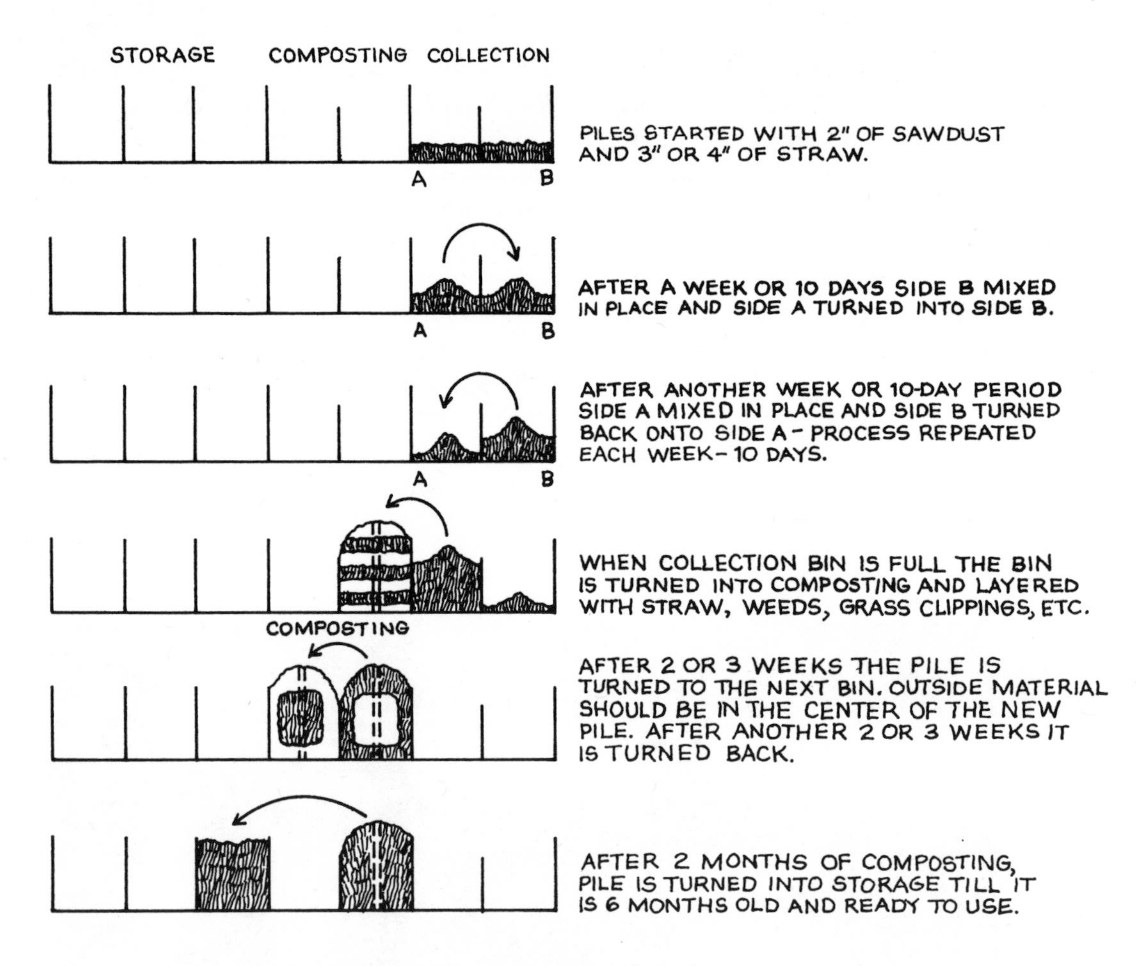

Turning schedule for the retrofitted, two-hole high-use-rate privy (10 to 12 people per day).

and room ventilation. Hand washing facilities are provided as is storage space for toilet paper and the high carbon add mixture. There is space enough for a woman's urinal and one is planned next spring or summer. The privy serves eight to 10 people and takes about two months to fill the collection bin.

The Farallones Drum Composting Privy

The drum privy is the simplest composting privy going and probably the cheapest. It's just an open-top 55-gallon drum with some perforated plastic tubes inside to admit air. A plywood seat with a sealed lid can be fashioned to fit tight on the drum, or the drum can be jacked up under a hole in the rest room floor with a squat plate or built-in seat with a tight lid. One drum is just about right for two people; it will take them about three months to fill it.

The drums are cheap (usually $3 to $5) and readily available. The ones with removable tops have a clamp that fits around to hold the lid on tight. They usually have a plastic or epoxy coating inside (look for ones that do) so they won't rust out. They should also have a flat black exterior and there should be a two-inch bung hole in its side near the bottom; this is where you're going to hook up the vent pipe.

Ventilation

The air-intake vent pipes should be made from two-inch ABS (acrylonitrite-butadiene-styrene) or PVC (polyvinylchloride) perforated plastic pipe and then coupled to a five-inch-long piece of two-inch ABS or PVC pipe that threads through the hole on the side near the bottom and is then fitted with a screened cap. The coupling should remain uncemented for disassembly. The cap should be fitted with a stainless steel screen so fine that even tiny fruit flies are kept from entering.

Another pipe fashioned as an exhaust stack must come up through the top of the drum to insure proper draft. It is not attached to the air intake pipes inside the drum; as a matter of fact, it should not go more than an inch or so into the drum, and the bottom should remain open. If it is set any deeper into the drum there is a chance that it will get clogged up as the drum fills up, and ventilation will be impaired. Since gas rises to the top of the drum the exhaust stack should be as high as possible, certainly higher than the seat. If

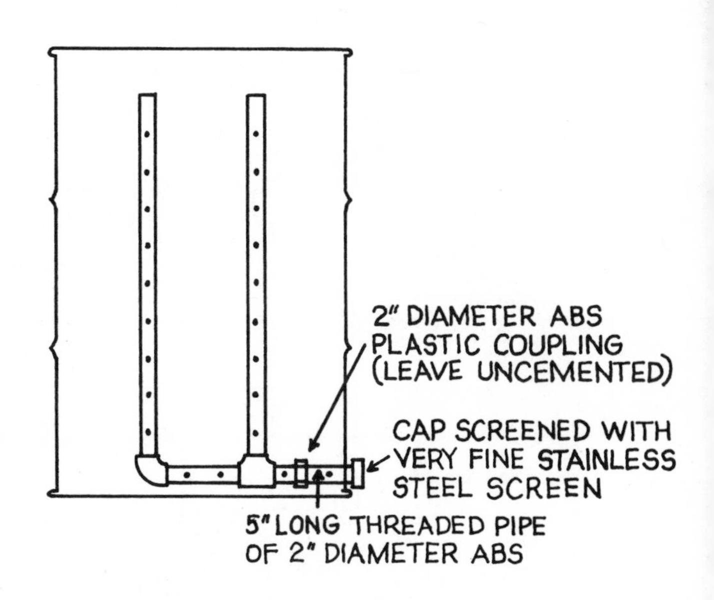

Cross section of a drum composting privy, showing the ventilation setup.

not higher than the seat the gas and its odors will escape through the seat and not the vent stack. Exactly how and where the exhaust stack is fitted into the drum depends upon how the squat plate or toilet seat is hooked up to the drum.

**The Simplest Setup:
Seat Right on Top of the Drum**

Basically there are two ways to do this. The simplest is to make the lid of the drum the seat as well. Cut a lid out of thick plywood that is slightly larger in diameter than the diameter of the drum. Saw two holes in it: one for the seat hole and one directly behind it for the exhaust stack. The stack hole should be slightly larger than the four-inch diameter of the pipe so that you can put a sealing strip between the circumference of the hole and the exterior surface of the pipe to form an airtight seal. (Weather stripping, neoprine foam, carpet padding, or any other flexible, spongy material should do the job.)

Make a circle with another larger strip of this sealing material on the underside of the wooden lid where it will make contact with the rim of the drum. This will insure an airtight seal between the lid and drum, even though the lid is not permanently fixed to the drum.

Cut yourself a lid for the seat hole out of plywood and put a handle on it. Attach

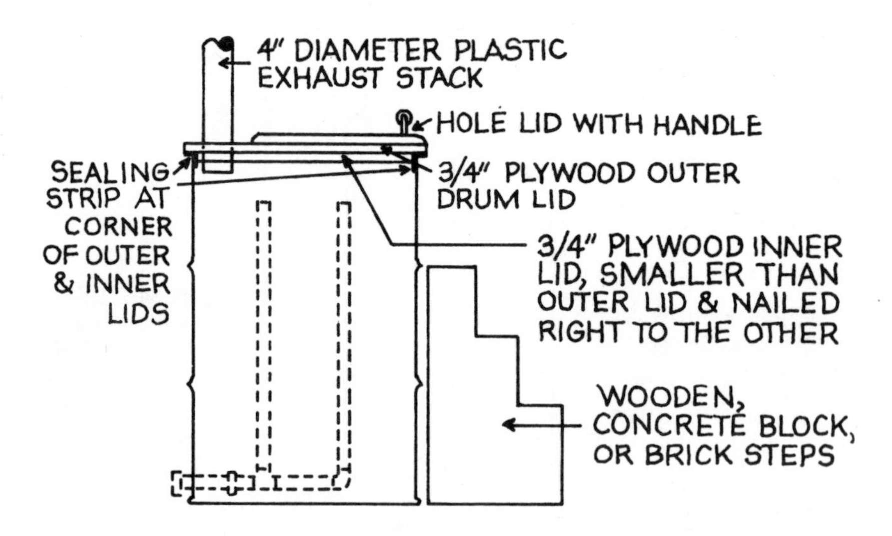

This is the simplest installation for the drum composting privy. The lid of the drum is actually the seat. When full, the lid comes off, the drum is moved to its storage place, and another empty drum with air-intake vent pipes inside takes its place.

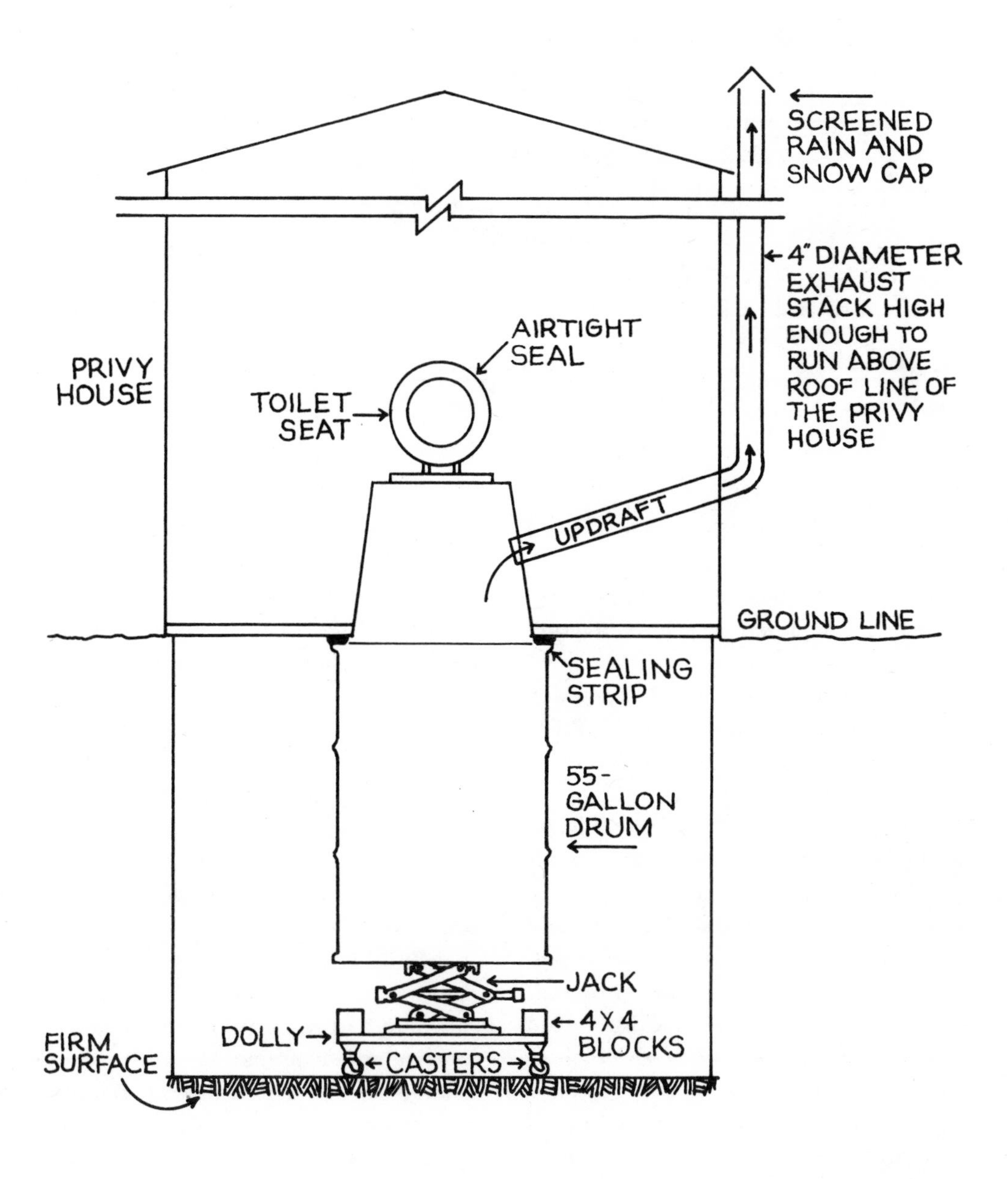

A drum privy can be held tightly against the privy house floor by a small scissors jack dolly.

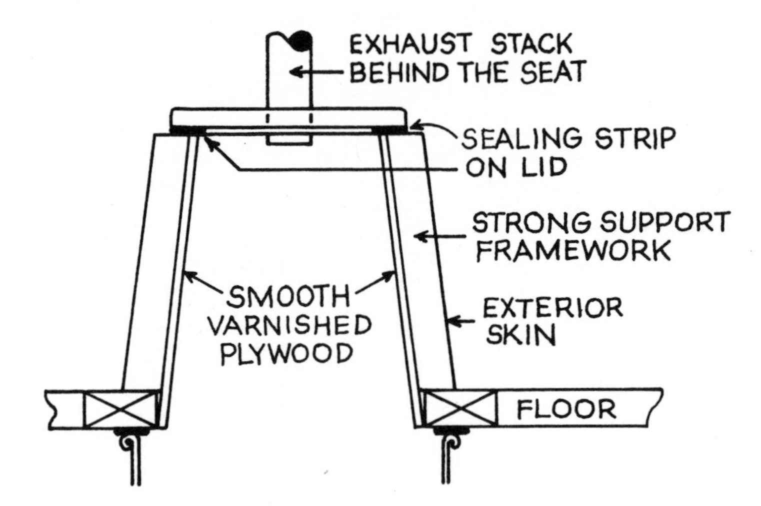

A cross section of a raised toilet seat on the floor directly above the drum.

the lid to the drum lid with hinges if you'd like. It's also a good idea to attach a sealing strip on the underside of this hole lid so that when it's closed you have a good seal.

The exhaust stack should be made from a four-inch-diameter piece of ABS, styrene, or PVC unperforated plastic pipe. If the drum privy is sitting out in the open, the stack need only be five to six feet high, but if it's enclosed in a superstructure it must be long enough to extend above the roof line of the privy house so that you get a good updraft. Run it up as straight as possible. Each bend or turn in the pipe will impede the draft necessary for evaporation of excess moisture, aeration of the drum's contents, and removal of odors. Fasten a rain and snow protection cover on top of the stack and screen it to keep flies out.

Set the drum where you want it, add the lid, exhaust stack, and hole cover. Then make steps out of bricks, concrete blocks, or wood so that users can get up to the top of the drum. Obviously, this setup makes squatting next to impossible; users will most likely have to sit right on the drum lid.

The Drum under the Privy House Floor

A trickier, but more aesthetically pleasing setup has the drum under the floor of the superstructure. The only thing you see when you use the privy is a raised toilet seat or squat plate in the floor.

Build your privy house with a space under the floor high enough so that the drum can stand upright in it and wide enough so that you can maneuver the drum out when it is full. You'll have to arrange some sort of setup that puts the drum right up tightly against the spongy sealing strip that's attached to the underside of the privy house floor.

The best way to position and support the drum is with a small scissors jack dolly. The drum is placed right on the dolly so that it can be raised up against the floor for use and then easily lowered when it comes time to replace it with an empty drum. The jack doubles as a cart to wheel the lowered drum out from under the house and into storage.

A permanent toilet seat can be built directly above the drum. The sides of the raised seat should slope inward near the top, and the inside should be smooth with no lip or inside members to catch feces or urine. Smooth polyurethane varnished plywood makes a suitable inside surface. The outside skin can also be plywood or made to match the decor of the room. Just make sure that the middle support is strong enough to hold heavy, as well as light and average people.

The toilet seat lid can be made from plywood and should be installed as shown in the illustration here. A sealing strip should be attached to the underside of the lid so that it closes tightly on the seat rim and doesn't allow air to enter there to interfere with the draft up the stack. It's a good idea to hinge the lid to the seat and attach a handle.

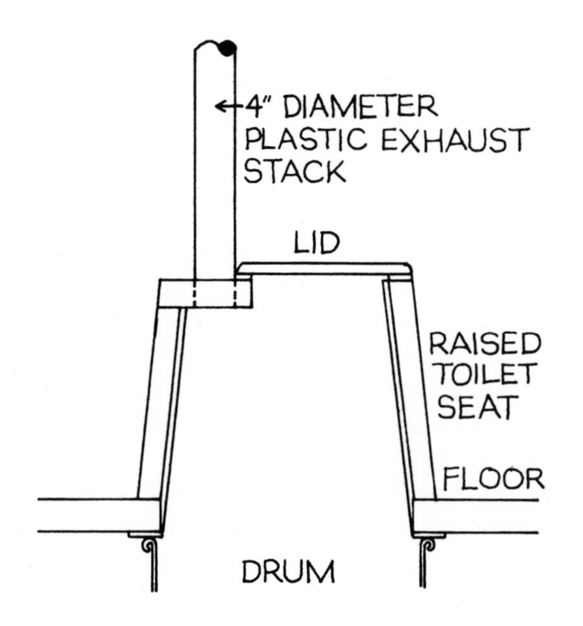

Raised toilet seat showing free-standing exhaust stack behind the lid.

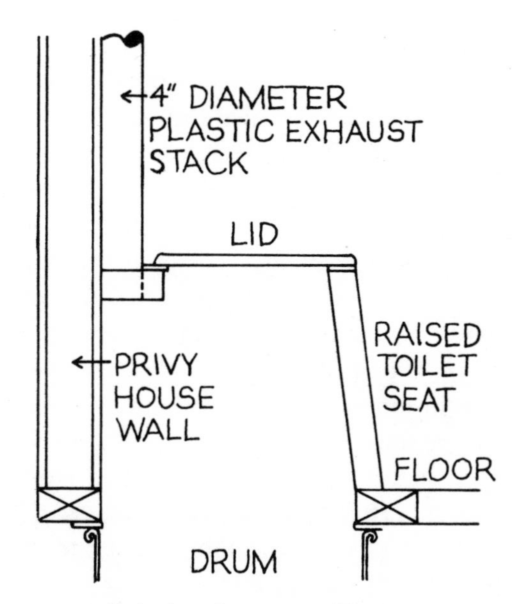

Raised toilet seat with exhaust stack running alongside one interior wall of the privy house.

The exhaust stack should be built as described in the earlier setup. It should be installed behind the toilet seat (be sure it's not in the way of the lid) and be long enough to extend out of the privy house above the roof line. See the illustrations here for two exhaust stack installations.

A squat plate can be used instead of a raised seat. See the illustration for details and alter the dimensions to fit the drum opening.

Using the Drum Privy

The drum privy is used much like any other composting privy. Before using, three to six inches of chopped straw, grass clippings, or leaves should be placed on the drum floor. Two cups of straw or grass clippings should be added after each use. Sawdust should be avoided because it doesn't compost well and doesn't contribute to a hot pile. Try to limit the amount of urine that goes into the drum by using urinals. Since the drum has no drainage hole, liquid can build up and cause the system to go anaerobic. Straw or peat moss can be added if you suspect there's too much liquid in the drum. It will absorb some of the urine and add bulk to keep the contents from compacting. Don't throw in garbage; it can cause odors.

It is a good idea to mix and aerate the pile regularly—every two weeks or so during regular use. Do this by removing the drum from the privy house floor or by taking off the lid/toilet seat. Then cover the drum with its original drum lid, seal it with the drum clamp, and just tip it over and roll it around a bit. There's no need to turn with a shovel and there's no spilling anything out while mixing. Turning schedules depend on use rates and individual initiative. More turning means better mixing and usually better aeration. If the moisture builds up at the bottom and causes odors from anaerobic activity, more turning will help distribute the moisture throughout the contents and alleviate this problem.

Storing or Composting for Further Decomposition

When the drum is close to full you can add some garbage to diversify the mass. Don't fill it to the top because you want some room in there so the contents can move around freely and mix well. Remove it from its use position and cover it tightly as described in the paragraph above. Roll it around to mix it well. There should be a two-inch-diameter hole in the lid. Screw a six- to 12-inch piece of two-inch metal or plastic pipe into this hole and top it with a rain cap. Then let it sit out for six months or a year to completely decompose and age, rolling it around once a month during this storage. A black drum sitting out in the sun is a pretty good solar collector and in moderate climates it will heat up well, aiding the composting and insuring pathogen destruction. Be sure to turn the drum around on occasion so all sides get that southern exposure.

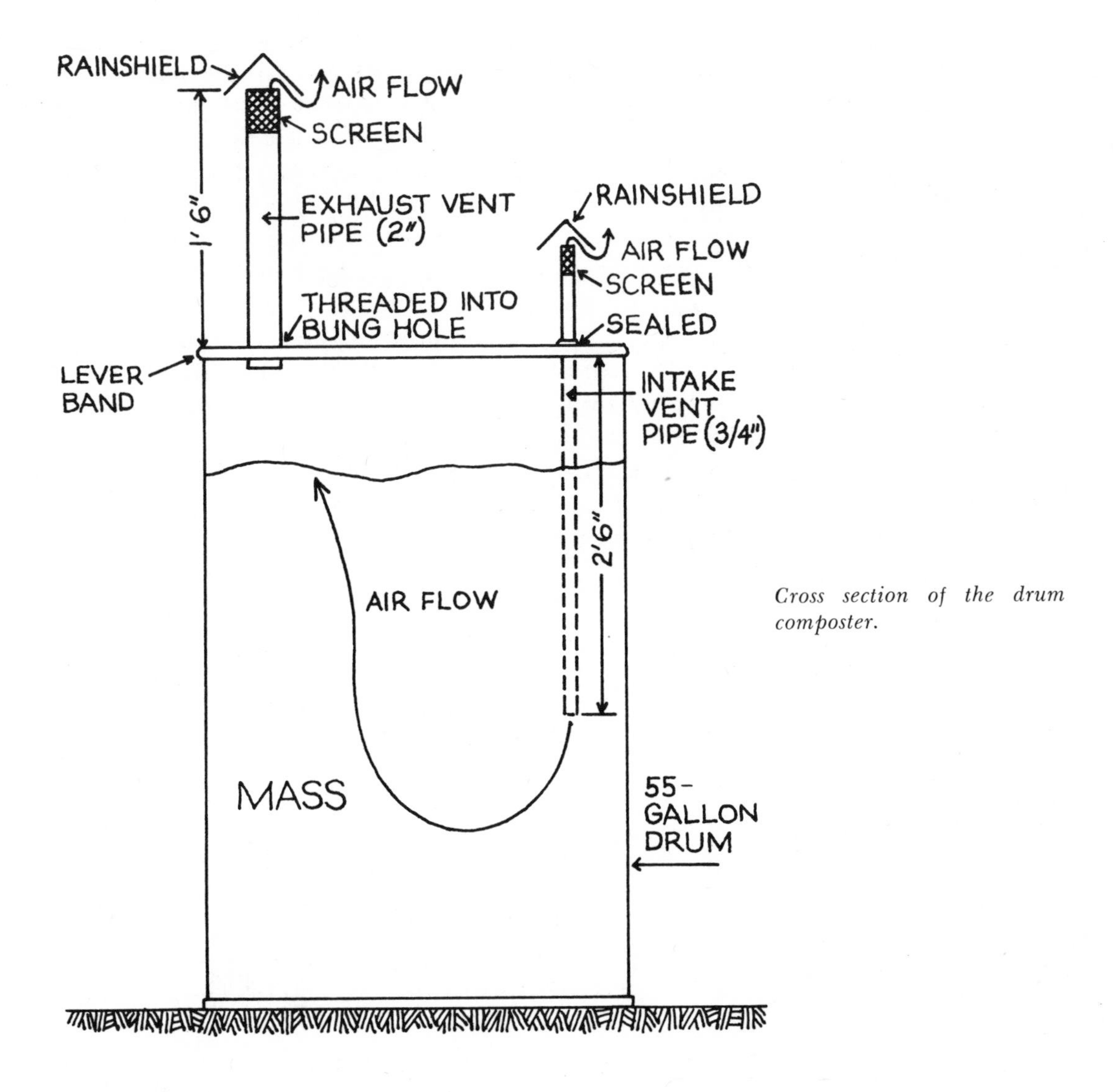

Cross section of the drum composter.

In a northern climate a black drum will lose more heat than it would gain from the sun. And of course it will lose heat all night. The mass of the pile is so small it will lose heat as fast as the microbes can generate it. For proper decomposition during such cold weather, store the drums in a warm cellar or shed.

One idea the Farallones Institute plans to pursue further is a simple solar oven large enough for about three drums. This oven is really just a well-insulated box with a good-sealing insulated door and a glass or fiberglass south wall with some sort of insulated shutter to be closed for nighttime. Several days of temperatures above 176°F (80°C) would be enough to pasteurize the material completely. Then it could be handled like any other agricultural manure—composted and used.

Instead of storing for six to 12 months in the drum, the contents can be removed and placed in the center of a large (more than three cubic feet) well-made compost pile. It is very important that it is placed in the center and not in the layers, because the center is the hottest spot, and high temperatures are needed to assure complete pathogen destruction. Build the pile on a slight slope to assure liquid drain-off. The pile should be situated on a base of rough straw, twigs, and rocks to permit aeration through the bottom. The pile should be covered with a six- to 12-inch layer of fine plant material or soil which is removed before turning and then replaced.

The pile should be turned after four days, and then every fourth day after that for 24 days, or until it is cool. Temperature of the pile should reach at least 140°F (60°C). It's recommended that you keep a thermometer inserted as deep into the pile

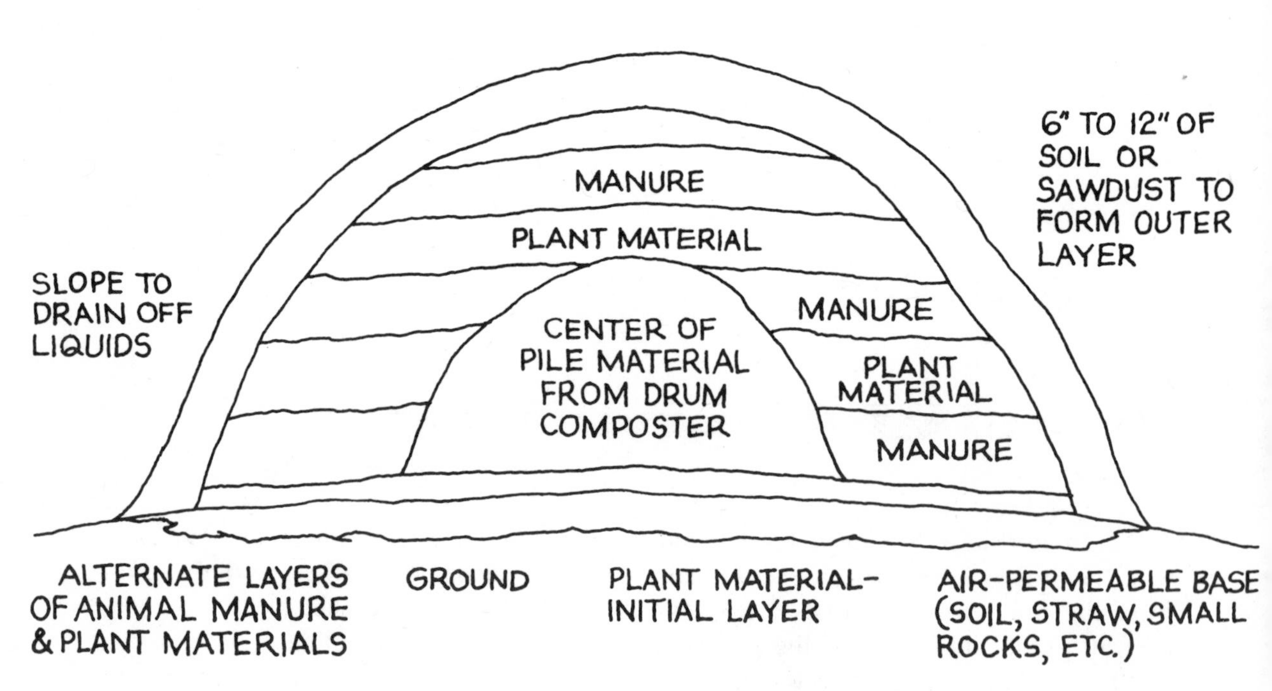

The contents of the drum privy are placed in the center of a large, well-made compost pile.

as possible throughout the composting period and check it frequently. The pile should be protected from soaking rains.

The composted excreta, which has either been stored in the drum for six months to a year, or composted in a hot pile for at least 24 days, should not be used directly on edible plants and especially not root crops. The Farallones people bury it around fruit trees, under a 12-inch cover.

The Backcountry Bin Composter

The Northeastern Forest Experiment Station, Backcountry Research Project, Durham, New Hampshire, has developed a sewage disposal alternative for use in remote forested areas where hiking and other forms of "primitive" recreation occur. This new system, called the "Bin Composter," is the result of work which began in 1972 on composting human waste. By this time it had become increasingly apparent that many tent-site and overnight shelter facilities were threatened by a combination of heavy use and soil conditions unsuitable for the traditional privy. In some cases, overflowing privies were suspected of polluting nearby springs and streams.

These conditions created a need for an alternative to the pit privy that was simple to operate, low in cost, and environmentally safe. Further constraints were imposed because many backcountry sites are at high elevations in remote regions. Conse-

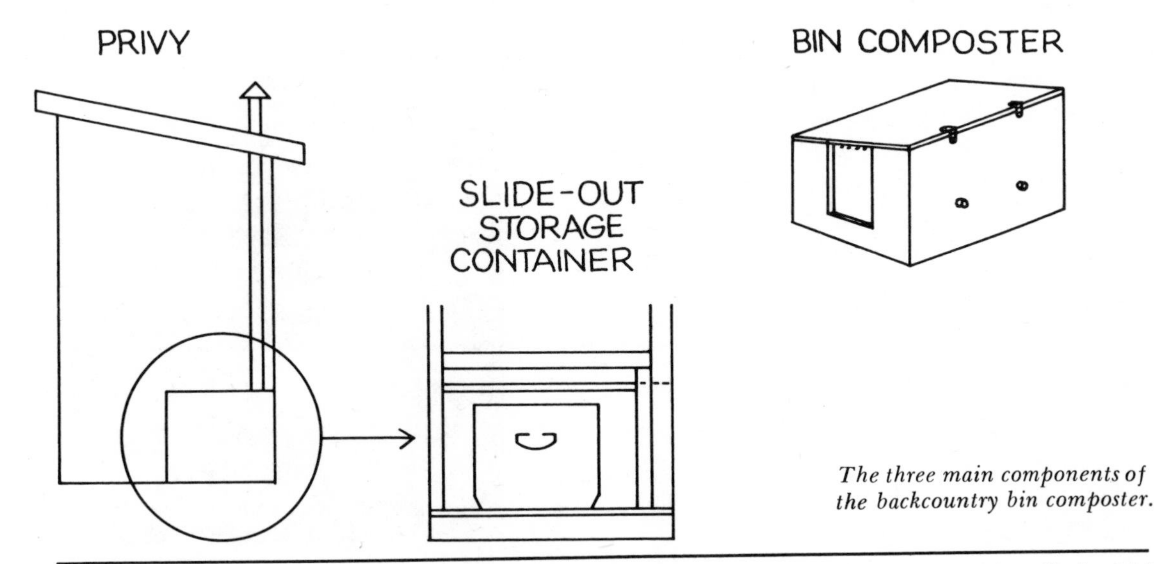

The three main components of the backcountry bin composter.

Further information is available by contacting the Backcountry Program, NEFES, USDA-Forest Service, P.O. Box 640, Durham, NH 03824. Staff member Stephen C. Fay is conducting all the field trials of these units.

A slide-out door at the lower end of the bin composter allows for easy turning of the compost pile. Note: This picture was taken before the unit was fiberglassed.

quently, any new system had to operate without water, electricity, or frequent technical assistance.

The bin composter was designed to operate under these conditions. It is lightweight enough to pack into the woods, and its leak-proof container makes it independent of site conditions. Composting human waste and ground hardwood bark is a simple process which requires only limited

Two 2-inch PVC pipes pass through the center of the compost pile. Holes drilled in the pipes speed the composting process by increasing the supply of oxygen and allowing excess moisture to evaporate.

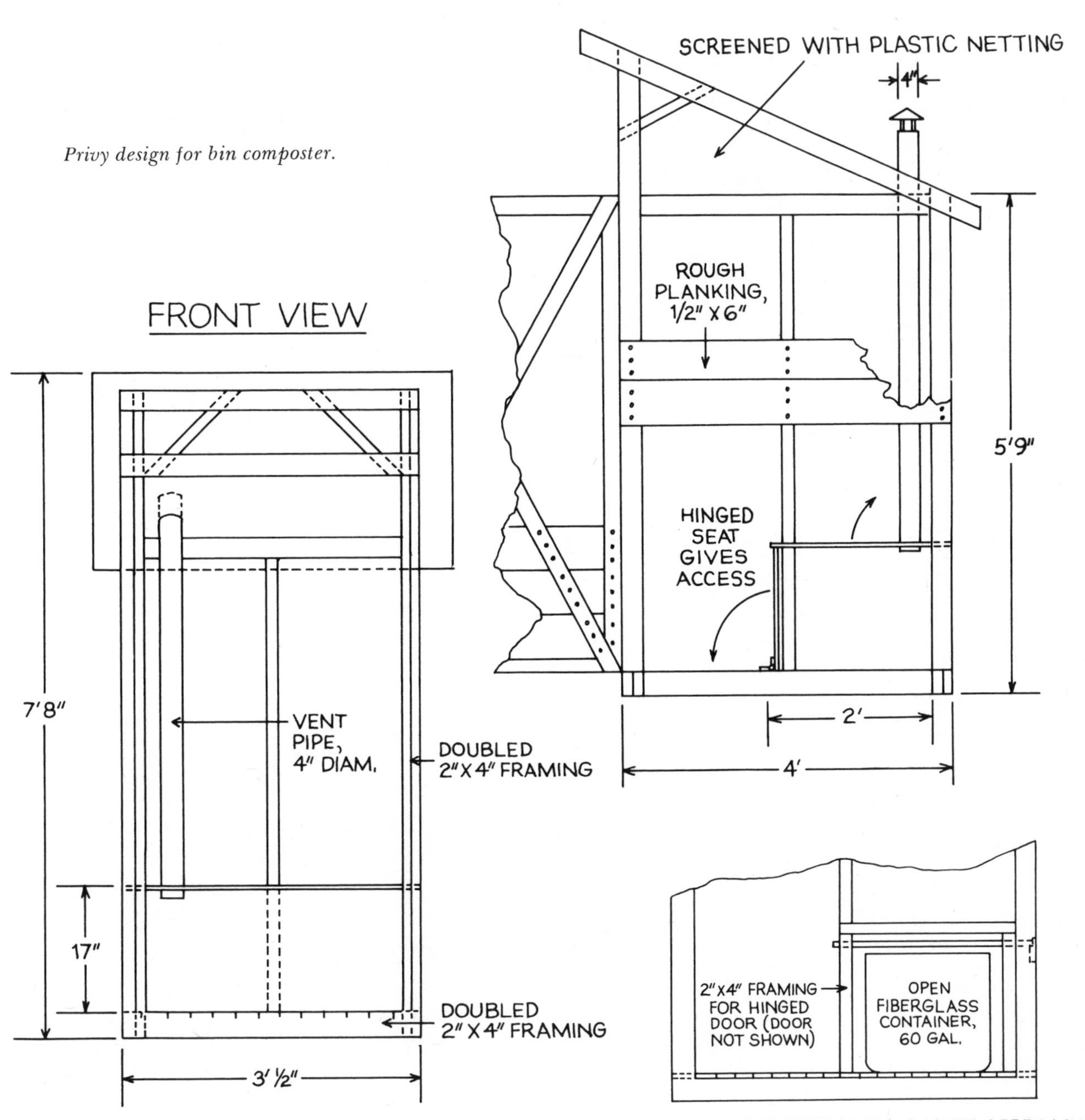

Privy design for bin composter.

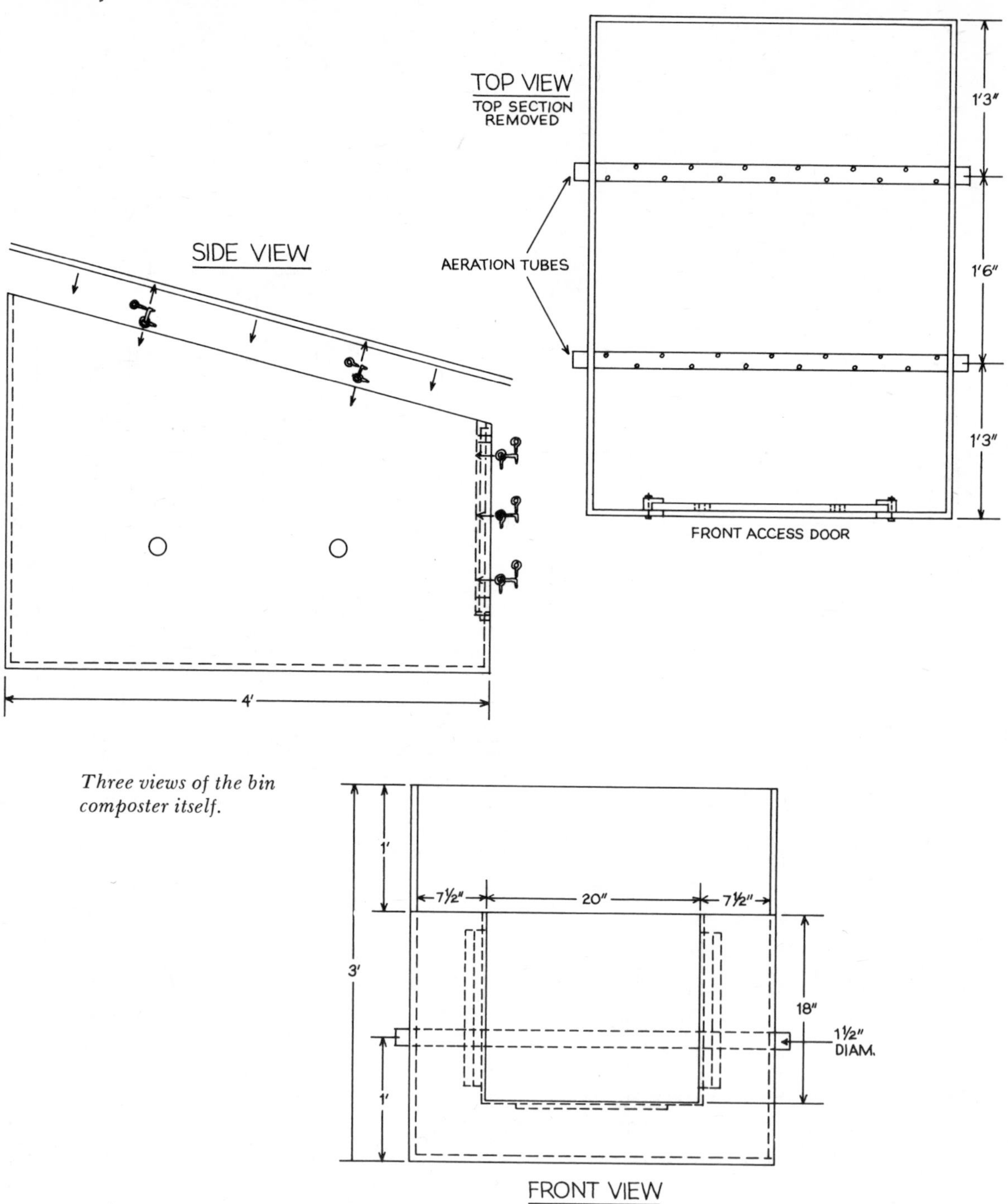

Three views of the bin composter itself.

maintenance. Also, in these times of increasing recreational use of backcountry areas, there are limitations to setting up an expensive waste disposal system with a set capacity. A bin composter, however, costs only about $125, and additional bins may be added without altering the most expensive part of backcountry waste disposal: the privy building itself. This makes the bin approach suitable for facilities receiving heavy use, as well as lesser-used sites where use may increase over time.

To operate the bin composter, human waste is first collected under the privy seat in a leak-proof fiberglass container. When a total of 10 gallons is collected, the container is slid out and the contents emptied into the bin composter. Forty pounds of ground hardwood bark are then thoroughly mixed with the waste to create suitable composting conditions. The bin is fully loaded when it contains 40 gallons of waste material and 160 pounds of bark. At this point, the bin contents should be allowed to compost with no further additions for two weeks; the pile should be turned inside-out after the first seven days. This increases the supply of oxygen—essential to aerobic composting. Temperatures inside the pile have been measured at 140°–149°F (60°–65°C).

Plans for 1977 are to install five units in the mountainous backcountry of Vermont. The objective is to test for logistical or operational problems in using the bin method along a trail/shelter system.

Chapter 5

Commercial and Owner-Built Composting Toilets

Moving up the technological scale, so to speak, from holding pits to compost piles to composting pit privies and composting privies, we come to composting toilets. They're called toilets instead of privies simply because these nice, tidy boxes that sit inside one's home resemble flush toilets more than they do traditional outhouse privies.

Unlike composting privies, composting toilets are a quite recent invention. It was in Sweden in 1939 that Rikard Lindstrom built the first self-contained indoor toilet that decomposed wastes instead of discharging them into the soil or nearby waters. The toilet, which was later named Clivus Multrum, was first used in Lindstrom's own house that sits on a rock slab on one of the fingers of the Baltic in Sweden. Although the unit seemed to work well at the time, World War II prevented this new toilet from gaining any real acceptance until about 1964.

Since then engineers and tinkerers in Sweden have come up with all sorts of variations on the original composting toilet, partly in response to Sweden's real need for alternatives to the flush toilet. Most all the likely spots (about 75 percent of the urban areas) have already been sewered, and what's left is outlying places that would just be too expensive to hook up to sewer lines. This country was scoured by an ice sheet, leaving very little soil, and because of this rocky terrain, septic systems are not considered good alternatives. What's more, the Swedes love to spend their summers outdoors along their thousands of lakes and on islands off the coast. Everyone who can afford it has a boat and a small summer cabin, but there is no way these vacation homes can be served with sewers or septic systems.

The Swedish government has been encouraging the development of composting toilets because they see them as valid solutions to their wastewater problems. Grants are available to communities who want to install composting toilets, and there are other government-sponsored incentives, like releasing homeowners from garbage collection and their costs if they compost and recycle their wastes. Because of such encouragement, Carl Lindstrom, son of

Clivus Multrum inventor Rikard Lindstrom, estimates that there are at least 20 to 30 different makes of composting toilets being manufactured in Sweden right now. Of this 20 to 30, about a half dozen are being sold in North America at the time of this writing. It's quite possible, though, that more are available now, as you read this.

All the toilets work on the same general principle: human wastes, when mixed with enough plant matter, like kitchen scraps and garden wastes, and exposed to enough air, will, in time, decompose and become nutrient-rich, humusy fertilizer. What goes on inside a composting toilet is very similar to what goes on inside a backyard compost pile.

Those who know their composting toilets divide the different designs into two general categories: the larger units that are rather big, bulky, and expensive, but have few moving parts, need little maintenance, and are emptied only every couple of years; and the smaller, less expensive units that are heater- and fan-assisted, may have automatic mixing units, and must be emptied at least once or twice a year.*

Larger Composting Toilets

The larger units (Clivus Multrum and Toa Throne, which are both available in North America, fall into this category) are what you might call "passive systems" in that they have few or no moving parts. They use no or very little electricity, and there is almost nothing to get broken or worn out. It is the design itself that effects two of the most critical factors of biological decomposition: heat and air.

The tank size above the Clivus Multrum and Toa Throne allows for a lot of surface area inside, insuring good exposure to surrounding air. The wastes spread out as they come down the toilet chute (and in the case of the Clivus Multrum, the kitchen chute as well) and spread out even more and mix with other organic matter put down the chute(s) as they slowly slide down the inclined tank. Small holes emit air into the bottom of the tank, and the updraft created by the long vent stacks draws air up through the air ducts in the Clivus Multrum and an air staircase in the Toa Throne. In some instances this updraft is assisted by a fan inside the vent stack or a wind-driven rotary turbine on top of the stack. There are small pockets scattered throughout the decomposition tank that don't get this air, but they generally don't stay anaerobic for long because they are likely to be converted and utilized further by surrounding aerobic bacteria in this or later stages of decomposition. Liquid accumulation which, along with compaction, is responsible for a pile going anaerobic, is prevented in the top part of the tank by its incline. It can, however,

* Some contend that these smaller units are misnamed. Because the piles are so small and air and heat are forced through them, they should more appropriately be called dehydrating toilets.

build up in the bottom part as the liquid slowly but steadily runs down the slope. This is most evident in piles that have compacted because of insufficient bulky material in the tank. The updraft is also supposed to draw odors up and out the vent stack and evaporate much of the liquid in the tank.

As was explained in chapter 3, the decomposition process generates its own heat, and as long as this heat can be held in the decomposing pile, the process will continue until there are no more raw wastes for the decomposing bacteria to act on.

Temperatures inside these larger units never get as high as 149°F (65°C), which is considered to be the ideal for good decomposition. The continuous circulation of air removes much of the pile's self-generated heat—just enough to slow down the decomposition, but not enough to stop it altogether. If the tank is insulated or kept in a place where there are

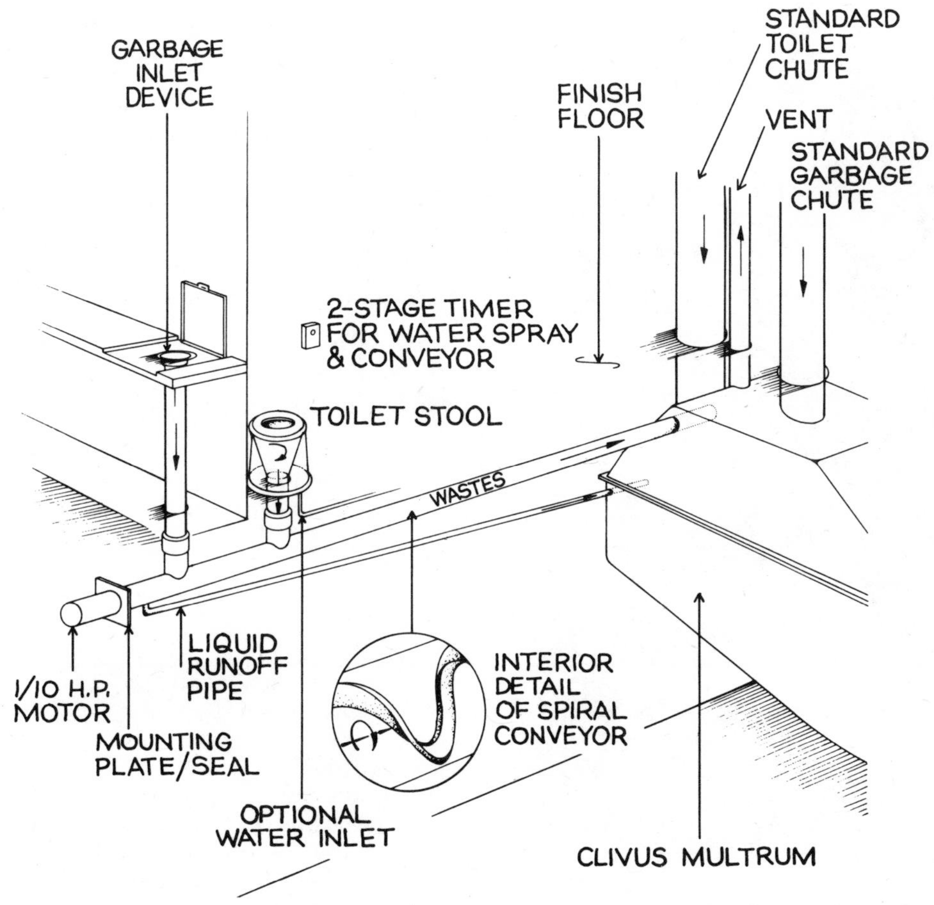

The Clivus Multrum's horizontal spiral conveyor that makes installation of the toilet easier, since the tank need not be located directly under the garbage and toilet chute.

air temperatures of at least 65°F (18.3°C), it should stay warm enough to decompose slowly—or molder, as the process is more accurately called.

Keeping the toilet in a moderately warm place is not usually difficult for home-owners, but it can present problems for outdoor installations and recreational areas. Boston-base composting toilet distributor, ECOS, in conjunction with the Appalachian Mountain Club, has designed a solar-heated shelter to keep the composting toilet inside warm throughout the year. Soltran is still in developmental stages, but because of its simplicity and low maintenance, it holds much promise and would make toilets in state and national parks, campgrounds, etc., feasible. One asset of a Soltran is that the sun shines hottest and the shelter works best during its period of greatest use—the summer.

Installation can be quite tricky because of their size, slanted bottoms, and the fact that they're designed to sit right under the bathroom floor so that the chute leading from the toilet to the tank runs vertically, with no even slight bends. Clivus Multrum now offers a horizontal spiral conveyor which makes it possible for the tank *not* to be directly under the chutes. But this pipe is a relatively new idea, and it has presented some users with problems. In a small, old house, or one with the bathroom on the first floor, installation of either of these larger toilets might be next to impossible. If you're not handy, and your distributor cannot help you, you may have trouble finding a carpenter who is familiar enough with one of these toilets to install it properly. Fortunately, both Clivus Multrums and Toa Thrones now

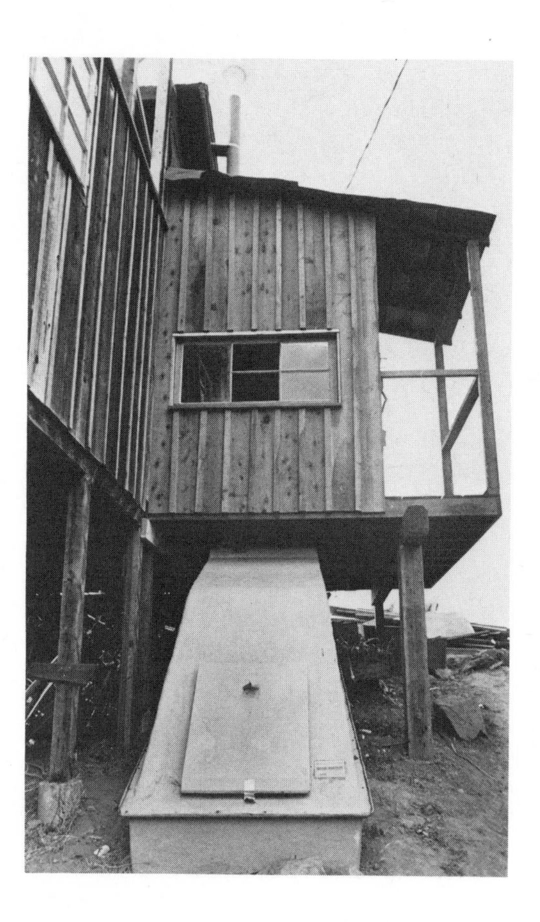

Since there is no basement in this pole structure, the Clivus Multrum was installed outside and under the bathroom itself. An outside installation like this is only possible in warm climates since cold air surrounding the unit can slow down the decomposition process and impede proper evaporation.

come with installation manuals to guide those unfamiliar with the design and the principles behind it.

Some Clivus Multrum owners we contacted paid upwards of several hundred dollars to retrofit the units in their houses by knocking out a wall, expanding a crawl space, or digging out and insulating a space for the tank outside the house. Others who planned to buy and install one in their homes decided against it when they found out what was involved, while still others who were building new homes had to redesign the whole thing and build the house around their toilet!

Hooking up the ventilation stack can be a bit of a trick, too. In order to get a good updraft, particularly in units unassisted by a fan, the stack has to be quite high. Clivus Multrum recommends that there be 15 feet of vent stack above the highest opening in the house (the toilet or kitchen opening), or longer. It seems reasonable to assume that the same would hold true for Toa Thrones.

How much longer depends upon the density of your neighborhood, the location of tall trees, the elevation of your land, etc. The stack should be high enough to get an uninterrupted air flow. A New York State family had their stack as high as the roof line and thought they were okay, but they actually should have had it higher than their chimney. The stack was close to the chimney, which apparently cut off some of the air flow. When they realized the problem and lengthened the vent stack they had no further problems with the draft.

The whole vent stack should be insulated to prevent water vapor from condensing on the inside of the pipe and dripping back into the tank, causing liquid buildup. Condensation will occur if the difference between the temperature of the pile and the temperature of the vent stack is larger than 4° to 5°F (2° to 3°C). But the stronger the draft, the more tolerant the system will be to such temperature differences. Theoretically, if there is a good updraft because of a strong fan or high winds outside, condensation will be less.

In principle, air is pulled up the stack, due to the Venturi effect, as breezes blow across the top of the stack. Unfortunately, this isn't always the case. Things like fireplace chimneys and wood stove flues can interfere with the draft, as can ventilation fans or opened windows in the bathroom, or toilets seats that don't seal tightly when the lid is closed. Many owners of these larger units had problems with liquid buildup and odors until they installed a wind-driven rotary fan at the top of the stack and/or a small electric fan in the stack itself. This simple, inexpensive addition seems to take care of the problem. All Toa Thrones and new Clivus Multrums come with the fan now.

The updraft, which is essential to the small toilets as well as the large ones, presents a problem to those concerned about home heat conservation (and who isn't?). All composting toilets pull a lot of hot air from the house through the toilet and/or kitchen chutes and up and out the vent stack. Those units that have a fan running all the time evacuate the most heat. The Clivus Multrum, without a fan, pulls 7 liters a second; the Toa Throne, 20 liters per second; the Mull-Toa, 9 liters; the Mullbank, 16 liters; and the Bio Loo, 18 liters a second. If you have a very well-insulated house, 50 to 60 percent of the heat in your house could go up this vent stack.

Some Clivus Multrum owners in Sweden have licked this heat loss problem by installing a heat exchanger in the vent stack. This device is a series of pipes designed to transfer heat from both the house and the decomposition process itself as it escapes up the stack to the air inside the pipes, which, once heated, makes its way back into the house. A prototype, which hopefully will be available in North America before too long, is now being tested in Clivus Multrum president Abby Rockefeller's home. It's projected that this heat exchanger will recover 50 percent of the heat otherwise lost up the stack.

With such a heat exchanger in it, the Clivus Multrum might be better renamed a bio-furnace. If people really want to use the Clivus Multrum not only to recycle their wastes but also to help heat their house, they should really save their extra wastes and let the compost pile remain small in the summer so that it slows down and produces little heat. Then, as the weather gets colder, this extra waste can be added to the tank to beef up the decomposing pile and allow it to generate more heat that can then be rerouted back into the house.

Before these large composting toilets can be used, several cubic feet of an absorbent organic matter like compost, peat moss, or humusy garden soil must be spread over the bottom of the tank. This starter helps to add bulk to the pile, absorbs liquid, and introduces the natural bacteria necessary to start slow decomposition.

If you're using peat moss, it's a good idea to open up the bag and leave it exposed to the air for a few days so that it can absorb moisture. Dried peat moss is not absorbent, as many gardeners know, and the liquid will just run around it but not through it. If you don't want to be bothered doing this, then mix your soil or compost and peat moss together well before putting them in the bottom of your tank.

Once installed and primed, the toilets are ready to use. The beauty of these larger units is that they operate pretty much on their own. They shouldn't need more than an occasional stirring, and they should never need watering. Because they are so big, they can withstand shocks, like urine overloads from beer parties, and several gallons of tomato skins dumped down the chute during canning season. Kitchen wastes like grapefruit halves don't have to be cut into small pieces like they do for the smaller units. The toilets are able to handle all kinds of organic wastes, including meat, dust and lint, the contents of vacuum cleaner bags, Kitty Litter, disposable diapers (without the plastic backing), sanitary napkins and tampons, *cold* ashes from fireplaces and wood stoves, hair, twigs, leaves, and grass clippings. Actually these last three, which are high in carbon, should be added from time to time to keep the carbon/nitrogen ratio close to an ideal of 30:1.

The variety of organic matter that can go into one of these large units, and the fact that the more added, within reason, the better, can present a dilemma to organic gardeners who view a garden compost pile as crucial to their garden's well-being. They soon find out that their compost toilet is in competition with their outdoor compost pile for all their grass clippings, leaves, straw, and other garden rakings. And the composting toilet, because it's so efficient, reduces wastes much more than does the pile out back. If these same people raise animals that live partly on kitchen scraps, they have another problem: who gets those carrot tops and lettuce trimmings—the pigs or the toilet?

Because of the long retention time, the Clivus Multrum people say that the contents which are eventually removed from the storage tank should be safe to use right on the garden. To play it safe, Toa Throne recommends composting its contents for six months first.

Despite the minimal amount of maintenance, you just can't treat one of these large compost toilets like a conventional flush toilet—flush and forget. You've got a living process going on inside there all the time, and you have to be sensitive to what's happening. Almost everyone we've talked to knew this would be the case before they bought a toilet, and they welcomed the involvement. We were struck at how tolerant even those who were having quite a bit of starting-up trouble were. Most folks felt that the benefits in the long run far outweighed the immediate problems.

Problems do arise from time to time, especially within the first year or two before the decomposition process stabilizes. For one thing, there may be liquid buildup, and this is affected by whether or not the tank is kept warm by the surrounding air or its own insulation, the ventilation, and the amount of bulky material in the pile that absorbs liquid.

The larger units also seem more prone to insect problems within the first year or so of operation, before the pile has achieved a balanced ecosystem where flying insects are held in check by their predators. Owners we spoke with complained of insect population explosions from time to time. House plant lovers can rest easy—the insects seem to be only those that feed on decomposing matter, not living plants. Some insects will come no matter what you do, like those that hover around all compost piles. But others can be kept at a tolerable level.

Fruit flies get introduced to the decomposing wastes by means of kitchen scraps, so don't leave vegetable and fruit trimmings standing around before you throw them down the chute. Dump them into the tank as soon as you can. Having a kitchen chute right next to the sink is an asset in such a situation because you'll be more likely to dump the scraps right away and not wait until it's convenient to take them into the bathroom. Grain moths can also give you trouble. Anyone who buys organic grains or flours will at sometime bring home some of these little insects. Dumping the infested grains down into the tank can cause a real population explosion.

Many composting toilet owners have found that they can alleviate such an infestation by throwing a handful or two of sawdust down the chute as soon as insects become noticeable, and keeping it up until the insects have gone. The sawdust effectively covers up the fresh material on which these insects travel. Peat moss or leaves will also work in the same way, but most people feel that sawdust is the most convenient and the most effective, and less likely to introduce new pests.

Sticky fly paper suspended in the chute has worked well for some Clivus Multrum families. A safe, organic insecticide like pyrethrum will also cut down on the number of insects. In emergencies, a no-pest strip can be hung in the chute, although it's important that the strip is secured in such a fashion so that it will not accidentally fall into the tank itself. If it does, the strong insecticides on the strip can destroy the delicate balance of insects living in the pile.

Mats Wolgast, a Swedish physicist, lives in a tight, superinsulated house that uses its Clivus Multrum not only as a waste composter but also as a heat resource or a "biological furnace." It has been calculated that the waste in the Clivus Multrum has an average heating value that is about the same as wood. The aerobic decomposition process gives off the same amount of BTUs that would be generated if the material were burned instead, but at a much lower temperature. Both composting and burning are oxidation processes. They end with the same end products: carbon dioxide, water vapor, and ashes. The well-insulated house needs a controlled ventilation and so does the composting process.

Both the house and the composting process generate heat. The house generates heat from people, lights, radio, TV, stoves, and other household appliances. The average amount of heat produced from these is about 3 kwh/day. The compost process gives off an average of 1 kwh/person/day. In the Wolgast house with five people this means about 1,800 kwh/year from the composting wastes, which can be compared with the 9,000 kwh/year of total heat demand for the house (about 20 percent). The heat exchanger which is necessary to recover this heat is hooked up to the composting toilet vent stack. The heat that is consequently recovered both from the house activities and the garbage is calculated to be worth about $150.

Carl Lindstrom

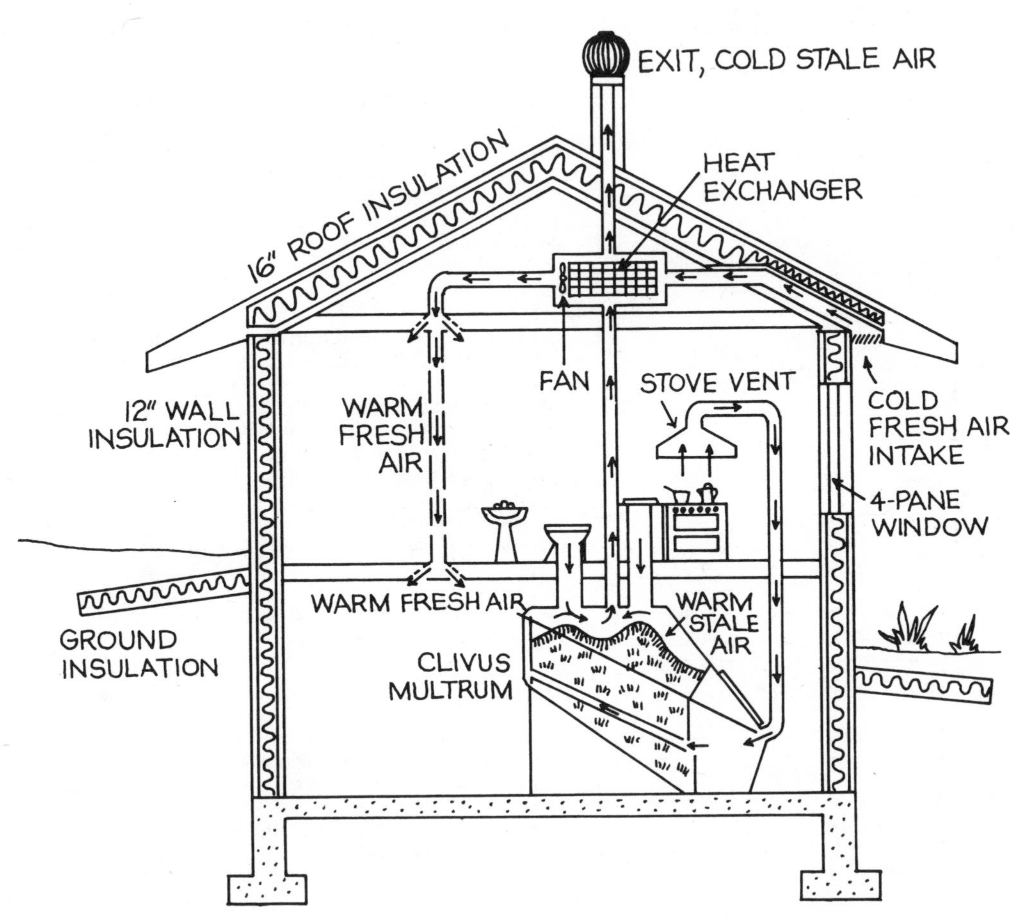

A heat exchanger built into the ventilation stack of the Clivus Multrum, as shown here, enables cold outside air entering the house to be tempered first by the rising warm air inside the decomposition tank.

CLIVUS MULTRUM

Problem	Possible Reason	Response
Odor comes out of the chutes as well as out of the container.	The air pressure in the house is lower than inside the tank.	Turn on the fan. If no fan is installed, order one from the CM office.
	More specifically, this lower pressure can be caused by competition from exhaust fan, an open fireplace, a wood stove drawing air from the house.	Allow more air to enter rooms (from outside the house) where such devices exist.
	Wind conditions cause a downdraft at the top of the ventilation pipe.	Put a rotary turbine on the top of the vent, if this has not already been done.
	The temperature in your basement is much lower than outside.	Insulate both the container and the vent pipe better.
	There are obstructions in the vent pipe such as spider webs, or the insect screen at the top of the vent is covered by dead flies, leaves, etc.	"Sweep" the vent stack with a rag on a string and clean the screen.
	The vent pipe is disconnected somewhere.	Follow the vent all the way up to check for such possible disconnections and correct them.
Odor can be detected even though the air goes properly down the chutes.	The chutes are leaking.	Check the seams in the chutes to see whether liquid can leak out at any point (especially the toilet chute(s)).
	The fan pushes foul air through air leaks in the vent stack.	Go over the vent pipes with duct tape and a caulking gun.
Odor is strong from the vent on the roof.	The process leans towards putrefaction (becomes partially anaerobic).	Increase the draft (i.e., turn on or increase the RPM of the fan).
Liquid builds up fast in the storage chamber and has an ammonia odor.	The compost pile is not firmly sitting on the bottom of the container. (Under certain conditions the material gets hung up on the air ducts, preventing the liquid from being absorbed by the waste pile.)	"Stoke" the garbage pile.
	There is almost no organic material in the area between the toilet wastes and the storage chamber.	Add at least a bale of peat moss as an absorber media for breakdown of the urine and evaporation.

TROUBLE SHOOTING MANUAL

Problem	Possible Reason	Response
Liquid builds up fast but doesn't smell.	There is enough peat moss in the bottom but the garbage pile does not absorb enough.	Stoke the garbage pile, especially if it is dry on the top. This can be a perfectly normal condition following a high input of liquid. NOTE: **No odor** indicates **no health hazard.**
	The input of garbage is small in relation to the toilet waste.	Fill the garbage section with peat moss, leaves, or sawdust.
	Water vapor condenses on the inside of the container and the ventilation pipe and drains back through the pile.	Insulate the container and the ventilation pipe along its entire run. If there is already some insulation you may still need to add more.
	It is just constantly too cold in the area where the Multrum is installed.	Install an electric light bulb in the storage chamber. Be sure there is no fire risk from overheating. **Consult an electrician.**
Insects are flying up the chutes.	Many insects orient themselves toward odor. If the draft is faint or reversed they may get out.	Increase the draft. Check to see if there is a liquid buildup with odor; if so, follow that instruction.
Insects are too numerous.	The prey/predator balance is offset.	Use an organic insecticide (like Pyrenone) according to the directions.* Rake up some grass and leaves and add this for a richer predator life in the composter.
The insect population temporarily gets out of control.	It is possible that an explosion of one particular species may occur.	Keep a no-pest strip accessible as an insurance for such conditions. Hang the no-pest strip in the chutes under the toilet and garbage chutes under the toilet and garbage lids. This will only kill those insects that stay in the chutes without affecting the microbes in the compost pile. The downdraft will prevent fumes from no-pest strip from entering home.

* Editor's note: Although not mentioned in the Clivus Multrum literature, we have learned of other mild insecticides that reportedly work in keeping down flying insects without seeming to interfere with the decomposition process. Baytex, available from some composting toilet dealers, is effective on gnats and midges; Pratt's White Fly Spray is available from people who sell greenhouse pest control products and Hargate (made from mineral oil, sesame oil, and pyrethrum) is effective on flying insects and can be gotten from Walnut Acres, Penns Creek, PA 17862. Mothballs have varied success in controlling grain moths. Some people we talked to say they do the trick, others feel they're not worth bothering with.

The Smaller Composting Toilets

If we call the larger units "passive" systems, then we have to call the smaller ones "active" systems because they have more active or moving parts. A minimum size for a compost pile is two cubic feet, but four cubic feet is better. These smaller toilets have piles less than four cubic feet. For this reason they can't rely upon heat generated by the pile itself or the natural ventilation of air flowing around the pile inside the tank to keep decomposition going at a relatively steady rate. Most of these toilets—the Mullbank (Ecolet), Bio Loo, Mull-Toa, Bio Toilet 75, Bio Toilets A and M—have ventilation fans and heaters. The Soddy Potty #2 is kept warm by a house heater or a solar collector. Since they don't have the inclined tank like the Clivus Multrum and the Toa Throne, they also must have some means of aerating, leveling, and mixing the contents—a rake, stirring arm, or a rotating tank.

These toilets were originally designed for campgrounds and vacation homes. They're easily transportable, less expensive than the large units, and are fairly easy to install. All but the Bio Toilet 75, however, have a more limited capacity. Theoretically, they can be used in year-round houses, and in many instances they actually are. They are popular with people who don't want to install a regular flush toilet for ecological reasons or can't install one (due to proximity to a water body, a soil of poor percolation, or a rocky surface under and around the house), but don't want the expense and possible installation problems of a Clivus Multrum or Toa Throne, or the competition these units give their garden compost pile. Some of the smaller units have been installed as second toilets in a house that already has a flush toilet, and it's quite possible that in the future one of these will be a second toilet in the house that already has one of the larger models as its main toilet. We know of one family of 10 that actually installed two of these small units for full-time use. As of this writing, they've had the toilets in for two years and have had no complaints.

As we said, one advantage of these smaller toilets is the ease of installation. Because they're approximately the same size as a flush toilet and the seat is right on top of the decomposition tank, all you have to do is put it in a room with a minimum temperature of 64°F (17.8°C), install the vent pipe, plug it in, and use it. The only exception is the Bio Toilet 75 which should be installed partly under the bathroom floor.

It's recommended that the vent pipe be installed and insulated just as it is in the larger toilets. Since there are fans and heaters in most of them, an air flow is not as dependent upon height of the stack, but it's still a good idea to extend it above the house roof line. Then, if the fan malfunctions for a while, the height of the stack should create a tolerable draft, at least for a short period of time.

As with the large unit toilets, the seat lids are made to seal tightly when closed so that air and possible odors go out the vent stack, not out the toilet seat. There should be no other exhaust fan in operation, especially in the bathroom, when the toilet is being used so that the toilet draft is not diverted. Enough air should *enter* the bathroom so that there is good warm air intake, and this air should preferably come from another heated area, like a hallway or a bedroom, not a cool attic or from the outdoors. Make sure the air intake holes near the bottom of the toilet are clean. At least one family we interviewed partially blamed the liquid buildup problem in their Mullbank on the dust accumulation on these vents.

To some people the advantage of the easy installation is offset by the constant reminder of their own wastes. These units, as we've said, sit right in the bathroom, in plain sight. The user sits right on top of the tank, and although there are seldom any odors, some find it disquieting to see their own human wastes and other garbage only several inches away when the lid on most of these toilets is raised.

In most of the units the fan is on all the time and the heater goes on and off automatically to keep the contents at a uniform temperature of 90° to 100°F (32° to 37.8°C). The Bio Loo's pasteurizing hotplate heats up to 158°F (70°C) to pasteurize the tank contents before they're removed. The Bio Toilet 75B has an additional heater (which is referred to as an electric evaporator) that goes on automatically when the liquid level in the chamber gets above a tolerable level.

The Mullbank, Mull-Toa, Soddy Potty #2, and Bio Loo have manually operated stirring arms that are controlled by levers near the toilet seats. The stirrers should be used frequently to level the contents and mix them around a bit. Two of the Bio Toilet models have more ingenious ways of mixing the contents. The tank of the Bio Toilet A, which is on the horizontal axis, makes one complete turn every time the toilet seat is opened, thereby automatically moving the contents about. The larger Bio Toilet 75 has a blade that rotates through the pile every time the toilet lid is opened. It's true that all these automatic devices may prove to be a convenience to many people because not too much is left to nature, but the electricity to keep them running will cost you about $3 to $7 a month in utility bills.

Actually, whether or not the heater, fan, and means of mixing really alleviate major problems in these smaller units is debatable. Some owners who are quite happy with their toilets say that they are just about as easy and convenient to use as a flush toilet. Others, however, feel that even with such devices the units are just too small to function well without regular attention. To paraphrase one owner, the smaller units have the same problems that small fish tanks have; they're just not big enough to handle stress. They don't have the buffering qualities of the larger ones. If you overdo or underdo something even just a little bit, you can throw the whole process off balance for awhile.

One of the biggest problems that a number of owners have been faced with is maintaining the proper moisture level. During times of heavy use, these toilets can get too wet, but some can get too dry during normal usage and must have water added to them periodically like the Mullbank. The Mull-Toa has a hygrometer that tells you if the moisture is too high or too low, and it also has a feature which lets you adjust the air flow to control the moisture. The Soddy Potty #2 has a humidity-sensitive damper which closes and stops ventilation when humidity dips below 50 percent. You can check the moisture level in the other units by looking down the toilet hole and stirring the contents, or by pulling out the storage drawer and checking the organic matter that has accumulated there.

All of these "active" toilets except the Bio Toilet (which can serve 10) and the Bio Toilet 75B (which some say can serve up to 20), are designed to serve ideally about four people and can accommodate occasional overloads. A number of people we interviewed, though, said that weekend guests or a party can overload their toilets very quickly and create problems—like anaerobic conditions—that take more time than they'd like to balance out. This is especially true in the toilet's first couple of months, before there is sufficient bulk in the tank. The Mull-Toa tank is especially small and therefore more susceptible to overloading. New owners are advised that the toilet will only be able to handle four quarts of urine every 24 hours for the first one to three months. The Mullbank literature also suggests limiting the urine during the first couple of months. With this in mind you'll either have to wait to have guests until the toilet is operating normally or ask all the friends and members of your family to urinate outside, as did one couple who has a new small toilet.

This liquid buildup problem can at least be partially avoided by adding carbonaceous matter like shredded newspaper, sawdust, or peat moss before, during, and after heavy use. This extra bulk may make it seem as if the tank is being filled up too quickly, but it won't be long before decomposition reduces the pile again to a more normal level.

If liquid buildup isn't the problem, not enough liquid may be the trouble. If the pile becomes too dry, the decomposition process will slow down and the pile can start to harden. In extreme cases, the pile will harden like a rock and no rotating, stirring, or even wetting at this point will make any difference. Dryness is caused by what you might describe as too much of a good thing—too much heat and too much ventilation, which both add up to too much evaporation. There are two ways to alleviate the problem: 1. periodically pour water directly on the pile as you mix it around, or 2. add more bulky materials to absorb liquid and hold it in the pile.

Adding bulk, as you can see, will help to keep the moisture at just about the right level. It will also keep the pile loose so that it can stay well aerated. Manufacturers don't stress this in their literature, but at least one composting toilet distributor suggests adding one to 1½ cups of carbonaceous material each day in addition to kitchen scraps. If you have a lot of kitchen scraps, and by this we mean two quarts or more a day, these will do. But if your family is small, you eat away from home a lot, or eat many processed foods so that your wastes are basically plastic bags or cans, you may want to put sawdust, peat moss, or leaves near the toilet so users are encouraged to dump some down before they close the toilet seat lid.

On the other hand, don't overdo the kitchen or garden scraps. The small units cannot handle piles of leaves, wads of newspaper or cardboard, or more than a couple of quarts of fruit or vegetable peelings at a time. Fruit rinds and corncobs will have to be chopped up, or put in an outdoor compost pile. Disposable diapers, Kitty Litter, tampons, sanitary napkins, grease, or any matter that is especially acidic or alkaline should never go inside these small units.

Flies don't seem to be as troublesome as they are in the larger units. This may be partly due to the smaller amounts of kitchen scraps put into the toilet, but it's probably also because of the constant heat and air movement and the faster decomposition rate. Fly problems, if they do occur, can be controlled in any way suggested in the "Clivus Multrum Trouble Shooting Manual." The Bio Loo comes with its own built-in container for insecticide, and Mullbank supplies its customers with a mild insecticide for temporary insect problems.

Because of their smaller capacity, these toilets are emptied more frequently than the larger ones. If it's regularly used year-round by at least two people, the storage area should have to be emptied about once a year. All manufacturers advise not using the organic matter around edibles, but burying it at least one foot beneath the ground level around flowers, shrubs, or trees. It could also be composted in a healthy pile for several months and then used in the vegetable garden. The Bio Loo is the only exception. If you have turned the pasteurizing hotplate on for six hours before emptying, the contents will be pasteurized and supposedly safe to use anywhere.

Don't think that just because these small units are emptied more frequently than the larger ones that they'll provide you with more useful organic matter. The tanks and storage areas are smaller and the amount of material you take out is also less. The heat and constant air flow effectively reduce the waste to a fraction of its original volume. It has been said that the finished product in the Mull-Toa, which has the greatest reduction rate, is only about 9 percent of its original volume.

Several of these toilets have to be reactivated when the storage container is cleaned out. Any good, humusy garden soil can be used, as can garden compost. The Bio Toilet and Mull-Toa models come with a special soil mix which can be reordered for a yearly addition to the tank.

Since these smaller units were originally designed for seasonal use, they can be unplugged if they will not be used for several weeks or several months. The decomposition process will stop due to lack of heat, moisture, and ventilation, but will start up again when the unit is plugged in and wastes are added. It's advisable to add some carbonaceous material and topsoil and pour a quart or two of water over the pile before you start to use it again. During winter, let the heater run for several hours to warm up the pile before use.

A Survey of the Commercial Models

Clivus Multrum

The Clivus Multrum, the best-known of all the Swedish composting toilets, is also the original composting toilet. The present unit has been installed in about 1,300 Scandinavian homes, and over 500 of these toilets have been sold in the United States since 1973 when the American Clivus Multrum company was established.*

Basically, the unit consists of a large plastic composting tank, with incoming toilet waste and kitchen waste tubes, an outgoing air vent, and an unloading door. New units have an aluminum porthole on the side of the chamber. It's handy for retrieving objects that inadvertently drop down the chute, checking the chamber for dryness, etc., or reaching in with a rake in case of temporary overloading. The unit operates by letting toilet wastes slowly slide downhill into the kitchen wastes; the two combine and further compost and slide into the storage area. Good ventilation and this sliding from one chamber into the next make turning of the wastes unnecessary.

There are two basic sizes: the smaller unit, which handles the wastes of up to three people with allowances for visitors, is seven feet six inches long and six feet eight inches high when propped up at the correct angle of 30°, and is just about four feet wide. The larger unit, which is good

---

* In 1971 Abby Rockefeller read about the toilet in the November issue of *Organic Gardening and Farming,* was attracted by its advantages, and the following summer flew to Sweden to see one for herself. She returned with two for herself and in the summer of 1973 got the American license to sell the Clivus Multrum in the United States.

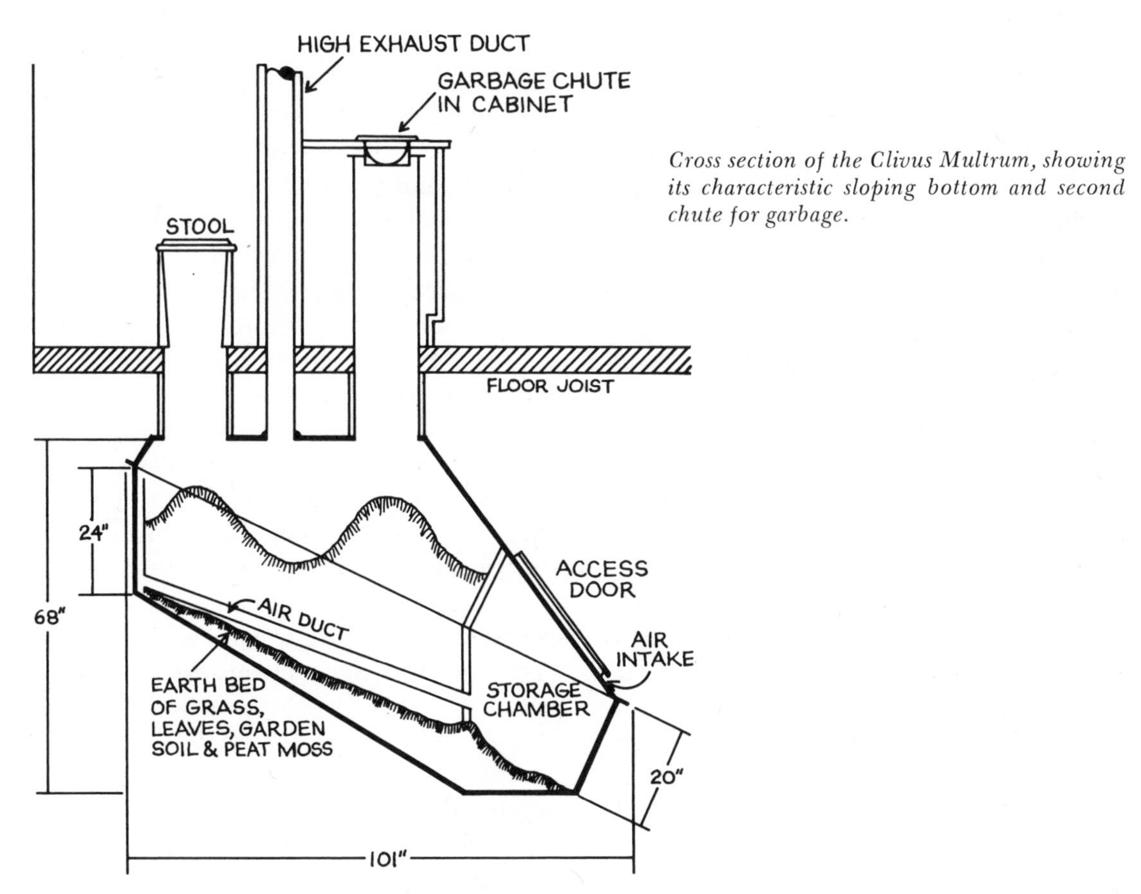

Cross section of the Clivus Multrum, showing its characteristic sloping bottom and second chute for garbage.

for four to six people, is nine feet five inches long, seven feet four inches high, and four feet wide. Both can be made to serve an additional three people by adding a mid-section.

Because of the tank's size and the importance of proper positioning, installation can be a problem. If the Clivus is going into a home that has not been built yet, this is no problem, but building it in an existing home is more difficult. Usual installation of the unit is in a basement, but it may be placed on the first or second floor of a home, in a partial basement, or in a crawl space, in which case the top of the tank may actually be above the bathroom floor and the toilet seat located right on the tank. It can also be installed outside as in a comfort station for campgrounds, or under an outhouse. However, precautions must be taken to prevent the unit from freezing during winter months. People who bought Clivus units in the past had to insulate them themselves, but now buyers can buy preinsulated tanks if they prefer.

With the unit in the basement of a home, the toilet pedestal would be located directly above the top end of a unit. The kitchen waste tube must be placed on the opposite side of the toilet wall, allowing the waste chute to be above the second chamber of the composting tank. This should not be inconvenient in most cases, as many houses are designed with kitchens and bathrooms sharing walls to reduce plumbing costs.

Clivus Multrum is experimenting with a horizontal spiral conveyor which is able to transport kitchen and bathroom wastes horizontally to the decomposition tank. It moves wastes by spiraling them along a horizontal chute. While it can help out people who have a problem installing their toilet and kitchen openings directly over the tank, this conveyor can create some of its own problems. The Wildwood School in Colorado has had some problems with solids stopping up the spiral. At an 8,000-foot altitude, temperature extremes have led to trouble and they had to insulate the pipe. The conveyor also broke down once. Clivus Multrum president Abby Rockefeller admits that it has some problems that haven't been worked out yet (only a few prototypes are in operation), but she says in principle it's a perfectly good solution and has some real advantages. For one thing, she says, it seems to be a good fly barrier.

Since there is no flushing method, the toilet pedestal is wider at the bottom than the top to prevent fouling of the sides. It also comes with a plastic liner. In the kitchen, there is an opening in the counter to drop kitchen waste through. Both openings must be kept covered when not in use because the unit creates a powerful downdraft.

A plastic roof-vent stack, with a snow and rain cover, creates suction to pull odors up through air holes in the storage chamber right under the unloading door, then through upside-down V-shaped air ducts in the chamber, and up and out the stack. In addition to pulling odors up and out, the ventilation system also provides the oxygen needed for aerobic composting.

The vent pipe must be kept warm to prevent water vapor from condensing and running back into the heap. All of the vent pipe that passes through areas that can get colder than 60°F (15.5°C) should be insulated to prevent water vapor from condensing on the inside of the vent pipe. This means all sections of the pipe passing through unheated attics as well as those sections exposed to the outside. (New units come with insulation for that part of the pipe that runs through unheated parts of the house. Owners must supply insulation for the pipe that's exposed to the outdoors.) And it is recommended that there be 15 feet of vent above the highest opening in the house (which is most commonly the toilet seat).

Before being put into operation or even before finishing assembly of the chamber, the Clivus must be primed. This involves spreading a minimum of three bales of peat moss over the entire bottom of the unit, followed by three inches of garden soil, leaves, and other garden rakings. The peat soaks up urine and its ammonia, and the filtering process of soil and garbage by the peat favors de-nitrifying bacteria leading to an increase in the range of species specializing in the breakdown of organic matter.

The composting or moldering process requires about a three to one ratio of cellulose to human waste, and it may be necessary to add more cellulose from time to time in the form of paper, sawdust, leaves, and such to maintain this ratio. The CM people say that once the equilibrium of the processes going on in the pile is reached (which should happen during the second year of operation) the unit will function with almost no maintenance at all.

Theoretically, the heat produced from the bacterial activity in the chambers should be sufficient to evaporate the liquid, but in practice this has not always been the case. A small heater or light bulb can be installed in the storage chamber to be turned on only if liquid starts to accumulate, which has happened in a number of instances causing an anaerobic process to set in, with its characteristically foul odors. As soon as the heater or bulb has evaporated enough of the liquid, it should be turned off. (The CM people caution that installation of either a light bulb or heater should be checked by an electrician, as an improperly installed unit can cause overheating and a possible fire.)

Abby Rockefeller is quick to point out that the brown liquid that can accumulate in the bottom of the storage tank is not urine but a sort of compost "tea." As we discovered on examining the liquid ourselves, it has no unpleasant odor. Abby recommends that instead of fooling around with a light bulb or heater, owners should simply remove the liquid when it begins to build up and use it to fertilize house plants or other ornamentals.

Clivus Multrum is now also recommending that a fan be installed in the top of the vent pipe when a proper downdraft cannot be achieved. Abby Rockefeller has installed a small wind-propelled rotary ventilator at the top of her vent pipe in her home in Cambridge. She told us that because of the density of her neighborhood, air currents are not what they should be, and the fan helps ventilate the moldering wastes. Such a fan is also recommended when the vent pipe cannot be of the proper height (see discussion earlier), or when the natural draft within the house might be diverted from the toilet unit by a fireplace chimney or woodstove flue. The CM people do caution new owners to make sure that an accumulation of spider webs in the vent pipe is not responsible for a poor draft. If spider webs do exist, getting rid of them is simply a matter of lowering a rag weighted down by a rock down through the top of the vent pipe.

The excrement and garbage chambers are never emptied. Only after the finished compost begins to appear in the storage chamber some two to four years after start-up does the unit need to be emptied.

While it is not necessary for a CM owner to be a soil microbiologist, he or she will have to learn how to live with a live and active biological process in the house. If you already have a thriving compost pile then you'll understand a bit better than other people how to get along with your composting toilet. The CM people make these recommendations in their owner's manual.

DO:

1. Keep the toilet seat and garbage inlet closed when not in use.
2. Keep the air inlet and outlet free of obstruction.
3. Put in organic kitchen wastes including all foods, grease, bones, lint from laundry machines, fireplace ashes (cool), vacuum cleaner bag contents, and other household dirt, cat litter, etc.
4. Put in organic outdoor wastes including weeds, grass clippings, leaves, seaweed, peat moss, etc.
5. Put in organic waste from industry and agriculture including sawdust, pressed apples from cider press, etc.
6. Put in all toilet wastes including sanitary napkins, tampons, disposable diapers (after removing plastic covers), etc.
7. Use only soap and a wet brush or paper towel to wash the inlets.

DO NOT:

1. Put in inorganic wastes including cans, bottles, plastic of all kinds, etc. These materials will not decompose and will take up space.
2. Put in inorganic waste including chemicals, antiseptics, drugs, antibiotics, other medicines, darkroom wastes, etc. These materials will damage the process by killing sections of the decomposing organisms.
3. Put in certain organic wastes including large volumes of newspaper, cardboard, or colored paper. Although these will decompose, they contain certain toxic dyes and inks that will harm the process.
4. Put in lighted cigarettes or warm ashes; if they're hot enough they could start a fire in the tank.
5. Put in large quantities of liquid (i.e., more than one gallon at a time). The unit is not designed to accommodate any liquid other than urine. Too much liquid will not be readily evaporated and the system will go anaerobic.

The unit is never completely emptied; only surplus is removed from the storage chamber. An average family of four should be able to use the unit up to ten years before having to remove any fertilizer.

This long holding period, at least two to four years before any material can be removed, is how the unit kills pathogenic organisms. Temperatures within the unit seldom go above 90°F (32.3°C), but the organisms are subjected to the competition and predation of other organisms for such a long period of time that, according to the CM people, there is no danger factor involved in using the finished product directly on food crops.

Toa Throne

The Toa Throne, like the Clivus Multrum, is a composting toilet that breaks down kitchen and toilet wastes through aerobic decomposition. But it's quite a bit smaller than the Clivus, it's easier to install, and supposedly breaks down wastes quicker than the Clivus.

The polyethylene composting tank looks much like the Clivus Multrum, but it measures only about 5½ feet wide by 4⅓ feet high by 3 feet deep and will service four to six people year-round. Unlike the Clivus, it has only a toilet chute; all kitchen garbage, as well as toilet wastes, go down this chute. The Toa Throne people claim that one chute is better than two because, since all the wastes go to the same spot, better layering (feces-kitchen wastes-feces), so important in composting, can be achieved. (It is only fair to mention, though, that the CM people say two holes are better because wastes spread out more inside the tank and have better exposure to circulating air.)

The unique feature of the Toa Throne is its "air staircase," a stepped, perforated, slanting bottom to the decomposition area. This staircase mixes the air that comes in from holes underneath the steps into the wastes. Semicircular distribution conduits running through the stored wastes from the bottom of the air staircase down to the emptying hatch help in aerating the pile and also in breaking up the incoming wastes to prevent them from packing together.

The unit comes with an electric fan. It is installed in the ventilation chute to maintain good air circulation up through the air inlets under the staircase and out the ventilation chute at the top of the chamber. As in all composting toilets, good ventilation is necessary to evaporate liquid, keep the system from going anaerobic, and draw odors up and out the vent chute rather than up and out the toilet opening. The ventilation chute does not come with the unit, and the installation manual suggests you make one from PVC or sheet metal tubing. This chute should have an inside diameter of four inches and, like the Clivus's vent pipe, should be insulated wherever it runs through an unheated or outside space to prevent outgoing evaporated liquid from condensing along the sides of the cold chute and dropping back into the chamber. The tank itself should also be insulated if you suspect the air sur-

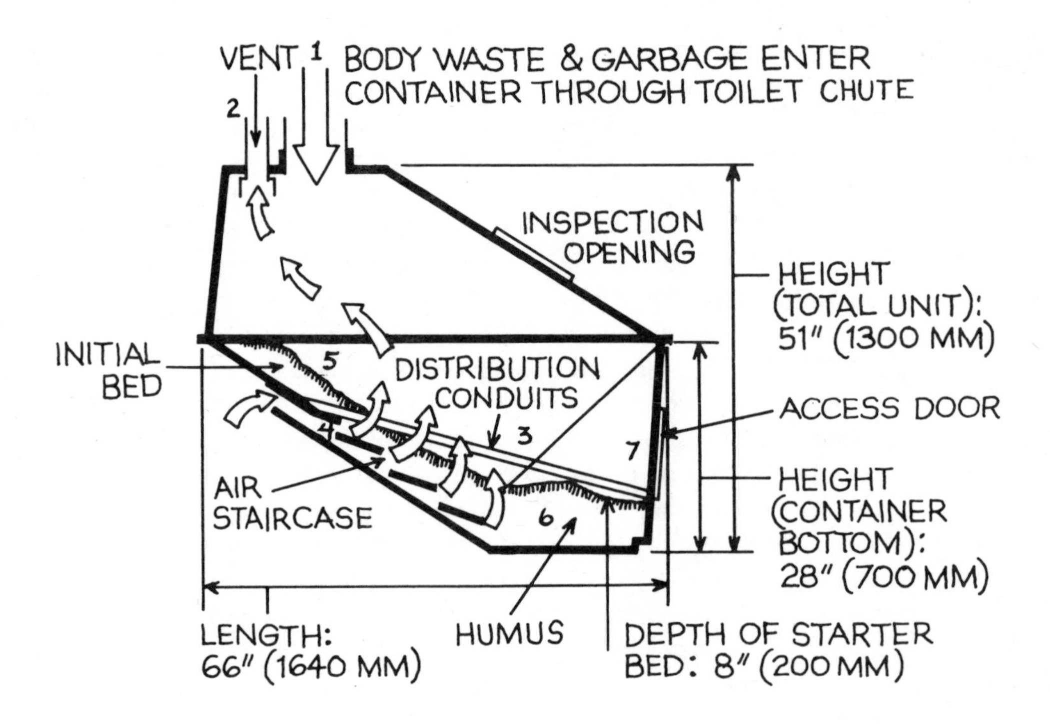

Cross section of the Toa Throne. The unit operates as follows:

1. Body waste and kitchen garbage enter the container through the toilet chute.
2. An electric blower draws air out, carrying evaporated liquid from the container. Where no electricity is available, convection created by heat from the decaying process is used for ventilation. Gases and the evaporated liquids are then conveyed to the outlet above the roof by the ventilation stack.
3. Longitudinal distribution conduits loosen up and prevent mass from packing together, as well as aerate the mass and drain the urine and other liquids to the earth and humus bed at the floor of the container.
4. The air staircase allows air to penetrate the decay bed from below. The stairs are inclined downwards at a 41° angle and cannot be obstructed. Thus, fresh air continuously circulates through the mass. After the air passes through the decaying matter, it is drawn off with the gases and the evaporated liquid through (2) the ventilation pipe.
5. After primary decomposition, the mass ends up on an initial bed of peat moss and compost soil or bark humus.
6. As the microorganisms decompose the mass, it is transformed to humus at the bottom of the container near the removal hatch.
7. The humus is removed through the access door from one to three times a year, depending upon the volume of usage.

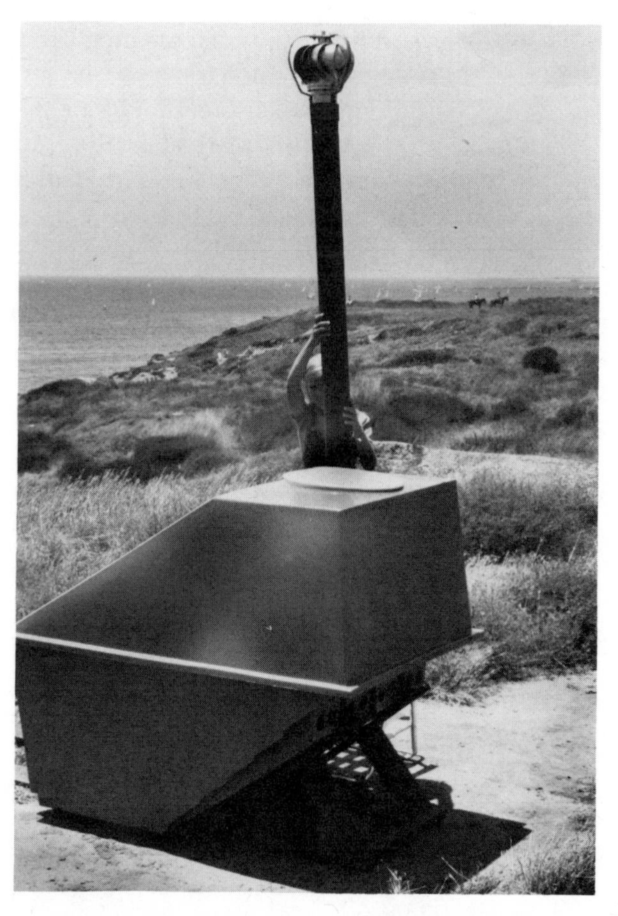

This picture of the Toa Throne shows the toilet seat in its simplest installation, directly on top of the decomposition tank. Most people, however, probably prefer using the toilet seat chute so that the tank can be installed out of sight, under the bathroom floor.

rounding the unit will drop below 66°F (19°C).

There is no heater; the Toa Throne people say that one is not needed because air circulation is sufficient to evaporate excess liquid. As a matter of fact it seems that the concern should not be excess liquid but occasional overdrying due to the effectiveness of the air staircase. In case liquid buildup does become a problem, though, the tank is designed so that a standard heating element can be installed on the floor inside the tank, under the air intake holes.

For household use, the tank may be installed below floor level, with the toilet above the floor attached to the container by a connection tube. For second floor installations, ideally the unit should be installed in the first floor in a closet, with the connection pipe leading to the toilet on the floor above. (Installation on the second floor itself is *not* recommended due to the weight of the loaded unit.) In outdoor installations, the Toa Throne may be set up with or without the toilet because in its simplest installation the seat and lid can be mounted directly on the container.

Like the Clivus, a starter bed must be put in the Toa Throne before it can be used. Six cubic feet of peat moss and three cubic feet of finished compost or bark humus should be spread loosely over the entire length of the staircase and the bottom of the tank. Plants containing tannic acid, like oak leaves, should not be part of the starter bed, and actually should never be added to the toilet. The Toa Throne people suggest that two pounds of lime be added to the starter bed and that a handful be thrown down the toilet opening from time to time every year.

No decomposed waste should be removed for 1½ to two years after the unit is in operation, and then decayed matter can be removed one to three times a year, depending upon the number of people using it regularly. The Toa Throne people claim that the unit will produce about 60 pounds of decomposed waste per person per year.

Mullbank (also known as Ecolet)

The Mullbank people describe the process that goes on inside their unit as decomposition "accelerated with an assist

End Product of Complete Aerobic Decay

The end product of complete aerobic decay consists of approximately:

Dry substance	45 %
Humus	30 %
Nitrogen total	1.9 %
Ammonium nitrates	.025%
Phosphor	.8 %
Calcium	1.5 %
pH	7-8.5
Carbon/nitrogen relation	15-20
E. coli (per gram)	0-100

Source: Chart is reprinted with the permission of Enviroscope, Inc. (P.O. Box 752, Corona del Mar, CA 92625) which compiled the data from the results of a test conducted under the auspices of the Swedish Health Department.

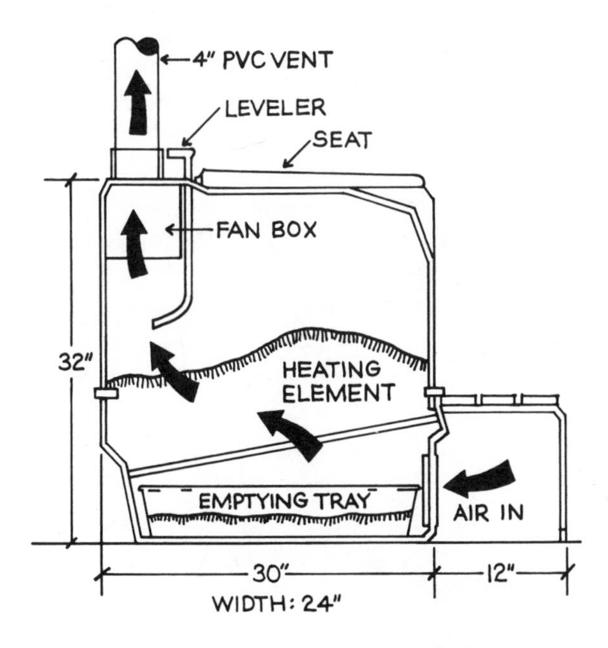

Cross section of the Mullbank, showing that air is introduced in the bottom of the unit, under the footrest.

from modern technology." What this means is that their self-contained toilet is fitted with a thermostatically controlled heating coil that keeps the temperature inside at 100°F (37.8°C). It also means that a ventilating fan works automatically to aerate the pile, remove odors, and evaporate the liquid; and that a manually operated stirring arm spreads and turns wastes in the tank.

Because the Mullbank doesn't have to rely entirely on natural processes like the natural heat generated by a composting pile of wastes and natural drafts, the unit can be small. And actually, it is quite a bit smaller than the Clivus or Toa Throne. It measures only 3½ feet long by 2¾ feet high by two feet wide. The manufacturer claims it will serve five to six people. The polystyrene chamber and toilet seat are one and the same: the unit is merely a storage chamber with a seat and lid on the top. A foot-high wooden platform allows users to step up to the toilet seat. The whole thing is designed to sit as is in any room of the house, so installation is quite simple. The most difficult part is making and installing the long ventilation pipe that should, like that of the Clivus Multrum and Toa Throne, extend above the roof line of the house and be insulated everywhere it runs into unheated or outdoor spaces.

The Mullbank must be primed by spreading 20 pounds of peat moss (that comes with the unit) over the heating coils in the bottom of the tank. The manufacturers give these guidelines for use and maintenance:

1. During the initial period of use, the amount of liquid coming into it should be limited since, at that time, the Mullbank only contains peat moss and there is not enough solid matter to absorb all the urine from three to five persons.

2. Always keep the waste pile moist. If the waste becomes too dry, the organic breakdown process stops and the capacity of the waste pile to absorb liquid diminishes. The rear part of

Like the other small units, the Mullbank, decomposition tank and all, sits right in the bathroom. Although the vent stack is not included in this picture, it is hooked up to the round opening, directly behind the toilet seat. The small handle behind the seat controls the leveler.

the waste pile easily dries out unless moist material is spread rearwards in the Mullbank. For this reason, water should be added to the rear part of the waste pile now and then.

The waste pile often becomes too dry if the Mullbank is used by only one or two people. Under such conditions, a couple of quarts of water should be spread over the entire waste pile now and then.

3. Make a habit of occasionally covering the waste pile with a little grass, a few leaves, or a thin layer of ordinary peat moss. You may also use natural topsoil. Paper also stimulates the organic breakdown process, so there is no reason to worry about too much toilet paper getting into the Mullbank. Paper has a good absorption capacity as well.

4. Check the collecting tray at least once every month. If there is a little liquid in the tray, this does not matter. You can add some peat moss or natural topsoil to absorb the liquid and break down the substances in the urine.

If the liquid does not evaporate or the amount of liquid in the collecting tray increases, then the Mullbank is either being overutilized or it is not working properly.

5. Spread out the waste pile, at least once every month, by rotating the distributing arm. This ensures that the whole surface area is used and organic breakdown is thus accelerated.

6. Do not allow the toilet seat lid to stay open unnecessarily. If the lid is kept closed, flies are kept out of the Mullbank and air is prevented from coming in the wrong way.

7. Keep the bathroom (or other room that houses the Mullbank) warm.

8. Small quantities of kitchen scraps can also be thrown into the Mullbank, primarily bits of fruit and garden vegetables as well as other refuse rich in carbon.

9. Certain substances must never be allowed to get into the Mullbank. These include inflammable materials (acetone, gasoline, etc.), burning or glowing materials (cigarette butts, burning or glowing matches, etc.), metal parts, plastic materials, glass, detergents, or other chemicals that can disturb the biological process or attack the plastic material. No disposable diapers, tampons, sanitary napkins, etc.

10. If your Mullbank is in a cabin or summer house that is left unoccupied for a short period (about one week), a couple of quarts of water should be spread over the pile of waste unless it is already very moist.

11. In the case of the toilet not being used for a long time (two to three weeks), the electric power should also be switched off before leaving the house. If there is surplus liquid present or if the pile of waste is very moist, the current should be left on (for a period of about three weeks) and no water should be added.

When the toilet is to be used again after being idle for a long time, a thin layer of grass, peat moss, and a handful of natural topsoil should be spread over the waste pile, followed by two to three quarts of water.

During the winter, the waste pile must be thawed out before the Mullbank is used. Have the power switched on about 24 hours beforehand.

The Mullbank should always be half full if it is to function properly. Emptying once a year is usually sufficient, but this will, of course, depend upon the use it gets. Emptying is a simple matter of removing the wastes from the collecting tray and scraping down a new layer from the tank into the tray so that the tank is half full. The removed waste should be put on garden ornamentals and not on food crops.

Mull-Toa (also known as Biu-Let, Bio-Mat, and Soddy Potty; formerly known as Humus H5)

This is another small-volume unit, that, like the Humus E, distributes the waste after each use. But instead of a rotation drum, the Mull-Toa has a rotor that levels the wastes. The slice-iron bar should be periodically used to rake down decomposed matter into the mold boxes.

The special feature of this toilet is its patented thermostat-controlled air-recirculation system that evaporates the moisture and maintains aerobic conditions. The heating elements in the air-recirculation pipes keep the temperature inside the unit at about 95°F (35°C). Humidity is monitored by the built-in hygrometer, and the air-outlet control can be manually adjusted to maintain proper humidity. There is a fan in the pressure chamber to insure a continuous air flow, and a diffuser installed on the upper part of the vent stack prevents unpleasant odor outside the house.

It should be started by distributing three to five quarts of compost or good garden soil on top of the mat of cellulose tissue inside the toilet. Once a year a pound or so of compost, good garden soil, or the specially prepared "humus soil" sold by the Mull-Toa people should be added. During the first one to three months or until the contents have reached up to the stirrer, the toilet cannot accommodate more than about four quarts of urine in each 24-hour period. After this period there is enough bulk in the tank to absorb five to six quarts each full day.

The mold box need be emptied only once a year. If the unit is only used on weekends or vacations, it probably will only need to be emptied every two years. The contents should be buried at least a foot below ground level around ornamentals.

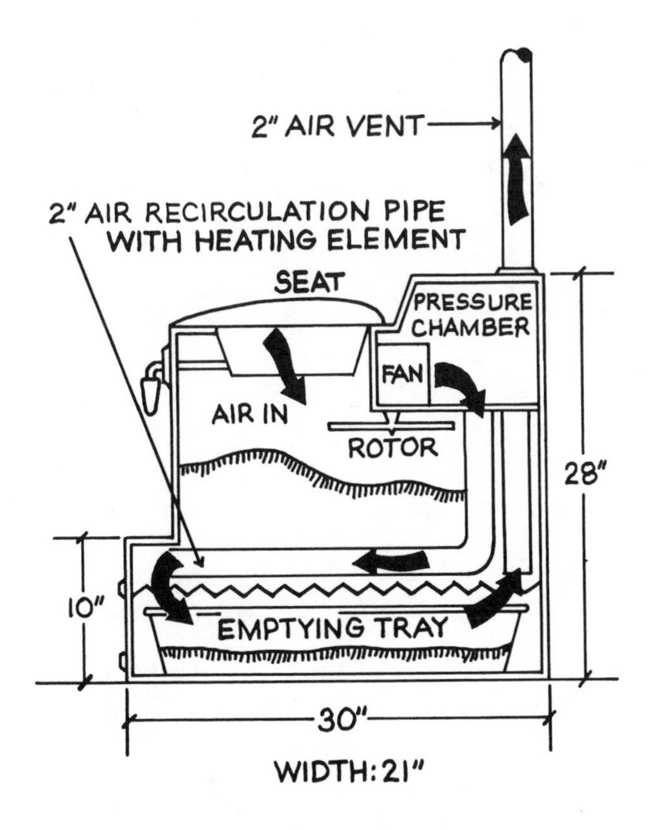

Cross section of the Mull-Toa toilet. The fan in this unit pushes the air through and out the tank, rather than pulling it out, as in the other toilets.

One company that sells the Mull-Toa (under the name Soddy Potty) has, just at the time of this writing, come out with the Soddy Potty #2, which is the Soddy Potty/ Mull-Toa without any electrical devices. It has a hand-operated leveling bar and no electric fan, just a wind-driven rotary ventilator at the top of the vent stack. Heat loss from inside the unit is decreased by the one inch of foam insulation that covers the exterior.

There is no heater for the unit; instead, PVC piping is hooked up on one end to the toilet and on the other to the house's wood stove, baseboard heater, hot air vent, the back of a refrigerator, or other convenient heat source. A solar collector (which comes in sizes of two feet by four feet or four feet by four feet and is available from the Soddy Potty people as an option) supplements the house heat during winter and takes over in the summer. The updraft pulls the heat from the house and/ or the collector through the toilet chamber.

The unit has a number of cleverly designed dampers. Most are thermostatically controlled with a mercury switch. These dampers only open when the temperature is 90°F (32.3°C) so that a lot of hot air is not drawn up the stack and lost from the house. The damper in the lower vent stack is humidity and ammonia sensitive

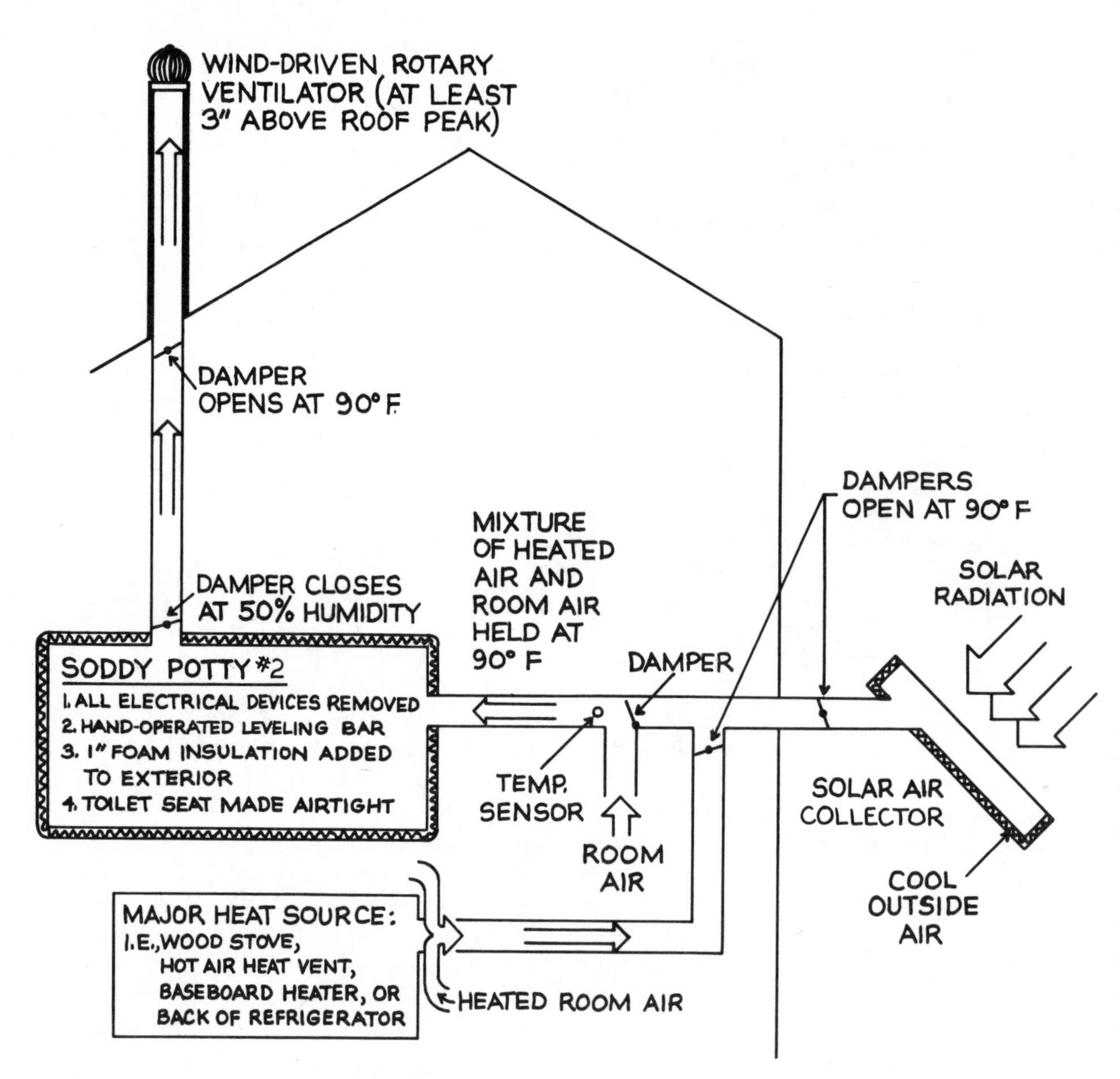

The Soddy Potty #2 is a version of the Mull-Toa toilet that is heated not by a fan, but by either a solar collector or the house's major heat source.

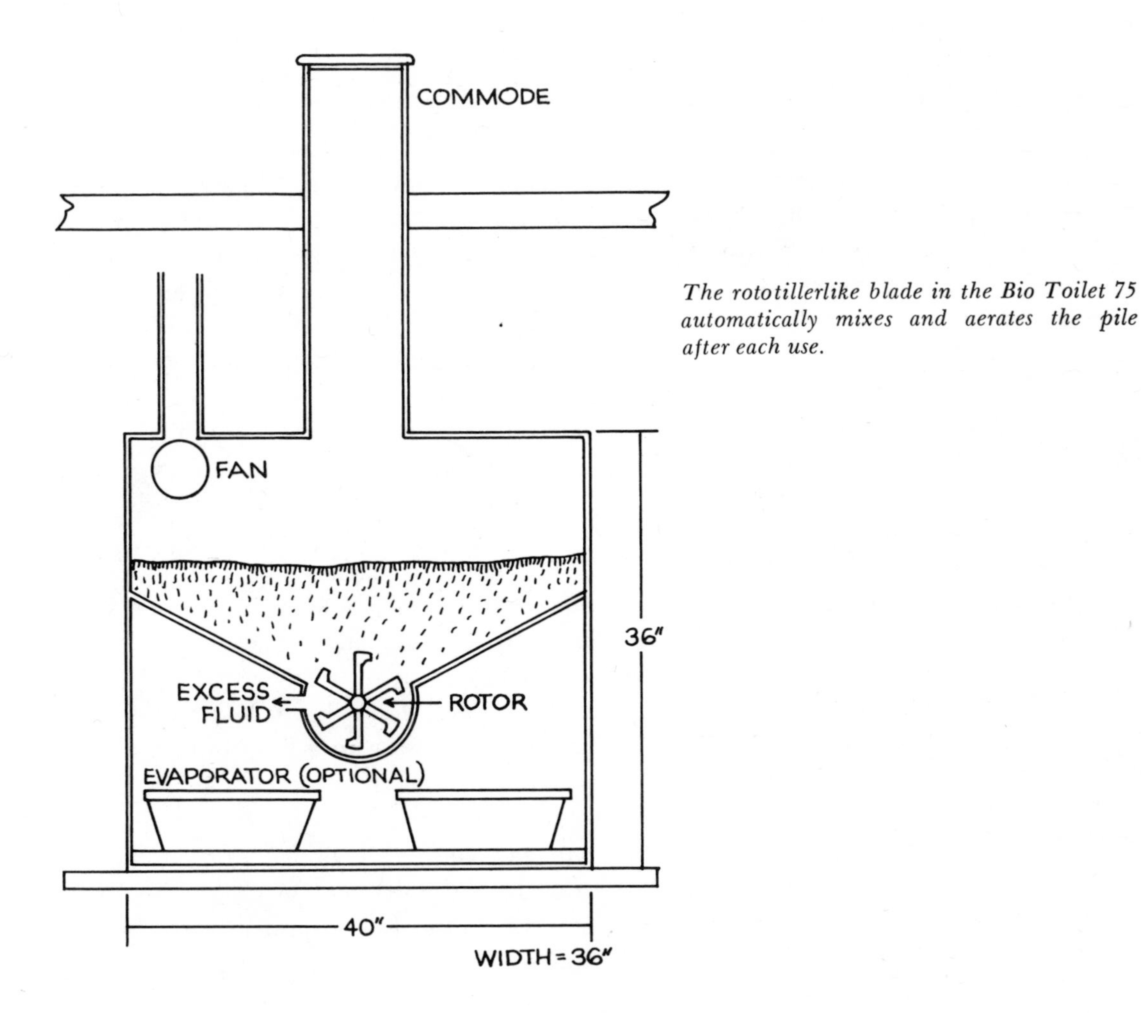

The rototillerlike blade in the Bio Toilet 75 automatically mixes and aerates the pile after each use.

and closes at 50 percent so as not to let the updraft dry out the decomposing matter too quickly and create compaction problems.

Bio Toilet 75 (and 75B)

This is the largest of the active units. The manufacturers claim that the Bio Toilet 75 can serve up to 10 people on a regular basis. The Bio Toilet 75B has a greater capacity because it has three compartments instead of one, and this supposedly makes for greater evaporation. (The removable, funnellike chute under the seat tends to soil more readily than a straight chute or one that widens as it goes down into the chamber.)

All wastes—bathroom and kitchen—go down the toilet chute. The decomposing tank, which is installed under the bathroom floor, is fiberglass and measures three feet wide by 3⅓ feet long by three feet high. It does not come insulated, so owners should either insulate the tank or keep it in a room-temperature area. Higher temperatures are not necessary because the pile is kept warm by air that is electrically heated and forced through the decomposing tank. A blade similar to that on a rototiller automatically mixes and aerates the pile every time the toilet lid is opened. There is an exhaust fan inside the chamber to insure proper draft, and a diffuser, as on the Bio Toilet A and M, on the upper part of the vent stack to prevent unpleasant odor outside the house. The vent stack itself should be insulated, as with the other composting toilets, and although the height of the vent is not as crucial as it is with the Clivus Multrum or Toa Throne because of two fans, it should go beyond the roof line of the house.

The Bio Toilet 75 should be started by spreading good garden soil (or the humus soil available from the distributor) on the floor of the chamber through the toilet opening.

It should not have to be emptied for at least one year of normal use (and less if used only on weekends or vacations). What organic matter you do remove should be buried at least a foot below ground level around ornamentals.

Bio Toilet A (and M)

One distributor of the model A describes it as "the cement mixer toilet." This is because there is a tapered drum inside the Bio Toilet A that rotates on the horizontal axis every time the lid is opened. This rotation both aerates and mixes up the pile as it slowly transfers the wastes from the toilet opening to the "mold box" where the partially decomposed contents are further composted and stored.

Another purpose of this rotating drum is to provide a barrier between the toilet opening and the decomposition tank. When the drum comes to a stop the opening is on the underside of the drum so that upon lifting the lid you see nothing but the solid topside of the drum. There is no black hole to look down, and this is important to those who are squeamish about seeing their own wastes, which would otherwise be pretty visible in such a small unit. Of course, the hole does rotate to the topside of the drum right under the toilet seat when the lid is pushed back all the way so that it can be used properly.

The Bio Toilet M resembles the model A in every way, except that it does not have this rotating drum. A manually operating slice-iron bar moves back and forth, forcing the decomposing material to fall through a grill down into the storage compartment. This slice-iron bar should be used every two weeks after the first two months of operation.

The Bio Toilet A and M are said to serve as many as six people on a daily basis. They are small units (22 inches wide by

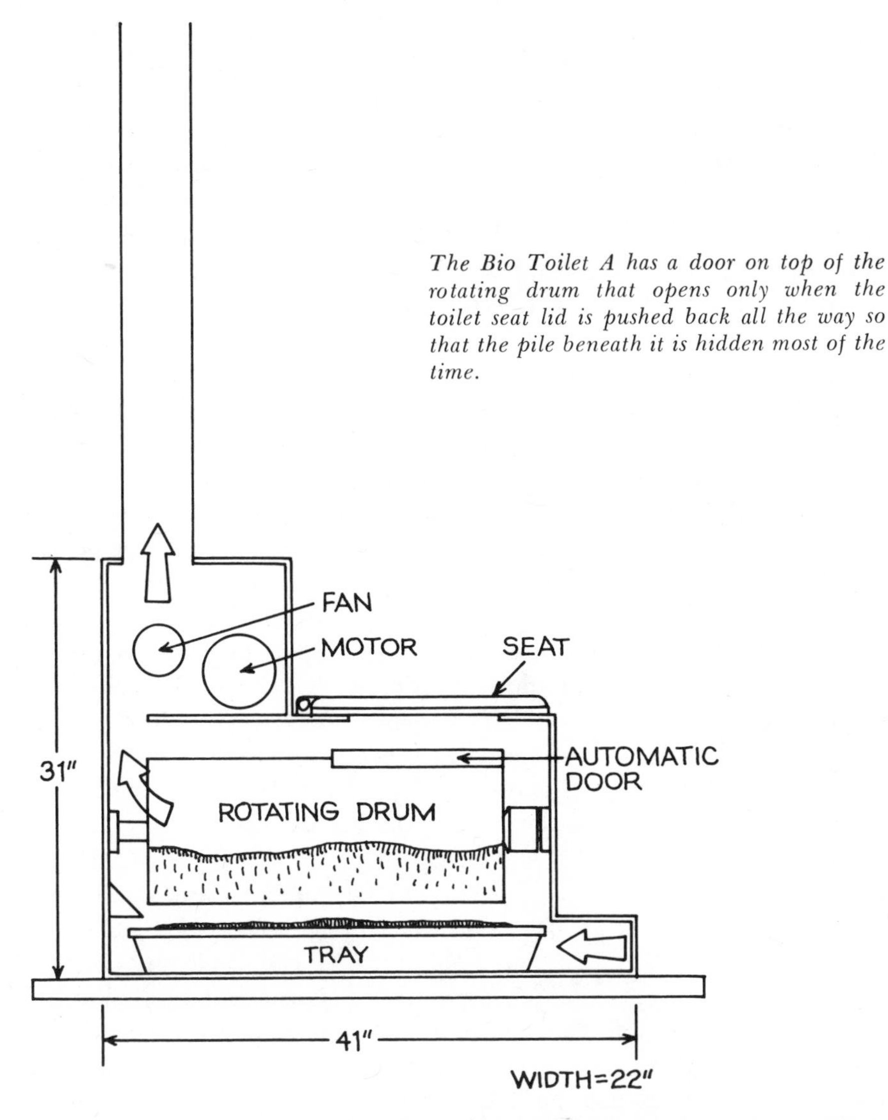

The Bio Toilet A has a door on top of the rotating drum that opens only when the toilet seat lid is pushed back all the way so that the pile beneath it is hidden most of the time.

41⅜ inches long by 31 inches high) and should be installed right in the bathroom, where a 65° to 75°F (18.3° to 24°C) temperature should be maintained when the toilet is in use. Like the other small models, the toilet seat is directly on top of the decomposing tank, and the user steps up a few inches to sit on the "throne." Installation is quite simple. There is an exhaust fan inside the unit and near the top of the vent stack there is a diffuser which is a perforated pipe that allows air to be sucked into the stack and mix with the air rising from the toilet. The manufacturer claims this diffuser will cut down on any foul odor coming up and out the stack. The air inside the tank is electrically heated and automatically controlled to maintain an optimal decomposing temperature of about 90°F (32°C) and the right humidity. The moisture will seldom be too high under normal use because of the heater and fan, but can be too low on occasion. If that becomes the case for more than a day or so, a quart or two of water should be poured evenly over the pile.

Start either model off by pouring the humus soil that comes with the units through the toilet opening. All that remains after a year of use is a drawerful of organic matter. It should be removed and buried a foot under the ground around ornamentals. Small food scraps and other organic matter can and should be added along with feces and urine, and it's recommended that one pound of specially prepared humus soil (or good garden soil or compost) be added once a year inside the rotation drum or in model M, on top of the grill.

Bio Loo

What sets this toilet apart from the other small units is the built-in heater which not only maintains a steady decomposition temperature in the tank of about 95°F (35°C), but can also be turned on high enough to pasteurize the organic matter in the stainless steel storage drawer. When the drawer is full (which will happen about every other month if four people are regularly using the Bio Loo) the heater should be turned up to 158°F (70°C) and left on automatic time for six hours. The Bio Loo people say the wastes are then safe to use anywhere. If you don't wish to use the pasteurizer, you can remove the contents of the drawer and incorporate them in a healthy compost pile for at least six months.

Such pasteurization or long composting is necessary in the Bio Loo because kitchen and bathroom wastes don't remain long enough in the unit to decompose completely. The Bio Loo has the shortest storage time of all the toilets we know—only two months. For people who want a good amount of useable organic matter pretty regularly (and are hesitant to put their own fertilizer in the garden without pasteurization) the Bio Loo may be just the thing. But for those who want to be bothered as little as possible with emptying the storage tank of their waterless toilet, the Bio Loo may prove a nuisance.

The polyethylene unit is about the same size as the other small units: 25 inches wide by 31½ inches long by almost 26 inches high. It's designed for a four-person household and should be able to accommodate use by guests from time to time. Small kitchen scraps should be added from time to time, as should a handful of garden soil and some grass clippings or other garden cuttings.

There is a built-in steel rotor for loosening up and aerating the pile. It's manually operated by turning a lever behind the toilet seat. The mass should prob-

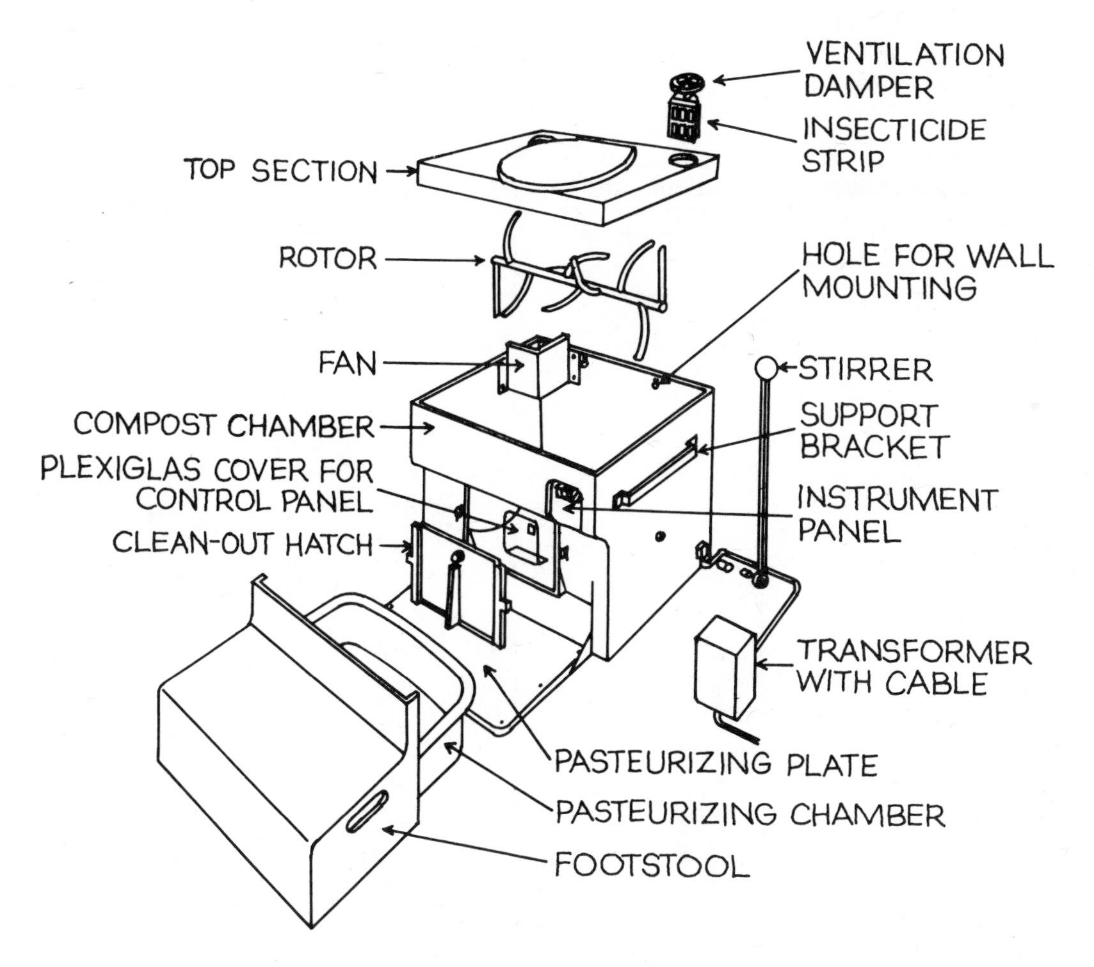

The Bio Loo is the only toilet with a pasteurization tray whose heating element brings temperatures up high enough to render organic matter in the tray safe enough for full garden use.

ably be stirred after each use. A fan inside the chamber aerates the pile, and like the other toilets, the air is drawn up and out a vent stack. An air outlet control can be adjusted in case the built-in hygrometer registers too high or low.

As mentioned earlier, the Bio Loo will have to be emptied about every other month. To reactivate the decomposition process after each emptying, a bag of specially prepared peat material (or good garden soil or compost) should be added.

A COMPARISON OF COMMERCIAL

Toilet	Cost as of April 1977	Size	Capacity
Clivus Multrum	$1,300–$1,685 depending upon accessories	45 x 101 x 68	3–10 persons depending upon the size unit and number of mid-sections
Toa Throne	$975–$1,190 depending upon accessories	39 x 66 x 51 (in cm: 99 x 164 x 130)	4–6 persons with overload

COMPOSTING TOILETS

Features	Advantages	Disadvantages
rotary roof ventilator fan 110V, 46W separate kitchen waste chute insulated tank optional horizontal spiral conveyor which makes it possible to put toilet and/or kitchen chute somewhere other than directly over composting tank	uses little electricity tolerates temporary stress like dips in temperature, large amounts of urine, plant wastes will handle a greater variety of wastes than most other units large surface area to encourage natural vent and material mixing large pile is relatively self-insulating; end product claimed to be pathogen-free is one of the most popular composting toilets and has benefited from extensive "field use" and testing	the most expensive its large size can lead to installation problems needs large amounts of wastes and will compete with garden compost pile long start-up period (1 year or more) before equilibrium is reached liquid buildup in the bottom section is not unusual and occasional emptying may be necessary
optional fan 110V, 20W	uses little electricity easier to install than CM because of its size it can handle temporary stress, as the CM will handle a greater variety of wastes than most other units large pile is relatively self-insulating air staircase encourages natural mixing and ventilation	air staircase can over-ventilate the pile and cause too much drying end product not claimed to be pathogen-free; should be composted for 6 months long start-up period (1 year or more) before equilibrium is reached has not been in use as long as the CM, not tested as extensively needs good quantities of wastes to function properly and can compete with garden compost pile

A COMPARISON OF COMMERCIAL

Toilet	Cost as of April 1977	Size	Capacity
Mullbank (Ecolet)	$736 inclusive	24 x 42 x 32	6 people with occasional overloads
Mull-Toa (Biu-Let, Soddy Potty, Bio-Mat, and previously called Humus Toilet H5)	$795 inclusive	21 x 30 x 28	4 people with only very occasional overloading

COMPOSTING TOILETS (cont.)

Features	Advantages	Disadvantages
heating coil 42V, 140W fan 42V, 21W manually operated distributing arm for mixing and leveling organic matter	most widely sold composting toilet easy installation extensive testing has been done on the end product and on the unit itself	kitchen and plant wastes should be cut into small pieces before adding to toilet scraping organic matter from between the heating coils and elsewhere into the collecting tray for removal can be inconvenient urine must be limited during start-up period end product not said to be pathogen-free; bury around ornamentals or compost for 6 months rear part can easily dry out unless moist material is spread rearwards and water is added occasionally pile can also dry out if only 1 or 2 people are using it
hygrometer air outlet control powered rotary arm that goes on automatically after each use and then levels the pile fan 110V, 250W thermostatically controlled heating elements	easy installation automatic leveling of organic matter greatest reduction of waste by weight of all the units (those who want plenty of humus for their gardens may view this as a disadvantage)	can be easily overloaded during the first 3 months urine must be limited to 4 liters/24 hours susceptible to drying out; water should be added in this case can handle only kitchen and plant wastes in small pieces raking down the contents into the collection boxes to avoid caking should be done once a week and can prove to be inconvenient and difficult at times

(continued)

A COMPARISON OF COMMERCIAL

Toilet	Cost as of April 1977	Size	Capacity
Mull-Toa (continued)			
Soddy Potty #2	NA	21 x 30 x 28	4 people with very occasional overloading
Bio Toilet A (and M)	(A) $795 inclusive (M) $550 inclusive	21 x 41 x 31	6 people with occasional overloads
Bio Toilet 75 (and 75B)	(75) $995 inclusive (75B) $1,395 inclusive	36 x 40 x 36	10 people (75B: more than 10 people)

COMPOSTING TOILETS (cont.)

Features	Advantages	Disadvantages
		when the rake gets stuck in the grate above the collection box noise of the fan, which runs continuously, can be bothersome
wind-driven rotary motor optional 2 x 4-foot or 4 x 4-foot solar collector insulated tank 4 thermostatically controlled dampers 1 humidity and ammonia sensitive damper	no electricity required dampers automatically control humidity and can prevent great amounts of heat from being pulled up and out of the house	installation may be difficult because unit must be hooked up to collector and/or house heat source particle size and quantity of kitchen and garden wastes must be limited one of the newer and lesser tested models
(A) power operated rotating drum 115V, 115W (M) manually operated slicer fan 115V, 23W 3 heating elements 115V, 110W each	(A) pile is mixed and aerated automatically after each use easy installation wastes are hidden from view when lid is opened; this barrier also discourages flies and odors from coming out the seat opening	(A) uses more electricity than most other units susceptible to drying out plant and kitchen wastes should be cut into small pieces before depositing end product should be buried around ornamentals or composted for 6 months one of the newer and lesser tested models
2 fans 115V, 32W ⅓ HP rotation motor 115V, 248W heat blower 110V, 1,200W	will tolerate overload urine overloads handled in the 75B model by an evaporator that goes on automatically when tolerable liquid level is exceeded	uses the most electricity of all the composting toilets uses more bathroom space than other units unless it's installed partially beneath the floor

A COMPARISON OF COMMERCIAL

Toilet	Cost as of April 1977	Size	Capacity
Bio Toilet 75 (continued)			
Bio Loo	$795	$24\frac{13}{16}$ x $31\frac{1}{2}$ x $25\frac{5}{8}$	4 people with occasional overloading

COMPOSTING TOILETS (cont.)

Features	Advantages	Disadvantages
	pile is automatically mixed and aerated after each use by rototillerlike blade	one of the newer and lesser tested models funnellike toilet chute soils more readily than those on most other units end product should be buried near ornamentals or composted for 6 months
fan 110V, 23W heating coil 110V, 30W pasteurization hotplate 110V, 160W manually operated rotor	easy installation pasteurized end product is safe to use anywhere	must be emptied every 2 months if 4 people are using it regularly rotor can get stuck or break if pile compacts plant and kitchen wastes to be deposited must be in small pieces pasteurizer uses more electricity than simple fan or heating element found in other models

Owner-Built Toilets

Because of the expense of the commercial models and the relative simplicity of their design, do-it-yourselfers have gotten into the act, just as they are designing and building their own solar collectors, windmills, and other low-technology hardware. Composting toilets seem easy enough to put together, and while, indeed, there are no highly technical mechanical parts to build, install, and keep running properly, they must be designed very carefully so that the composting process, once it gets going, can stay in balance.

In our pursuit of good, homemade units we learned that one should not be fooled by their outward simplicity. We heard several sad stories of homemade toilets that worked terribly right from the start, and of others that seemed to function well enough initially, but after a few weeks began to clog up, attract flies, or create odor problems that got so bad that the units had to be removed or rebuilt.

What we present here are three designs that are some of the best we've seen. Their builders warn us that their toilets are still in experimental stages and have not been in operation long enough to withstand the test of time. We pass this caution on to you; should you wish to build your own using any of these as a model, plan on doing some experimenting.

Before designer-builders discuss their own units, we'd like to share with you some tips we got from Zandy Clark, who has had much experience in building composting toilets. Although they can be built in many different shapes and sizes, there are some basics that apply to all types.

Some Tips for Building Composting Toilets

The best way I know to give guidelines for building a good unit is to talk about some potential problems and how to prevent them from happening. And the best way to explain some of these problems is to relate the experiences a friend of mine had with the composting toilet he built himself.

Mike, the builder, is a knowledgeable and skilled person who had access to drawings of some brand name toilets. His toilet has a good draft due to a straight eight-inch vent pipe of blue fiberglass sewer pipe, two stories high, insulated above the roof and assisted by a blower. The box is plywood coated with fiberglass resin. Mike's design looks very much like the Norwegian Kombio composting toilet. There are two layers of perforated four-inch plastic pipe running up the slope, connecting the bot-

Zandy Clark heads up Maine Natural Systems in Brunswick, Maine, a group that specializes in alternative energy and wastewater systems. He is also publisher of Compost Toilet News, *Star Route 3, Bath, ME 04530.*

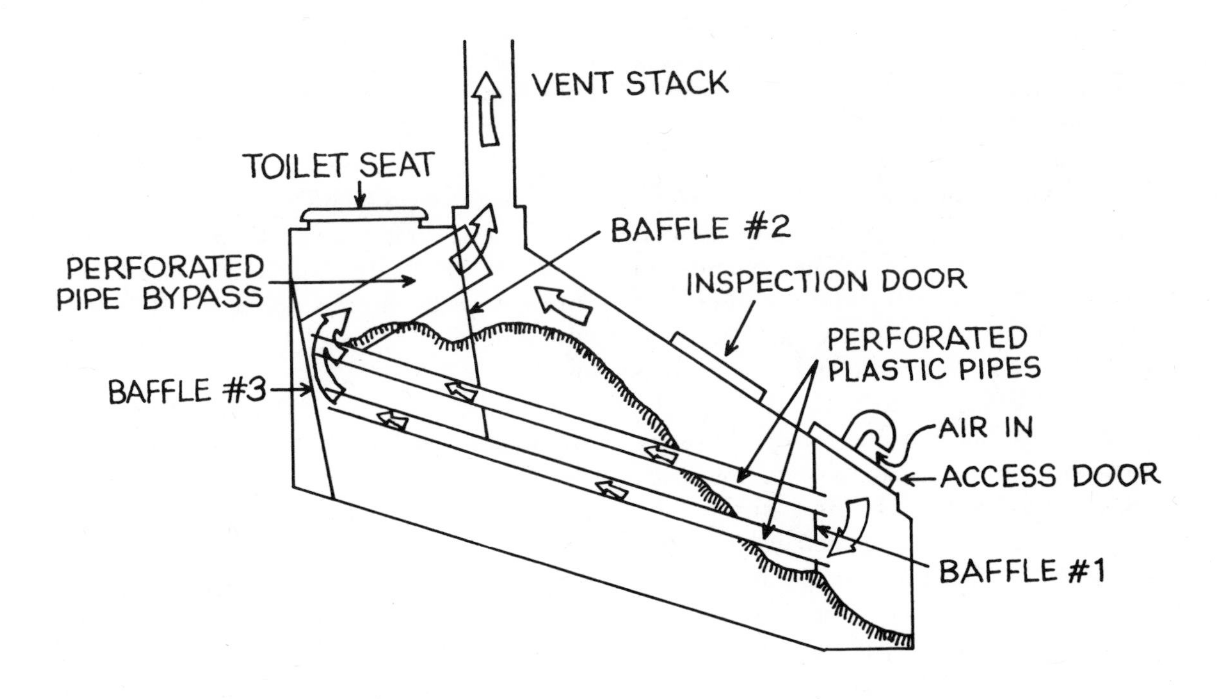

The original Kombio toilet with two perforated plastic pipes running the length of the decomposition tank.

tom chamber to the upper baffle #3. These pipes pass through all three baffles through four-inch holes cut in the baffles. They are screwed in place to the baffles so they don't slide down with the compost and become dislodged. There are three of these perforated pipes in each layer, making six altogether. They are spaced about a foot apart, to allow the compost to pass through and down them.

This baffle and pipe arrangement allows air to penetrate the pile. It also allows most of the air to pass unhindered directly out the vent, via the bypass pipe. This second function is essential, because if no air can pass through directly, evaporation will proceed at a very slow rate, due to high relative humidity in the container. The toilet must evaporate two pounds of moisture per person per day, and compost only three ounces of dry solids per person per day. Obviously, it is the two pounds of moisture that can cause the problems.

Some of this moisture will not be evaporated by the time it seeps through the bed, so it will pool up in the bottom chamber. Here, it must especially be exposed to moving air. Wind dries clothes hung on a line even in winter and even when the sun is hidden. The baffle and pipe systems allow a small wind to move through the bottom chamber, into the perforated pipes, and through the bypass pipe to the vent. This causes evaporation from any pooled water in the bottom, as well as from the bed itself. Much more evaporation takes place in the bed because it is warm, and because each small particle of cellulose (peat moss, sawdust, or grass clippings) acts as a wick, exposing moisture on its surface to the surrounding tiny air pockets. If enough peat moss, sawdust, or grass clippings are added *daily* to keep it aerated, water is spread to every portion of the pile by capillary action and evaporation takes place uniformly from most of the bed.

Mike's toilet ran well for the first six months or so, because he added at least a quart of peat or sawdust daily, and because, unlike most homemades, his pipe and baffle system was allowing free air movement through the pile. He did have occasional pooling in the bottom chamber, but this is normal in the early months of operation, before the bed is relatively full. Because of moving air, this occasional buildup evaporated, and Mike did not have to bail it out, though he did try to limit urine input for a few days when he noticed buildup. A few small leaks occurred near the access door, probably because the box got wrecked during installation. But so far no great problems showed up. True, the pile did not seem to be moving down at all, but he figured this might be normal until it really heated up, or until it was time to remove some compost from the bottom.

Then one day, Mike noticed some odor when he opened the seat. Turning on the vent blower helped at first, but the odor got worse as time went on. He realized that the bypass pipe, which was perforated according to the drawings he had, was now covered and clogged due to the perforations. Also there were some flying insects now, probably from the garbage he occasionally added. He knew that poor evaporation and excessive nitrogen caused this, but he could not continue to add peat and sawdust to remedy the situation because the pile was getting too high. He also knew that fine ashes, superphosphate, or lime dusted on the pile discourages larvae from growing, but he did not know that various insecticides for ornamentals and greenhouses could have controlled these pests. These insecticides contain pyrethrins and rotenone, both short-lived poisons derived from plants. In small amounts they do no harm to the compost.*

Of course, Mike tried to rake out as much compost as he could after a year, and then he and his family cut down their use of the toilet as much as possible. But the

* Editor's note: See "Clivus Multrum Trouble Shooting Manual" earlier in this chapter for specific recommendations.

pile had been blocked so long that it would not move at all. The middle baffle, #2, was too deep, and the six perforated pipes compounded the problem. Since they continued to use the toilet without the added peat or sawdust, this compaction only worsened. After two years and a period of little use, Mike braved the task of cutting a hole in the second baffle near the top where he could reach it, and then removed the top layer of perforated pipes. Then he knocked through the pile under the seat with a 2-by-4 from the seat opening.

This spring, when the toilet will be three years old, Mike intends to let the toilet dry out, empty it completely, and

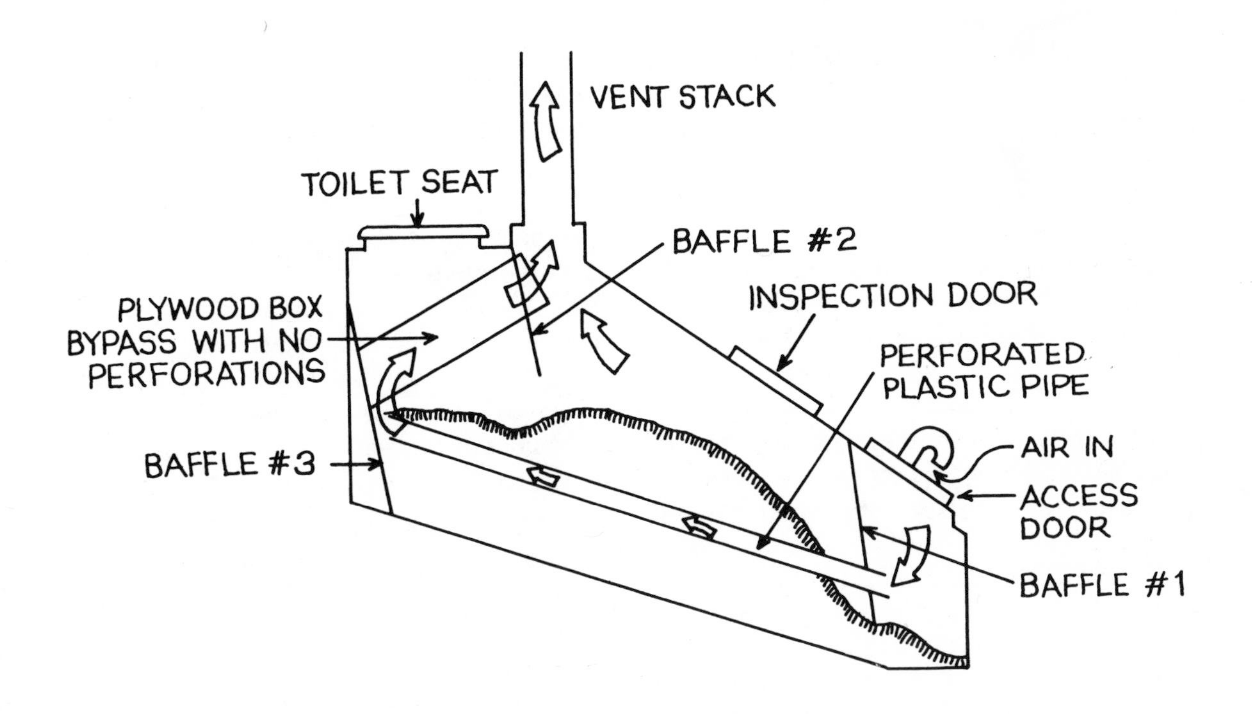

The redesigned Kombio, now with only one perforated plastic pipe and a plywood box bypass instead of the perforated pipe bypass under the toilet seat.

remove baffle #2. Then his unit will look like the present Kombio design.

Mike's unhappy experience points out the danger of obstructing the flow of compost through the unit, as well as the extreme compaction and wetness that result from not adding some form of fine cellulose fiber (peat or sawdust) daily. Other problems, relating to these two, include condensation in the stack, flooding, odor, and leakage.

Perforated pipes can clog due to the finer particles entering suspended in water and settling in the bottom holes. After these holes clog up, any water that enters the pipes can run directly to the bottom of the container without being partially absorbed by the bulky material and partially evaporated by the draft. U-shaped, inverted channels work better than the perforated pipe. You can use four-inch PVC Schedule 40 pipe with a three-inch slot cut out of the circumference lengthwise. These are supported where they pass through baffles #1 and #3 by passing through four-inch holes cut in the baffles, and screwed to the baffles. The three-inch slot faces downward, so that the channel forms a tunnel through the bed, allowing air to seep into the bed and also travel through the channel and out via the upper baffle and bypass pipe.

With this construction, water seeping down around the channels is prevented from running down as quickly, since the bottom of the channel is compost, which, being relatively dry, absorbs the moisture. The object is to make it as hard as possible for moisture to get to the bottom chamber. Don't worry about the bed getting too wet and going anaerobic. Sloping box types drain themselves very well, and moisture evaporates better in the bed than from a pool in the bottom chamber.

These channels should define the top of the initial starter bed. That is, the starter bed should just cover the channels. The starter bed consists of topsoil forming a dam against baffle #1, then peat moss or mixed peat and very well decomposed compost up to the channels. The channels should not be closer than a foot apart, which usually means you need only three of them.

The single most common cause of *liquid buildup* and the compaction that generally follows is the failure to add cellulose material—like peat moss, sawdust, and grass clippings—daily.

Another cause of liquid buildup is condensation in the stack. When condensation occurs, the water that has evaporated and is travelling as vapor up the stack hits the cold sides of the stack, condenses, and runs right back into the toilet. This can be avoided by making the vent stack as straight as possible and using only 45° rather than 90° elbows if it must jog around something. These 45° joints create less turbulence, and the slope between them allows any condensation to run back, rather than collecting in the low places of the horizontal pipe.

Insulate the vent wherever it passes through cold attics and the outdoors, just as you would a stove pipe. Surround the stack with fiberglass pipe-wrap and keep it dry by putting galvanized pipe over the fiberglass. The galvanized pipe should be sealed on top with a six- to four-inch reducer or with duct tape to prevent rain and snow from wetting the insulation, which would destroy its R factor.

Install a raincap, preferably a rotary wind turbine, which increases natural draft when the wind is blowing better than four miles per hour. These turbines can be used even if you have an electric blower or fan. Cut a piece of window screen a bit bigger than the pipe size, place it over the vent pipe opening, and insert the turbine over it. If it is too snug, saw slots down about an inch vertically in the vent pipe, and try again. Don't cement this joint; you want to be able to remove the turbine to clean the screen every few years.

Flooding can also be the result of poor evaporation. Insulate the box, especially underneath where cold can conduct most easily into it. Excess moisture that collects here must be warm to evaporate well. In fact, the ambient temperature of any toilet's space should be at least 65°F (18°C), allowing large, unheated types to run at about 75°F (24°C) inside, which is a minimum for good decomposition and evaporation.

Slow air flow causes poor evaporation also. This is caused by air leaks in the seat cover or inspection hole. Natural draft will take its air at the highest possible point, so seal the covers with neoprene gaskets or silicone so that suction will be maximized at the air intakes. These intakes should be no larger in total diameter than the diameter of the vent pipe. Make them a bit large, and tape them off until natural draft is maximized. Make sure to bypass the bed via pipes and/or baffles so that all the air does not have to strain its way up through the compost.

Odor at the seat when it is open is caused by placing the vent opening too far away horizontally as well as vertically from the seat. Draft is temporarily overcome when the seat is opened suddenly, and can back-puff if the vent is too far away to recover. The closer the seat is to the vent, the better. Don't make a separate chute for garbage, as the more chutes you have, the more secondary chimneys you create, especially if there are any air leaks at their joints or covers.

Leaking can be avoided by waterproofing with resin-based coatings, preferably reinforced at the points with glass cloth. Concrete must be waterproofed with coatings, not just sealers, and care must be taken that the footings are deep enough to prevent settling and cracking of the walls or floor. Do not use any butyl liners or coatings, as they break down when exposed to urea. I caulk the clean-out door after each opening with silicone caulk until a gasket is formed.

Install a blower if you can, because it may be necessary to prevent odor when the seat is open on low-pressure days, and because it aids evaporation. It should be hung below the vent opening on nylon cords, so that it is aimed up the vent, but does not impede natural draft when it is off, which will be most of the time, hope-

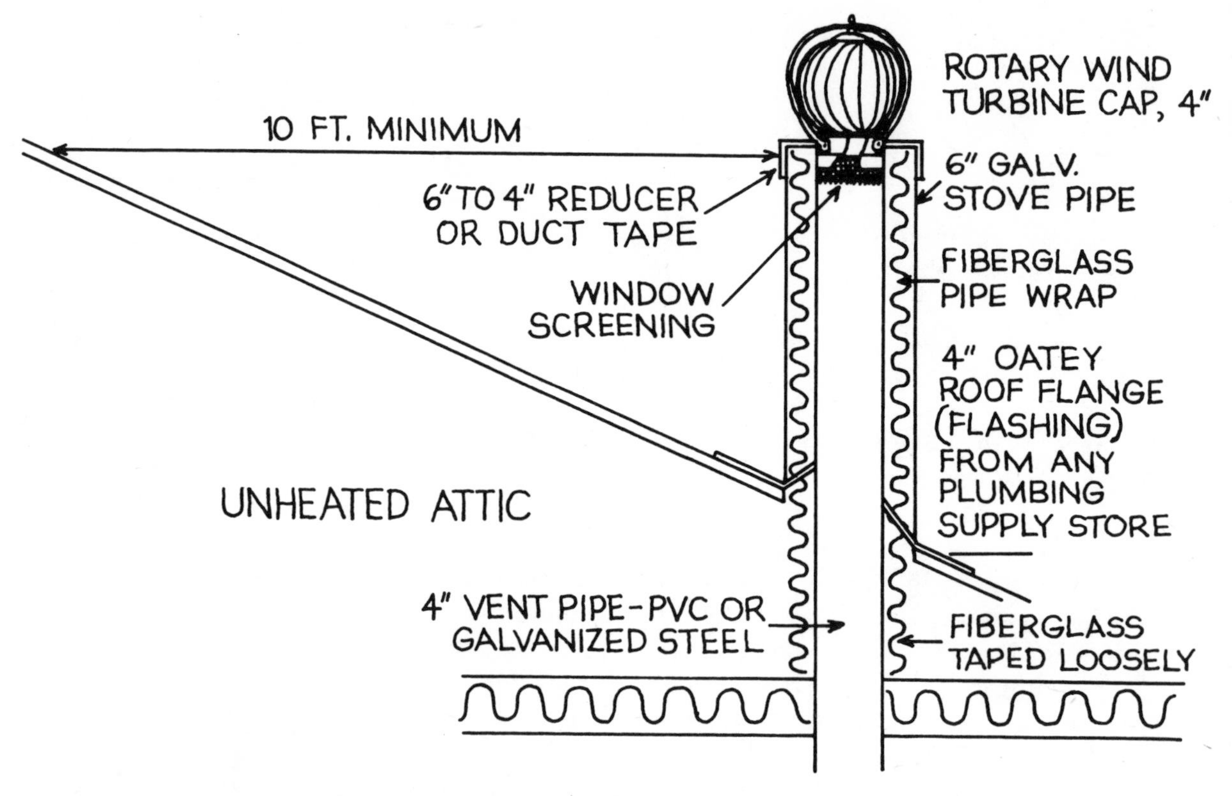

Cross section of a good vent pipe installation, showing insulation around the pipe, and proper distance from the roof peak.

fully. Similarly, fans installed in the vent pipe itself should be encased in a section of pipe larger than the vent diameter by at least an inch, so that air can pass by easily when it is off. A very small squirrel-cage blower, such as Dayton's smallest, is fine. Blowers or fans rated at 15 CFM are usually big enough and 50 CFM is usually too big and noisy.

Don't try to heat the compost directly with resistance heaters such as heat tape. The commercial toilets that have such arrangements are highly engineered. Heat the air instead, if you must, or use hot water in a coil, being careful that the coil does not impede the flow of compost. Small types are the most difficult of all to build, because they must run hotter—about 90°F (32°C) on the bottom—and can more easily flood.

Consider installing a monitoring device, such as an indoor-outdoor thermometer at the air intake, with the bulb in the decomposing material. An hygrometer in the access door or beside the vent helps, too.

Don't get scared off by what I've just said. I said it not to discourage the prospective builder, but to dispel the myth that composting toilets are simple devices that just anyone can build properly. Before you begin to build, study up on the composting process, study some of the commercial designs, and, if possible, talk to some people who have built their own toilets.

The Evans's Coprophage #3

Three years ago I went looking for an ecologically sound and aesthetically acceptable waterless toilet. After a year of testing and examining commercial models I came to the conclusion that none met my requirements. Apparently most had been designed by engineers with a minimal practical understanding of the process of decomposition, engineers who knew little or cared less about the chemical or physical characteristics of human excrement.

Because of my failure to find a suitable commercial model I decided to build my own. I wanted my toilet to have the following characteristics if possible: It must be visually acceptable by totally cutting off the view of the waste mass. It must be without odor at any time. It must be a biologically sound aerobic composter with little (or no) inputs of auxiliary energy. It must be large enough to serve the waste disposal needs, both toilet and kitchen, of an average family on a permanent basis. For hygiene purposes, feces should be subjected to high composting temperatures for at least a month and then stored, without the necessity of handling, for a year. Major cleanouts of aged materials should be limited to once per year. Regular servicing should be handled on a remote basis. The toilet must be foolproof and easy enough to use so that any member of the family can operate it.

The unit presently residing in my basement is called "Evans's Coprophage" model #3. The first two models were pretty good but they were so small that they had to be serviced every two weeks, and the stirring mechanisms were not up to the task. While I am pretty happy with #3, I am not under any illusion that it is a manufacturable item; much more design and engineering work has to go into it to produce a production model. However, I do believe that the basic concepts may provide a step towards a better composting toilet.

Description of Evans's Coprophage #3

In the bathroom there is a small ply-

John W. Evans is Associate Professor of Biology, Memorial University of Newfoundland.

wood box fifteen inches by twenty-one inches by fifteen inches high. A toilet seat is attached to the top of the box. The toilet seat has been slightly modified by placing circles of vinyl weather stripping under the rim of the two portions of the seat. This seals the seat to the box as a deep freeze top seals to the cabinet. When the seat is opened all that can be seen is an inky black void. The gentle hum of a ventilation fan can be heard. The slight negative pressure provided by the fan prevents any odors from emerging from the toilet. This ventilation makes the toilet more pleasant to use than a standard flush toilet because no odors result, even during use.

The bulk of the toilet is located in the basement directly below the seat in the bathroom. My basement is completely unheated, so that the air ventilated through the unit is cold. This does not seem to interfere with the heating action of the toilet.

The dimensions of the basement cabinet are six feet by four feet by six feet three inches high. In part, at least, the size was determined by the most efficient utilization of four-foot by eight-foot sheets of plywood. The cabinet is composed of two parts—a top and a bottom. The space within the cabinet is not subdivided.

The functions of the bottom are: (*a*) A second-stage waste storage container. After passing through the first stage, the mostly

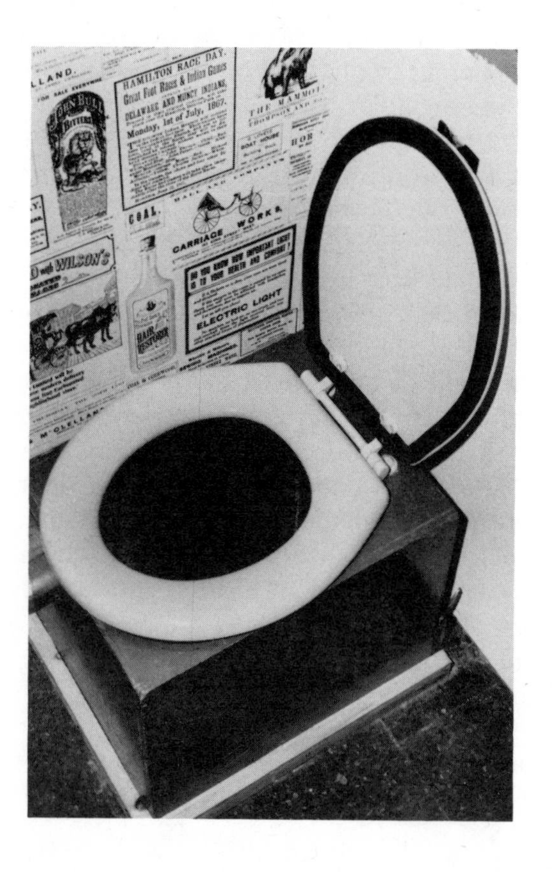

In Evans's toilet, a standard seat is mounted on a plywood box which opens directly into the large tank in the basement. The inside of the toilet is blackened so that nothing can be seen down the hole.
(Photo by John Evans.)

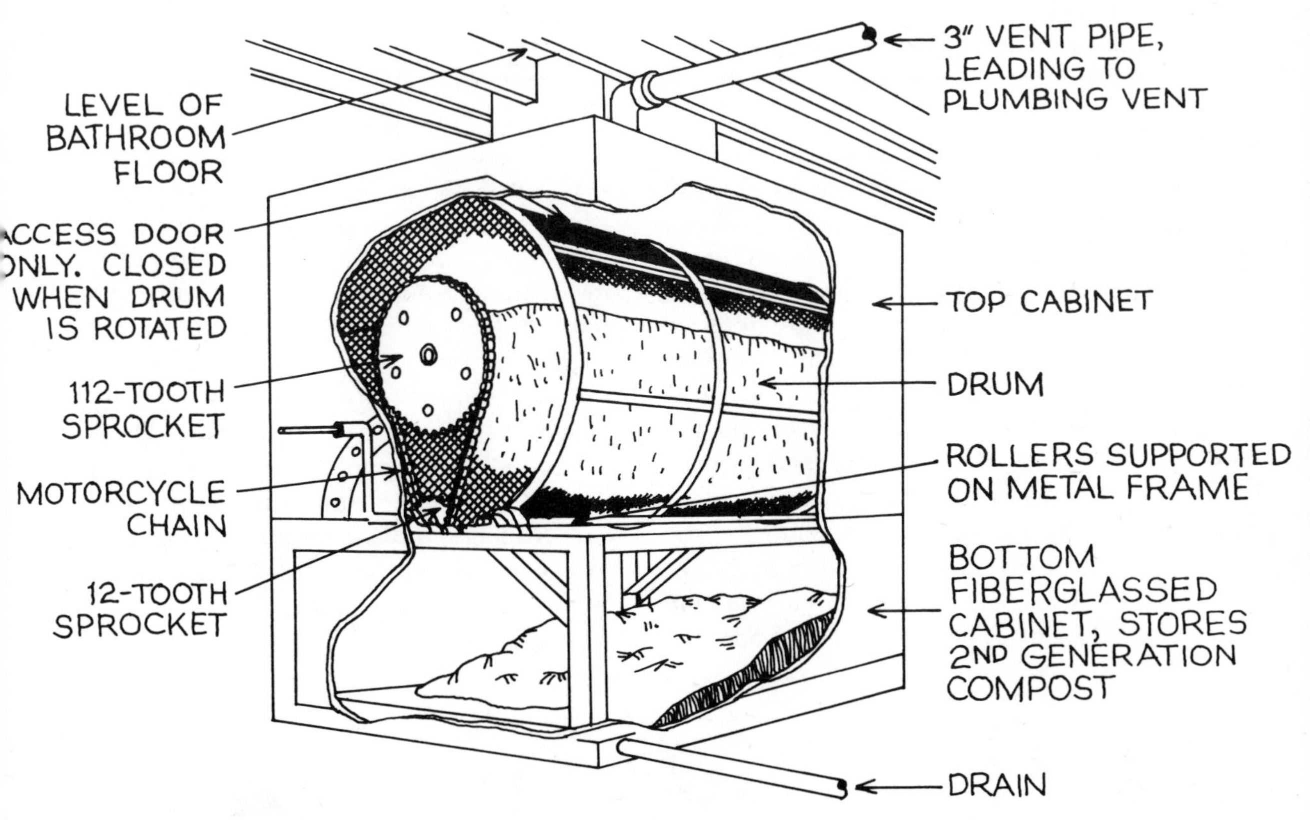

Cross section of Evans's decomposition and storage containers. Note the drain hole at the bottom which makes collection of any excess liquid easy.

decomposed waste is left undisturbed for another full cycle. (*b*) The floor of the bottom is gently sloped so that any excess moisture will collect in the front where it can be dealt with as necessary. (*c*) When clean-out time arrives only old, well-decomposed material has to be handled.

The bottom is fiberglassed with a dark pigmented resin. The front of the bottom is fitted with a closely sealed access port 2½ feet by one foot. This port has a window so that visual inspections of the lower section of the toilet can be made. The bottom is six feet by four feet by two feet three inches deep at the front and two feet deep at the back. As a result, the floor is slightly sloped towards the front end of the cabinet. The bottom is held off the cement floor by a low stand which is constructed of two-inch-thick lumber. The front of the stand is three inches high and the back is six inches high; this slope offsets the sloped floor of the bottom so that the whole cabinet is upright. A drain hole has been installed in the bottom cabinet so that if excess liquid accumulates it can be easily removed.

The top of the cabinet is four feet by four feet by six feet long. It has no floor because it is continuous with the bottom chamber. The roof of the cabinet meets an extension of the toilet seat box from the floor above. The top is firmly screwed together but not glued so that access may be had to the mechanical part of the toilet when necessary. The interior of the top and the access box from the bathroom is well painted with water-resistant black paint. This has the effect of reducing the light within the cabinet to such an extent that the waste mass cannot be viewed from the bathroom. The front end and one side of the top have viewing ports so that conditions within the rotary drum can be observed. No attempt has been made to seal the cabinet. Here and there are several small access and ventilation holes which allow for the free flow of air. As long as there is a slight negative pressure inside there is no odor because all air flows into the cabinet and out the household plumbing vent.

The unique feature of Evans's Coprophage #3 is the rotary drum which is horizontally suspended in the top part of the cabinet. The drum is 50 inches long and 37 inches in diameter and is supported by two pairs of rollers. The rollers rotate in bearings which are attached to a welded metal frame that rests inside the bottom of

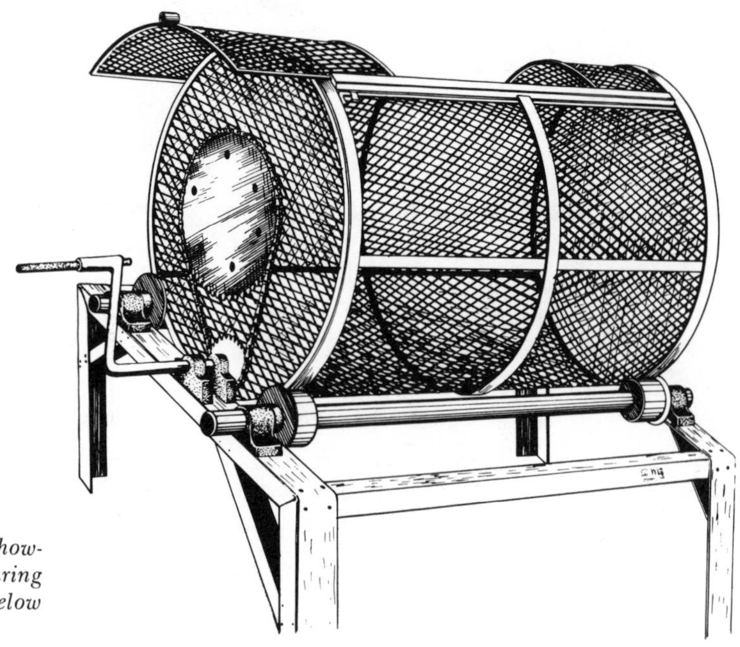

Close-up of the drum cage itself, showing access port which is open during use. The port is located directly below the toilet seat.

the cabinet.

The drum is constructed of a welded framework of flat iron which is covered by galvanized and painted expanded metal. The only regular access to the interior of the drum is via a 12-inch by 18-inch opening in the curved side. This access opening can be opened or closed at will by a sliding door. This door is operated with a pipe which is inserted through a small hole located high on the front of the cabinet. Most of the time the access door is open and the opening is located directly below the toilet seat. When the toilet is "flushed," that is, when the drum is rotated so that the fresh waste is buried, the door is closed so that no material falls out of the drum.

Since the wastes and filler material occupy the lower one-half to three-fourths of the drum, it is not possible to turn the cylinder directly because the weight of the mass is too great. A 9⅓ to one mechanical advantage was obtained by a chain and sprocket device attached to the end of the drum. A 112-tooth sprocket was bolted to the front end of the drum and a 12-tooth sprocket was bolted via two bearings to the metal frame. The two sprockets are joined by a motorcycle chain. The axis of the small sprocket extends out through the front wall of the cabinet where it is attached to a crank handle. Nine and one-third turns of the handle rotate the drum once. The handle is designed so that it can be fixed in place when not in motion. This keeps the opening of the drum in perfect alignment.

The accumulated waste is buried about once a week; the whole job, if carried out in a relaxed way, requires about one minute. The procedure is as follows: A light inside the toilet is turned on so that you can observe the process and check on alignment, and a pipe is pushed through the access port to close the door of the drum. The crank is turned until the drum has been rotated one or two times. The pipe is used to open the door again and the light is turned out.

At this writing the "coprophage" has been in continuous use for about three months and there have been no complaints from family or visitors. It looks nice, it doesn't smell, and it is easy and convenient to operate. However, if I were going to build it again I would make a number of alterations. (*a*) The drum would be constructed of finer mesh; a lot of fine sawdust is lost to the lower chamber when the drum is turned. (*b*) If I were building a toilet for a new house I would put a large (six inch to eight inch diameter) vent pipe straight up through the roof. This way I could probably eliminate the exhaust fan. (*c*) The problem of excess moisture is a vexing one if you let it bother you. This fluid looks like strong black coffee and smells like slightly rank fruit cocktail. About two gallons have accumulated in three months. After having attempted to design several fan gadgets to increase the rate of evaporation within the toilet, I have decided that it would be much simpler to drain it into a bucket four or five times a year. I'll bet the rose bushes will really love it.

The large mesh drum rotates on four small rollers under the two end rims. The crank turns the drum (when the access port is closed as shown here) indirectly via a chain and sprocket. In the upper left corner of the upper cabinet is an exhaust fan which forces smelly fumes out via the household plumbing vent. Old waste can be seen accumulating in the lower chamber. Ventilation holes in the side of the lower chamber allow fresh air to enter.
(Photo by John Evans.)

Steven Travis's Solar Composting Toilet

My home-built composting toilet is an integral part of my "Natural House for Northern Maine." Like the house, it's not dependent upon a utility company to keep it going. It uses no electricity; the sun heats it and the wind ventilates it.

This toilet is not the first I've built. I've made a good number of toilets before for myself and other people that ranged in performance from awful to good, and improvements have come pretty much through trial and error. The first was built from a 55-gallon barrel and worked at room temperature. It composted very slowly, took several months for one person to fill it, didn't smell, but eventually overflowed with urine. I added an 80-watt heat cable to the next one, resetting the thermostat

Steven Travis has designed and is in the process of building his "Natural House for Northern Maine," which integrates alternative energies and a waste-water system. Plans for Steve's house are available by writing to him at R.F.D. #2, Box 82H, East Holden, ME 04429.

to 95°F (35°C). It worked twice as fast, handled two people, evaporated faster, but still overflowed. On very cold nights, the vent would freeze solid, and odors backed up. Next, I built two wooden composters lined with polyethylene. These were easier to build and maintain. The box shape held twice the volume of the barrel for the same floor space, and it raised the seat higher from the pile. The drawer leaked some and still filled with urine. I varnished the next one with urethane and added wicks. Then I built one with a window and reflectors, lined the drawer with fiberglass, and added much more garbage. This worked well.

The solar model I describe here is my most sophisticated by far and represents all the past experiences I have had building and using waterless toilets and privies.

This toilet is supposed to sit right in the bathroom, provided the bathroom has a southern wall. My collector is placed vertically to attract winter sun, and is actually attached to the toilet itself, but if one would, for whatever reason, wish to have it a few feet away, no harm would be done. The pipe hooking the collector to the composter would just have to be insulated. The collectors draw solar heat from behind my homemade insulating panels which are a variation of Steve Baer's "Skylids." They pivot open during the day to admit the sun and close at night to hold the heat. They work automatically by sunlight: The sun heats a tank of Freon mounted on the sun side of the panel, the Freon begins to vaporize, and the pressure pushes the liquid Freon up a tube to a tank on the other side. This weight shift tips the panel open. At dusk, the room-side tank is warmer, so the Freon flows back and the panels close.

The toilet is made from plywood, coated inside and out with epoxy. Even the metal hardware nails are covered with epoxy to prevent corrosion. All joints are covered with hardware cloth to keep flies out. My unit isn't insulated, but if it had to sit in a cold room I would definitely insulate it—perhaps with one-inch Styrofoam panels.

The four-inch-diameter plastic pipe vent stack is located immediately behind the toilet seat opening, and like all commercial units, runs up above the roof line. It's capped with a screen to keep out flies. The screen is secured only with a rubber band so that it can be removed in winter to prevent freeze-ups. A damper in the vent stack automatically regulates air flow up the stack. A bimetal coil pulls open the Styrofoam damper as the air warms, and a counterweight closes it as the air cools. A knob sets the temperature range. Once set, I tape the knob securely so that it stays in place.

A thermometer next to the vent stack lets me know how warm the pile is. Although it can't damage if frozen, it won't compost unless it's about 65°F (18°C). It decomposes fastest between about 85° to 105°F (29°–40°C).

The toilet works like this: Wastes pile up on the pipes in the toilet. Sun-heated air enters the unit from the solar collector

(continued on page 157)

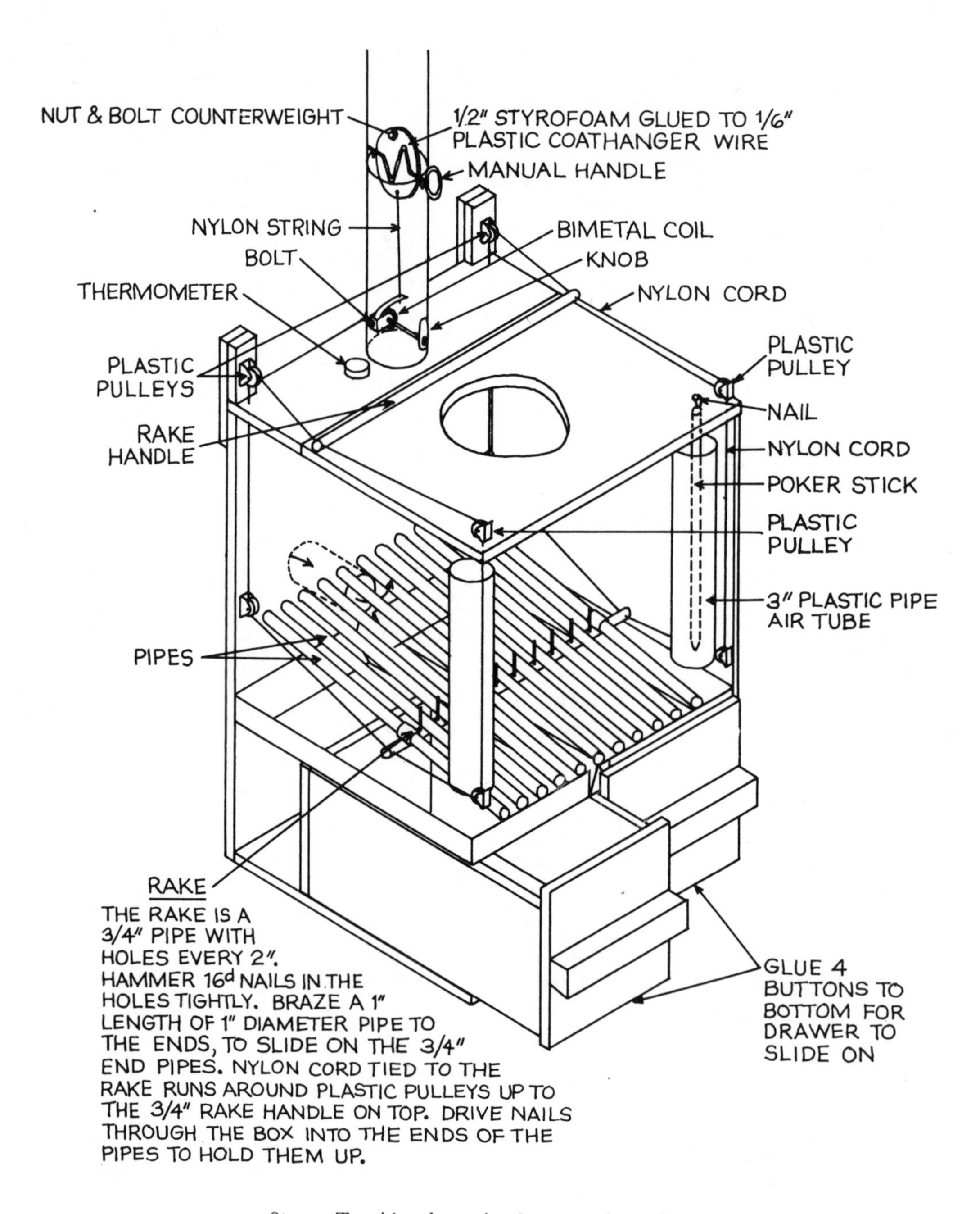

Steven Travis's solar-assisted composting toilet.

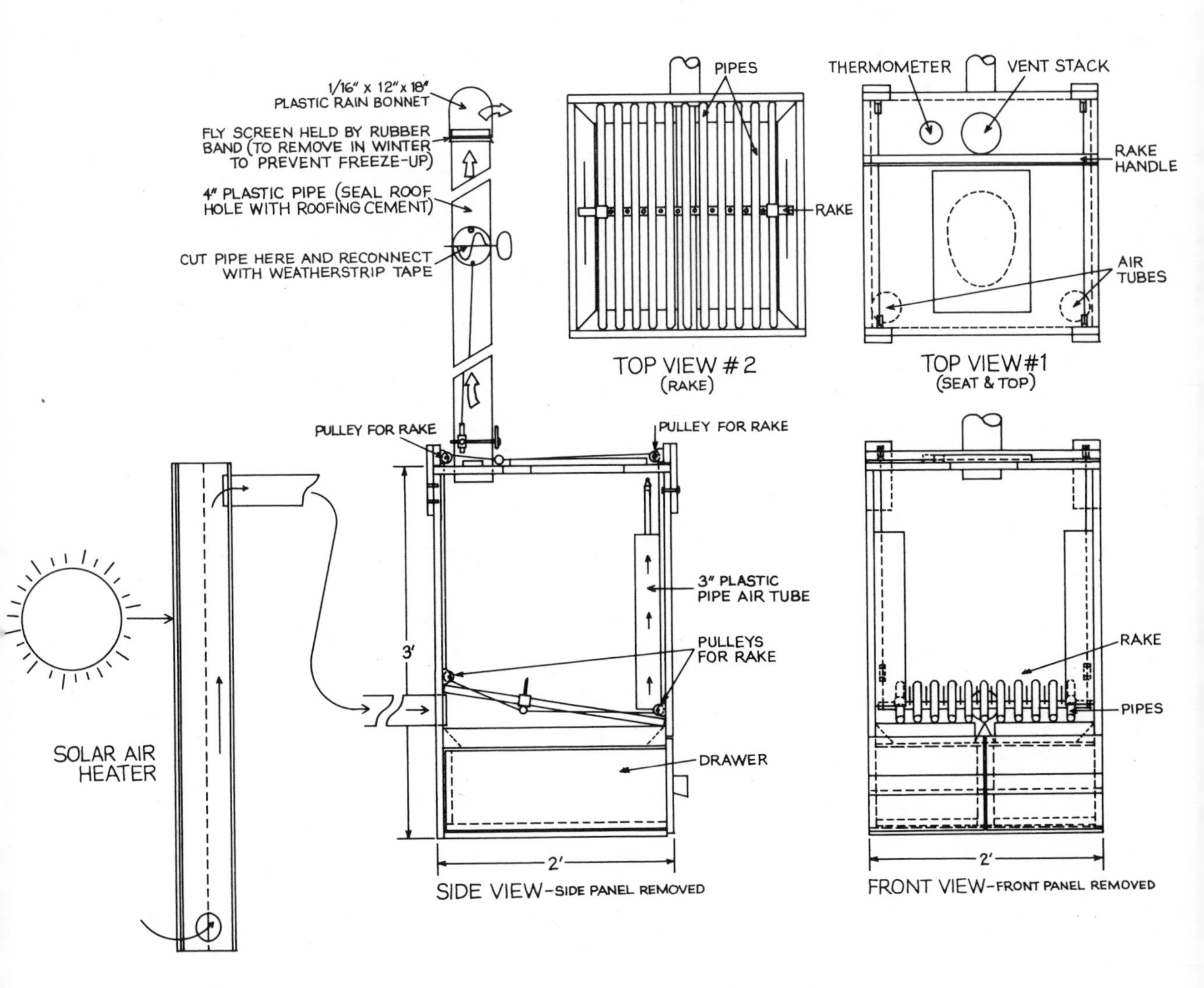

MATERIALS:

3/4" x 4' x 8' PLYWOOD
1 GAL. EPOXY
3/4" x 24' IRON PIPE
4' x 4' x .040" FIBERGLASS
1" x 4' x 8' STYROFOAM
4" x 10' PLASTIC PIPE
18' NYLON CORD
1" X 8" X 16' BOARD
8 PLASTIC PULLEYS
SPRING THERMOMETER
BIMETAL COIL
1" x 6" x 16' BOARD
2" x 2" x 12' BOARD
5 FT.² FIBERGLASS CLOTH
2 LB. 16d NAILS
1" x 20' SPONGE WEATHERSTRIP
3" x 3' PLASTIC PIPE
3/4" X 4' DOWEL
4' x 4' BLACK CLOTH
1/2" x 1" x 16' LATHE
2-1" HINGES
1" x 10" x 1 1/2" BOARD
1/16" x 12" x 12" PLASTIC
8 BUTTONS
1" x 2" x 3' BOARD
CAULKING
3/8" x 3/4" BOLT & NUT
6" x 12" FLY SCREEN
STAPLES
COATHANGER

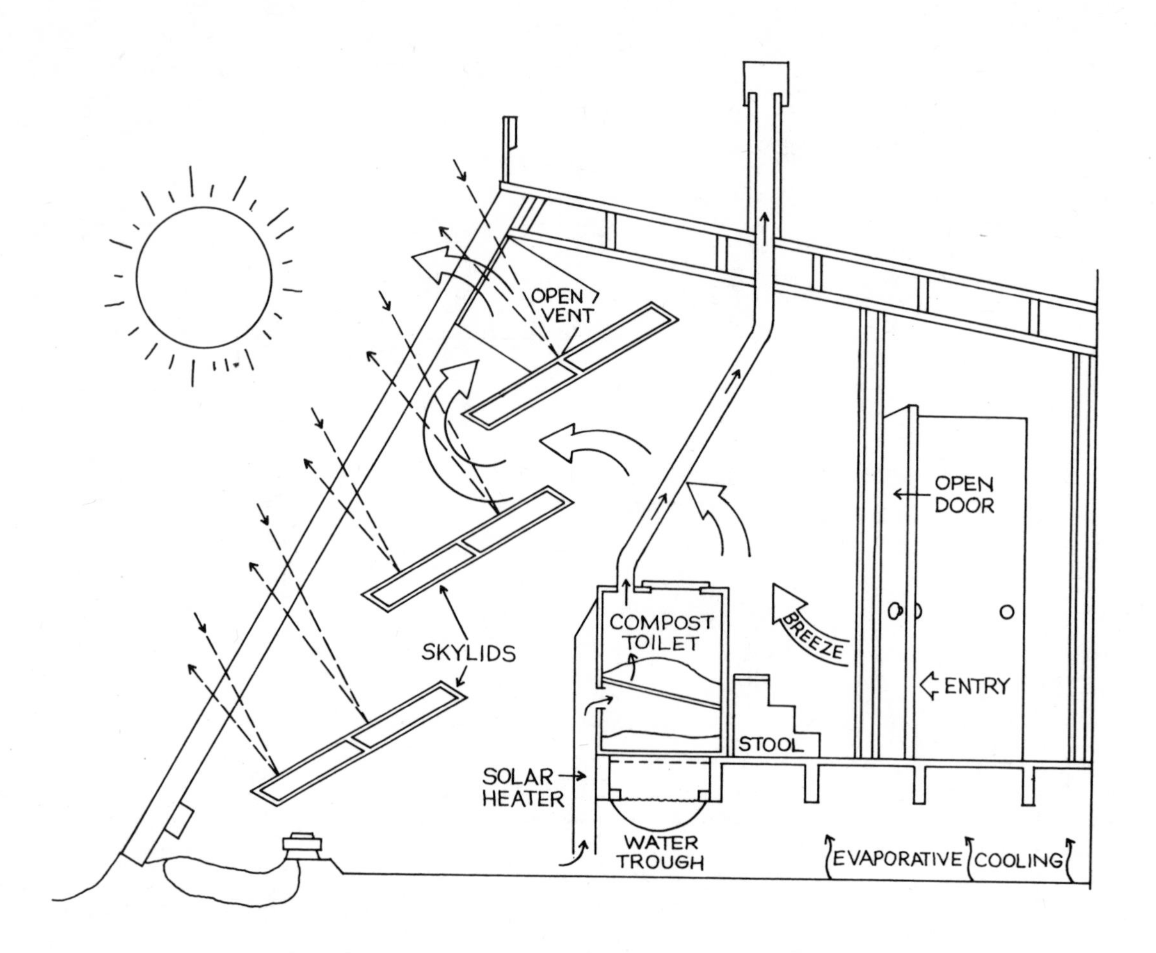

Cross section of the Travis house. Skylids, installed behind the glass on the south side of the house, open during the day to allow the sun's warm rays to shine on the solar heater which is attached to the toilet. The Skylids close up again at night to prevent heat inside from escaping through the glass. Vent at the upper left of the house creates a natural updraft which takes any odors up and out of the bathroom.

and is drawn up through the pile and also up through the two air tubes in the two front corners immediately above the storage drawers. These two air tubes are particularly important when the pile covers the pipes and effectively blocks air passage around them, because they permit good ventilation and evaporation to continue. If the air tubes should become clogged with decomposing material, the poker is there to manually push the material down through them into the drawers below. As the pile decomposes, it presses and drains between the pipes to the drawers below. A manually operated rake also works it down. The warm air passes through the drawer, through the pile, and up the vent, drying the compost to as little as one-tenth of its original volume. Most all the odors go up and out the vent.

So far there has been a little liquid buildup, and instead of trying to alleviate the problem with more heat or ventilation, I'm going to put on a urinal and cut down on the amount of liquid getting into the toilet in the first place.

I started the toilet off by spreading three layers of cabbage leaves on top of the pipes. I don't think the cabbage leaves are a must, they just happened to be handy and big enough so as not to fall between the pipes. Then I added two shovelsful of fresh soil over the leaves—and the unit was ready to use. I add as many kitchen and garden scraps as I can get and a shovelful of fresh soil monthly to balance the nutrients, build up the pile, and prevent compaction. I rake and poke at the pile whenever I'm in the mood or whenever I feel it needs leveling and mixing.

The Clivus Minimus

We had been concerned, at the Minimum Cost Housing Group, with developing an on-site design for a one-household toilet that would use little or no water and could be used in high density settlements. It was our intention to produce a design which could be adapted to slums and squatter settlements in underdeveloped countries, and which could be realized in a variety of materials.

We started this work in 1972 with a survey of existing alternative sanitation systems, which grew into the publication, *Stop the Five Gallon Flush,** still being updated and enlarged. It seemed to us at the time that composting would be the ideal basis for our work. This had been a traditional system for handling human waste in south Asia for centuries, in the rural areas. It usually consisted of a hole in the ground into which excreta and organic waste were thrown. When the hole was full, additional organic matter would be added and the hole covered up, to be dug up some time later. The humus was used as fertilizer in the fields. This avoided

Witold Rybczynski is Director of the Minimum Cost Housing Group at the School of Architecture, McGill University, Montreal.

* Witold Rybczynski and Alvaro Ortega, eds. *Stop the Five Gallon Flush* (Montreal: McGill University, 1976).

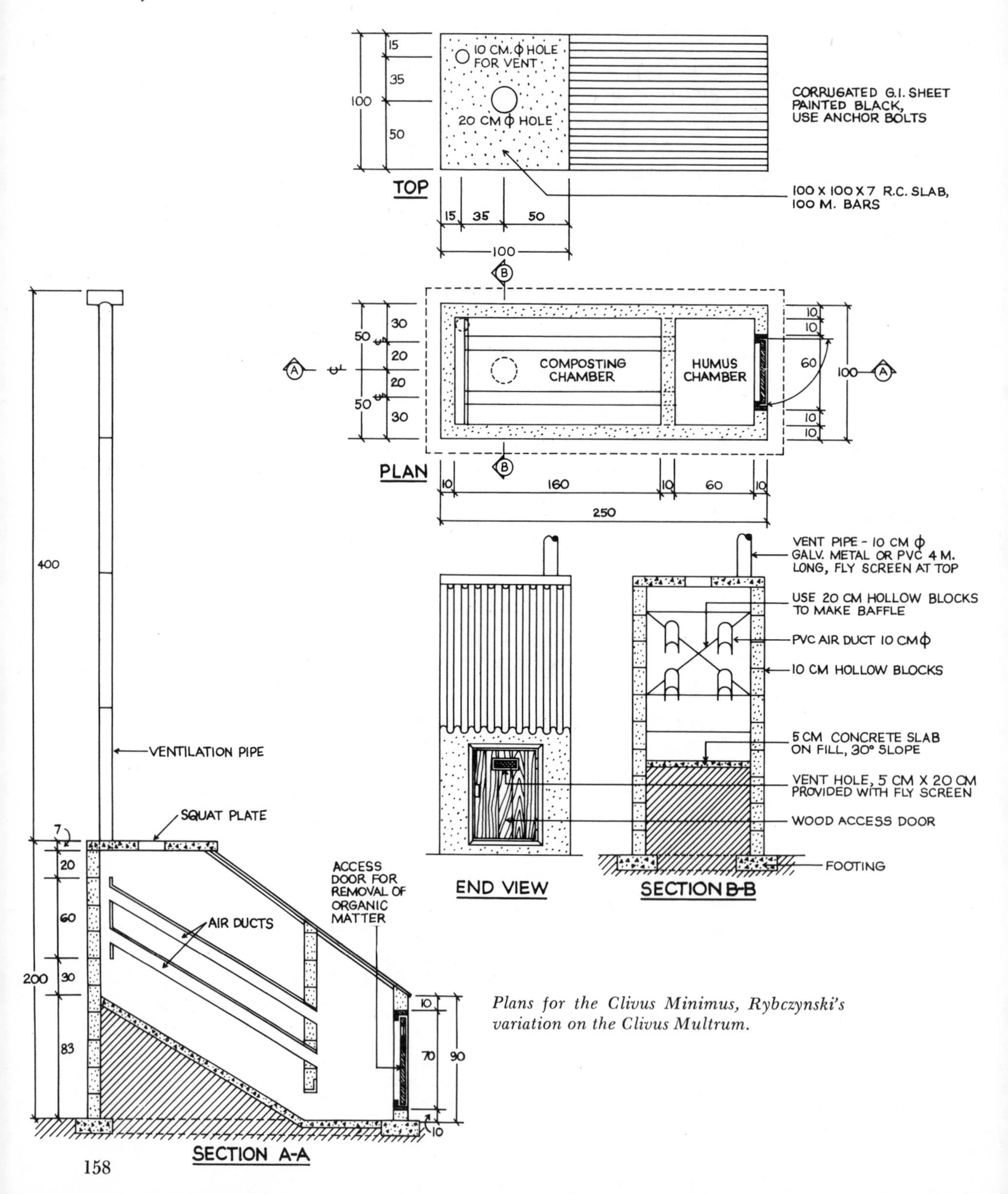

Plans for the Clivus Minimus, Rybczynski's variation on the Clivus Multrum.

the handling of fresh fecal matter, which was the cause of so much disease in countries such as China and India (see chapter 4).

The question was, "Could composting be 'urbanized' and used in high density settlements?" We found that in fact someone had already begun work in this direction. The system developed was called the Clivus Multrum. We became convinced that the Clivus Multrum was the most promising of the "composting toilets," a number of which were being marketed both in Europe and the United States. The two problems that had to be resolved were: (1) Could the Clivus be built in an on-site version at a low enough cost, and could it be made out of locally available materials in underdeveloped countries; and (2) Would the Clivus operate in various tropical climates?

The Clivus Multrum is made out of fiberglass, a much too expensive material for poor countries. The first prototype we built in 1974 was out of asbestos-cement, factory molded. This proved to be still too costly, and extremely heavy. We next built two on-site toilets out of cement blocks in owner-built houses in Québec in 1975. These have been operating successfully since. A third unit was just completed in a house north of Ottawa. All of these Clivuses, now christened the Clivus Minimus, were in cellars.*

The first installation in an underdeveloped country was completed in Magsaysay Village in the Tondo area of Manila. The Tondo is a squatter settlement of 160,000 people, three-quarters of whom have no waste disposal facilities at all. The prevalent system is euphemistically called "wrap-and-throw"! The Minimus chamber is built out of cement blocks, plastered inside, and has a concrete bottom. The vent pipe is galvanized metal, and the air ducts are PVC. The total material cost (not including labor) was U.S. $55. The construction time was six man-days. Half a dozen Clivus Minimus toilets have been built, both in the Tondo and in a resettlement area outside Manila.†

At the moment it seems that, for the Philippines at least, the cost of construction is acceptable. We will be trying the Minimus in India and adapting it to local materials and conditions. The Manila Minimus is built above ground, as a high water table and seasonal floods make burial too expensive. In other situations the Minimus would be buried, and humus retrieved through a "manhole."

The operation of the Minimus in hot climates remains to be seen, and the Manila models will show the problems, if any, of ventilation in hot/humid situations. It is too early yet to tell, though the fact that hot temperatures will certainly speed up the composting process itself is promising.

It should be stressed that there is no one "design" for the Minimus. It must be adapted to meet local climatic conditions, available building materials, local skills, and conditions. The application of

* The Minimus incorporates suggestions of Carl Lindstrom, with whom the author has been in contact since 1973.

† The installation of the Minimus in Manila was done as part of a United Nations Environment Program project.

This picture of the Clivus Minimus was taken in the Philippines. The decomposition tank was installed above ground because of the frequent flooding in the area.

composting sanitation technology to developing countries cannot be on a piecemeal basis. It must be done on a community (not individual) scale and integrated with social and educational development. It was precisely in such a way that rural composting toilets were introduced to North Vietnam during the years 1961–1965. As a result of this program of rural sanitation it has been reported that over 600,000 tons of fertilizer were produced annually in this fashion!

It is for this reason that the Minimum Cost Housing Group has not produced do-it-yourself plans for the Minimus and does not contemplate doing so in the near future.* The use of the Minimus represents a significant shift in cultural patterns and habits and must be clearly understood by the users in order to be successful.

* Editor's note: Just as we were about to go to press we learned of another group, Maine Compost, that is now selling do-it-yourself plans for a toilet very similar to both the Clivus Multrum and the Clivus Minimus. The Maine Tank, as it is called, has a concrete tank, metal pipes, and double wall metal vent stack. There are chutes for both toilet wastes and kitchen scraps. It can be made to a number of sizes and because the concrete is poured in place between forms, it can be fitted to almost any space.

The engineering plans, which cost $10 at this time, come with details of form construction, and instructions for pouring the tank and installing the upper parts of the system. Peggy Hughes of the Maine Compost group figured that materials for her unit, not counting the forms, came to about $500.

If you live in eastern Maine, the group will build the complete system for you for $800. The price includes the necessary cubic meter of working compost to get the system started, as well as at least one maintenance check within a month of installation.

Their address is Maine Compost, Deer Isle, ME 04627.

Chapter 6

Dealing with the Greywater

If you are considering the use of a compost toilet, or have one, you probably already realize that separate handling of toilet wastes in an ecologically sane manner is only part of the water system in a household. While the 30 to 40 percent of the water used is no longer needed for flushing toilets, the other 60 to 70 percent of the wastewater normally discharged from homes still needs to be treated. This wastewater is called "greywater"—as distinct from "blackwater" from the toilet. Because greywater could possibly be as "dirty" as combined sewage, adequate treatment is a must; however, its different characteristics and volume allow for use of alternative systems for disposal or reuse. Unless you are converting a home with an existing septic tank system, the use of such alternatives—which are usually cheaper—may determine whether or not you can afford a flushless toilet.

What Greywater Is

Greywater is all the wastewater produced in a household other than toilet wastes. It comes from the kitchen, dishwasher, laundry, shower/bath, and bathroom sink. It contains grease, food particles, soaps, detergents, hair, dead skin, bacteria and viruses, and a whole host of various other items that get thrown down the drain. Although greywater once was thought to be fairly uncontaminated when compared to blackwater, recent studies have shown that household greywater can contain substantial amounts of physical, chemical, and biological pollutants. These constituents point to the need for adequate treatment.

Before designing a system to treat your greywater, you will need to know what is likely to be in it, where it comes from, how much you will have, possible reuses for it, and the soil, bedrock, and ground water conditions on your land.

Patricia M. Nesbitt is a specialist in wastewater treatment. Presently a staff consultant for the Institute for Local Self Reliance in Washington, D.C., she has been a consultant for many environmental groups and government offices.

Volume

The volume of greywater produced ranges between 24 and 36 gallons per person per day, or 96 to 144 gallons a day for a household of four people. This is based on a dwelling that has an adequate supply of running water, and includes normal use of a washing machine and dishwasher. It does not include use of a garbage disposal. It is possible, of course, to reduce the amount of greywater produced even more, and figures of 50 to 75 gallons a day for a household of four are also frequent in the literature. It is hard to generalize; you must consider that the range is anywhere between 15 and 50 gallons a day per person, all depending on different habits and water conservation measures. (For more on water conservation see chapter 8.)

Thus, in designing your system, it would be best to determine the average daily amount of water used as measured over at least three to six months. Simply put a meter on the main water line, or figure it out by checking how much use your pump gets. Alternatively, you could actually measure the amount of water used in your appliances, bathing, and sinks. The table here gives the average household water consumption. When you determine

Composition of Average Household Water Use

Fixture	Average Household Use (gallons/person/day)	Percentage	Design Flow for 4 People
Kitchen sink/dishwater	3.9	9	23.4
Laundry	10.0	23	60.0
Bath/shower	13.5	30.8	80.0
Toilet	16.2	37.2	97.2
	43.6	100%	260.6

Source: Data is an average of data presented by Ligman, "Rural Wastewater Simulation"; Laak, "Home Plumbing Fixture Waste Flows and Pollutants"; Cohen and Wallman, *Demonstration of Waste Flow Reduction from Households;* Bennett, Lindstedt, and Felton, "Comparison of Septic Tank and Aerobic Treatment Units"; and Witt, "Water Use in Rural Homes." (See complete references in Bibliography.)

your own average greywater production, subtract out the toilet wastewater volume, and multiply the remainder by 1.5 to determine the capacity needed in the treatment system. This extra is to allow for overflow and surge loadings.

Remember that these are average volumes and do not necessarily take into account the higher amount used on weekends when more people are usually home, and the unevenness of the flow. The volume will probably also change with seasons, as people tend to take more showers and do more laundering in the summertime. Loading is likely to be intermittent, with some surges when great quantities of water are dumped quickly, followed by relatively little for several hours or even days. Variations within a day are great and unsystematic, and different families have different patterns. Surge loadings, however, may not contain as much pollution as smaller and more regular flows from the kitchen—for example, bath waters usually are very diluted and may not require much treatment.

Quality

Kitchen wastes contain food particles, grease, dirt, soaps, and dishwashing liquids, and frequently, cleansers, ammonia, and bleach. Bathroom greywaters carry teeth cleanings, shaving wastes, shower/bath waters, soaps, toothpastes, and mouthwashes. Laundry contributes body oils, dirt, detergents, softeners, and other contaminants.

In addition to these wastes, greywater also contains a significant amount of bacteria and viruses. Meager data suggest that these organisms could be pathogenic, although the potential for pathogenic bacteria is much greater in toilet wastes. The likelihood of transfer of disease would be

(continued on page 165)

The Standard Rule for Sizing Conventional Leach Fields

For homes with no water conservation measures, the conventional rule that health officials use to size the leach field is 150 gallons per day (gpd) per bedroom. Two occupants per bedroom contributing 75 gpd each is assumed. This guideline is used rather than a gallon per person per day figure because the number of occupants in a house could change, either through people moving in or out or through selling the house. This 150 gpd bedroom figure also allows a good margin of safety for unusually large flows. Of course, these flows include use of a water-flushing toilet, and greywater receives no special handling.

Average Pollution Loads from Rural Households

Source of Wastewater	BOD_5[a] (grams/person/day)	SS[b] (grams/person/day)	N[c] (grams/person/day)	P[d] (grams/person/day)
Greywater:				
Kitchen sink	8.3	4.1	0.4	0.4
Automatic dishwasher	12.6	5.3	0.5	0.8
Bath/shower	3.1	2.3	0.3	<0.1
Laundry	14.8	10.9	0.7	2.1
Greywater subtotal	38.8	22.6	1.9	3.4
Garbage disposal	10.9	15.8	0.6	0.1
Toilet	20-28	27	10.5-17.7	0.7-1.9
TOTAL	69.7-77.7	65.4	13.0-20.2	4.2-5.4

Source: All data except for toilet wastes are from R. Siegrist, "Segregation and Treatment of Black and Gray Waters" (Small Scale Waste Management Project, University of Wisconsin, Madison, WI, 1976; see R. Siegrist, M. Witt, and W. C. Boyle, J. Environmental Engineering Division, ASCE, *102,* No. EE3, June 1976). Data for toilet wastewater is from Rein Laak, University of Connecticut, Storrs, CT, 1977.

a. BOD_5 (Five-day Biochemical Oxygen Demand) is, technically, the quantity of oxygen used in the biochemical oxidation of organic matter for five days at 20°C. It is a general indicator of how much organic pollution is in the wastewater.

b. SS (Suspended Solids) is the amount of solid material suspended, as different from dissolved, in the wastewaters.

c. N (Nitrogen) is the total amount of nitrogen, in all forms, in the wastewater.

d. P (Phosphorus), usually present as phosphates, is the amount in the wastewater. Nitrogen and phosphates is the amount in the wastewater. Nitrogen and phosphorus are nutrients for aquatic organisms and contribute to water pollution.

increased if someone in the household were sick, but physical contact with sick people would pose more danger than exposure to greywater. Let us not forget that trace amounts of fecal matter and urine from the shower may harbor pathogens, and if you have a baby, rinsing dirty diapers will obviously add to the biological contamination of greywater.

Nevertheless, greywater is usually cleaner than blackwater.* Removing human wastes from the waste stream simplifies the treatment needed, since the highly soluble chloride and nitrogen compounds from urine which are not treated well in soils are kept out of the waste stream with the use of a waterless toilet. The primary causes of soil clogging in the leach field, the feces and toilet paper, are also eliminated. This reduction in pollutants also changes the nature of organic matter remaining in the greywater. Greywater contains about the same amount of oxygen-demanding compounds, and there is evidence to suggest that these compounds are greater and more easily stabilized than combined wastes.† This means that treatment of greywater in soils is much more likely to be complete, thus reducing the possibility of ground water pollution and increasing its reuse potential.**

Treatment

Greywater management is a relatively new area of interest, and as a result there has been no long-term research conducted to establish sound treatment technologies for separate domestic greywater. Today experimentation is going on in several areas of the country, but there are still more untried ideas and initial experiments than tried and true methods. Short of using the "Mexican drain," otherwise known as "throwing it out the window the way my granny used to," no widely practiced method has been found. In some regions, this will still work, though it is likely that granny only had some 30 gallons a day to throw out, even with her huge family. So unless you have no running water and your water consumption is limited by what you can carry in and out, this is not a wise choice of treatment method.

This evaluation of current experiments is biased toward low-technology solutions which rely on a minimum energy input, a minimum of resource consumption, and a maximum recycling of all wastes. The goal of domestic recycling

* Except for grease, greywater contains fewer pollutants than blackwater. As with all sewage disposal methods, efforts should be made to use a grease can or some other method of disposing of it, burying it. Grease should never be washed down any drain.

† However, many reagents are more complex than sanitary wastes. Caution should be taken to avoid dumping chemicals such as paint thinner, lacquers, or automotive oils down the drain, as they most probably will interfere with biological treatment.

** Private communication with Rein Laak, May 6, 1977.

should be to close the circle between wastewater and clean water with the help of the natural regenerative capacity of the soil. The "living filter" of the soil is the only filter which can do this cleansing job adequately on a regular, reliable basis. In this country, it has been only in the last hundred years that waste disposal has been separated from food production, and the disruption of this natural cycle has caused massive pollution and has forced us to rely on other, more expensive means of fertilizing our food crops. By using the organic filter of the soil, we can reconnect wastewater treatment to food production and can cut down on the costly use of petroleum-based fertilizer substitutes.

This notion of recycling the wastes carried by wastewaters is not new in the field of sanitary engineering. Today land application of partially treated sewage is being utilized in both small rural communities and in metropolitan areas in the United States, and it has been used in several foreign countries for decades. Of all the practical sewage treatment technologies that meet federal water pollution regulations, land application is, where soils permit, the most cost effective, most reliable, and least energy-intensive tertiary treatment method available. Here we are interested in a scaled-down version of this technology.

What follows is a compilation of the experiments as they have progressed so far. We caution you not to adapt any of these experiments (except those noted as proven) for your home use *unless* you can afford to experiment. Many of the ideas here "sound nice" but have not been developed in the field for more than six months, if that long. For this reason, one engineer we spoke with has recommended that folks should start small with the easy greywaters from bath and laundry. Tackling kitchen wastes adds whole new dimensions that might discourage the beginner quickly. Additionally, because greywater treatment technology is underdeveloped, waterless toilet owners might have to continue relying on methods that are designed to treat combined wastes for a while longer.

Methods of Pretreatment of Greywater

Conventional Design

The standard for the pretreatment of combined wastewater is a septic tank. Commercially sold septic tanks can be easily used for greywater, although very recent research * has suggested that greywater is sufficiently different from combined sewage that a specific pretreatment tank is warranted.

Septic tanks usually have a capacity of 750 gallons or more, but since the toilet waste is 30 to 40 percent of ordinary sewage, a tank of 500 gallons could easily be

* R. Laak, "A Gray Water Soil Treatment System" (unpublished, 1977), and J. H. T. Winneberger, ed. *Manual of Grey Water Practice, Parts I and II* (available from Monogram Industries, 100 Wilshire Blvd., Santa Monica, CA 90401, for $1 each).

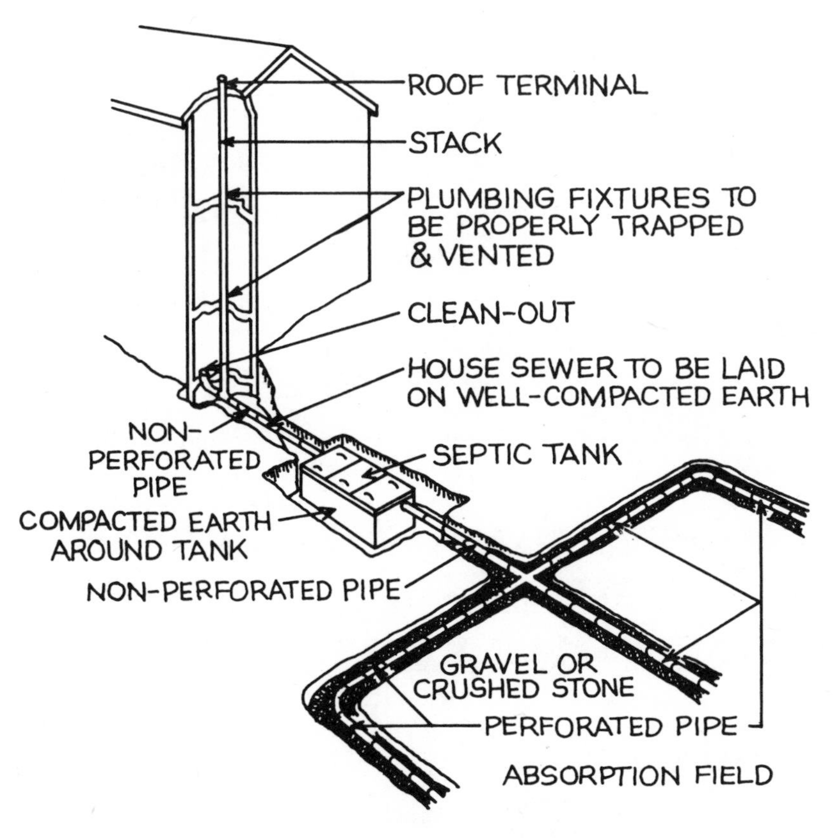

Septic-tank sewage-disposal system.
(From *Manual of Septic Tank Practice,* U.S. Public Health Service, 1972, p. 25.)

used if just greywater were to be discharged into it. If we assume that greywater will discharge at a faster rate because of fewer solids, an even smaller tank could be used. To size the tank generously, double the daily flow of greywater and add 40 percent to this volume for storage space for sludge. This will allow for the proper 24-hour retention and at least a year's accumulation of sludge and scum. Thus, if your average daily greywater flow is about 75 gallons, a tank with a capacity of 210 gallons will do you just fine. However, tanks usually come in standard sizes, and for the money, a larger tank is worth it. A larger tank will allow for longer retention with better settling and better grease removal, and thus probably will save on clogging problems

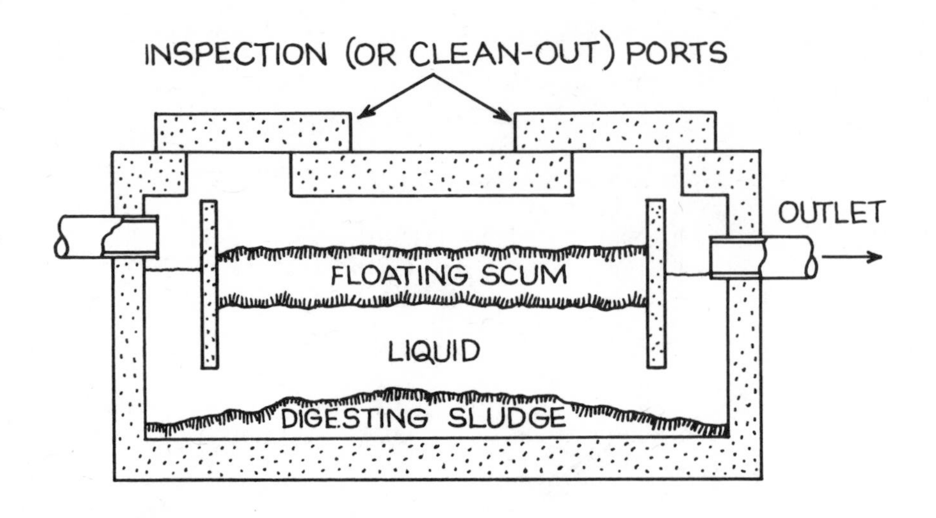

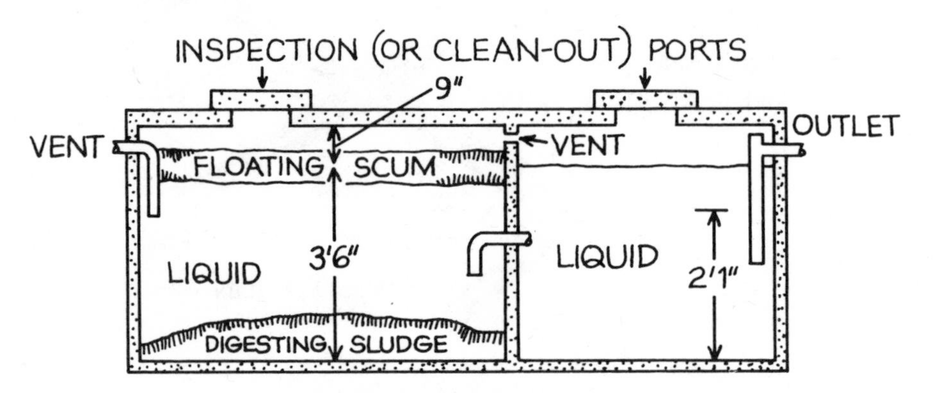

Two septic tank designs.
(From Victor D. Wenk, *A Technology Assessment Methodology, Volume VI, Water Pollution, Domestic Wastes,* p. 4. Mitre Corporation, McLean, Va. June 1971. Distributed by National Technical Information Service, U.S. Department of Commerce, Springfield, Va. [#PB–202 778–6].)

down the pipe in the discharge field.

The design of a greywater pretreatment tank takes into consideration that greywater is typically 30°F hotter than combined sewage. Because this heat helps to keep grease suspended in solution, a pretreatment tank should be designed to give off heat. Cooling the greywater will emulsify the grease and allow it to float better. Consequently, a greywater tank should be longer and narrower to provide increased wall contact with the soil and should be constructed out of a poor insulator like steel. Researchers have also found out that a two-compartment tank is better than one because it gives better settling.

The septic tank functions to trap grease and oils, settle solids, accommodate surge loadings, and perform some pollution reduction. In it, light substances float to the top forming a scum layer, while heavier substances settle to the bottom forming sludge. These sludges need to be pumped out every three to five years (depending on the capacity of the tank), and then can be put onto compost piles for recycling.* In the tank pollution reduction is dependent on how long the water is retained, although generally most of the treatment is received after discharge from the tank.

The effluent from a septic tank is anaerobic, and thus the water contains little or no oxygen when it flows out for treatment. This makes life difficult for aerobic bacteria and other oxygen-demanding organisms in a garden soil, but does not interfere with adequate treatment in conventional leach fields that are properly designed and maintained.

Recently, methods of aerating the septic effluent have been developed for use in failing septic systems. One commercial venture is to insert an electrical pump into the tank which blows air into what normally becomes the septic soup. This changes the wastewaters to an aerobic system and allows for more rapid and more complete decomposition of some wastes. Some contend that it can also minimize problems with soil clogging. While the aerator has been used primarily as a remedial action in septic tank failures,† aeration tanks are now being marketed for initial installment. Disadvantages of these include high initial costs, high maintenance, problems with shock loadings, and failure due to blackouts or interrupted electrical supply. Other pretreatment methods which provide for an aerated effluent seem to offer more promise.

Filtration

This is another common method of pretreatment in which a filter is used to strain, aerate, and partially treat the wastewater. A variety of media can be used for filters, either singly, in a series, or in sev-

(continued on page 172)

* The compost piles need to be very hot in order to handle these wastes. They should be distributed to a couple of piles and mixed in well. It should be noted that most states prohibit the reuse of this sludge and have strict requirements for its disposal. The usual practice is to dispose of it in some central facility such as a landfill. Before the turn of the century, however, it was common for "night soil" trucks to dump their cargo onto local agricultural lands (see chapter 1).

† See footnote on page 32.

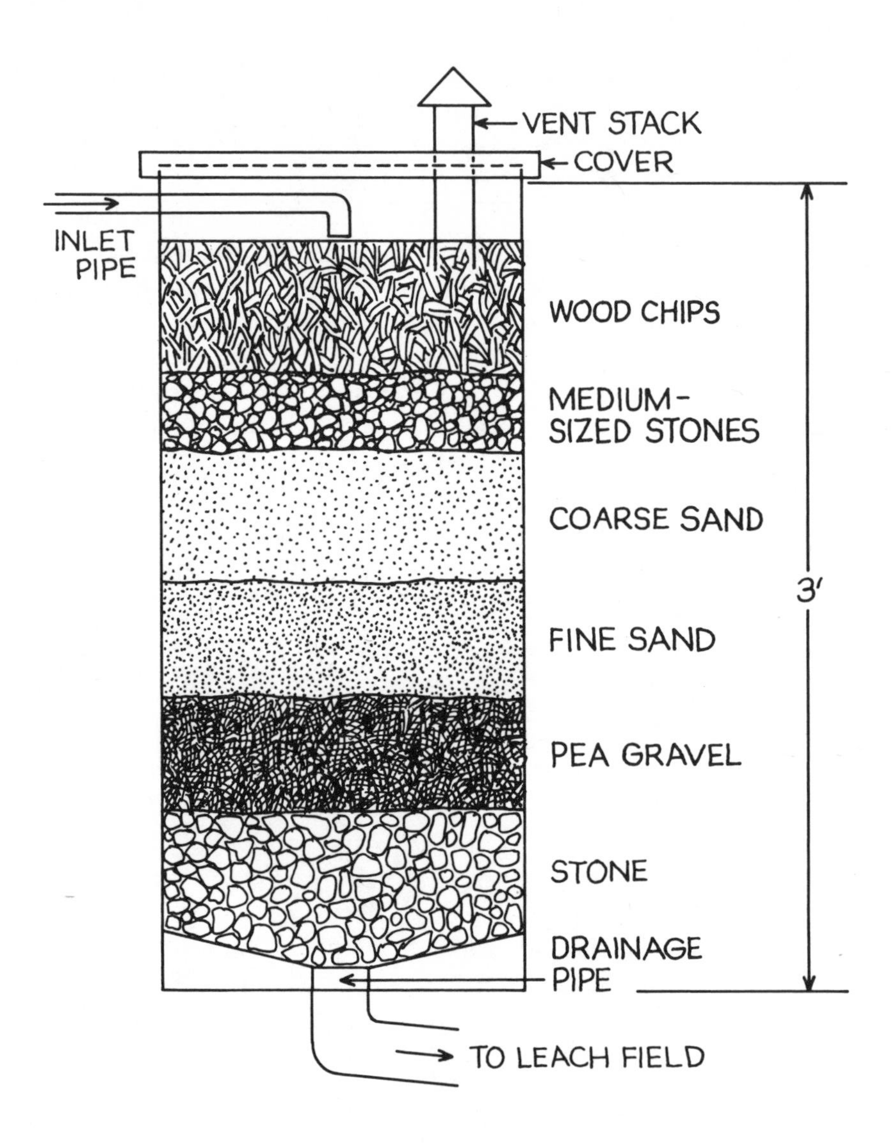

Simple, multi-media filter.

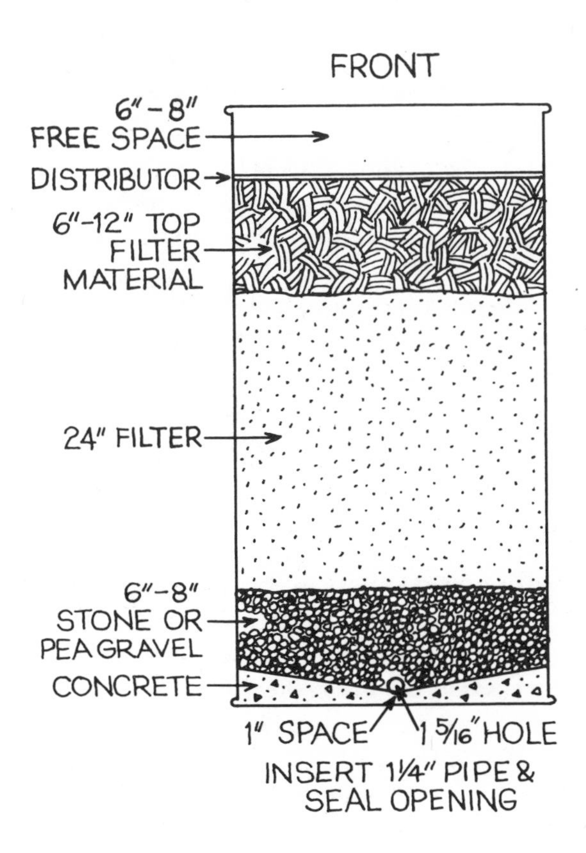

Oil drum greywater filter. To design a filter, a 55-gallon drum can be used. If you are lucky enough to find one with a removable top, use it. Otherwise, cut off the top in such a way that it can fit down inside the drum with a ½-inch or less clearance. This will be your distributor. Merely drill ½-inch holes in it, spaced 1 inch apart on center with an epoxy coating which will protect it from corrosion.

For the underdrain system, basically a funnel is needed to shunt the water into the drain pipe. For simple construction, make a V-shaped funnel out of concrete, with the bottom of the V laying across the diameter of the drum. The sides should roughly match each other, at any angle between 30 and 45 degrees. The idea is to build up two sides, gently sloping to the bottom where you can place your drain pipe. While the cement is still slightly workable, drill a 1$\frac{5}{16}$-inch hole in the side of the drum at a level just above the cement line inside. Now insert the drain pipe through the hole and let it lie along the length of the bottom of the V. For drain pipe, use 1¼-inch plastic pipe, capped at one end, and drilled with ¼-inch holes facing upwards. Seal the opening around the pipe on the drum to prevent leakage. When the cement is completely dry, backfill carefully with about 6 to 8 inches of pea gravel or stone to come up to the sides of the concrete funnel. Make sure to wash the gravel first to free it of dust and dirt that could later clog the system. This underdrain gravel will support your filter and keep the filter medium out of the underdrain. Now lay in your filter medium. If you are using mixed media, put the smallest on first. The filter should be at least 2 feet deep with an additional 6 to 12 inches more which can be removed as scum builds up. If you are using a sand filter, this top 12 inches could be wood chips as a preliminary strainer or just more sand. If you are not using a septic tank or some kind of settling basin to screen out big food particles and grease, this top layer will need frequent cleaning or replacement. Wood chips would be best in this case. There should be about 6 to 8 inches of free space remaining at the top of the filter which will hold the water as it perks through the filter. This filter is obviously limited to filtering small amounts of water at a time. Its holding capacity is less than 20 gallons.

eral layers within the same containers. The most commonly used media are sand (0.2–0.4 millimeters effective size), pea gravel, river bottom gravel, and crushed rock. Sand is an excellent medium for catching viruses and parasite ova, and it should be considered an integral part of any mixed media filter. Less traditional media are organic filters such as wood chips, straw, grass clippings, and compost, but these have not been tested on a long-term basis. An advantage is that they could be moved to the compost pile when they become clogged.

Wood chips the size of bark, commonly sold as mulch, seem to offer several theoretical advantages, as their irregular shapes would prevent packing and keep the system aerobic, their long fibers provide good surfaces for sites of biological activity, and their variable sizes would act as a good strainer. Redwood chips have been sold in California for this purpose, and it is worth checking if local woods could also work. Cedar and fruitwood chips are thought to be good filters also.

Filters can be used with or without a grease trap or small septic tank. They need to be insulated in the winter time, or they can be put in the basement or buried underground with an insulated cover above ground for surface maintenance. They should be covered and vented to protect against odors.

In principle, filters can be used as a pretreatment strainer or as the basic treatment method. But for greywater management, where disposal is in the soil as nutrient-laden irrigation water, the filter is designed to be a protector of the leach field, not a purifier. As the water trickles over the filter medium, the pollutants in the wastes are eaten by the bacteria living on the surfaces of the gravel or rocks, thus decreasing the biological oxygen demand of the water. The irregularly shaped granules of the medium support the column without packing and allow for direct diffusion of air into the filter. Water is distributed over the top of the filter so that all parts of the filter are dosed regularly and maintain the biological film on all particles of the medium. Distributors can be homemade, using a simple T-shaped distributor, a tilting bucket, an automatic syphon, nylon bags, or an inverted garden sprinkler. Alternatively, if one is using a 55-gallon drum to contain the filter, the top of it can be cut out, perforated at one-inch intervals with ¼-inch holes and laid on the top of the filter. The water will flow evenly over this plate. Filters should be screened and covered to discourage flies and prevent rain from entering. They should be vented at the top to allow gas to escape.

Placement of the filter is dependent on several considerations, including space availability, soil characteristics, desired access to its effluent (for testing or diversion), and most importantly, location of the disposal or reuse site.

There are several experimental filters now being tested for greywater.

Slow sand filters are being tested at the Small Scale Waste Management Project of the University of Wisconsin in Madison for the main treatment method. In their

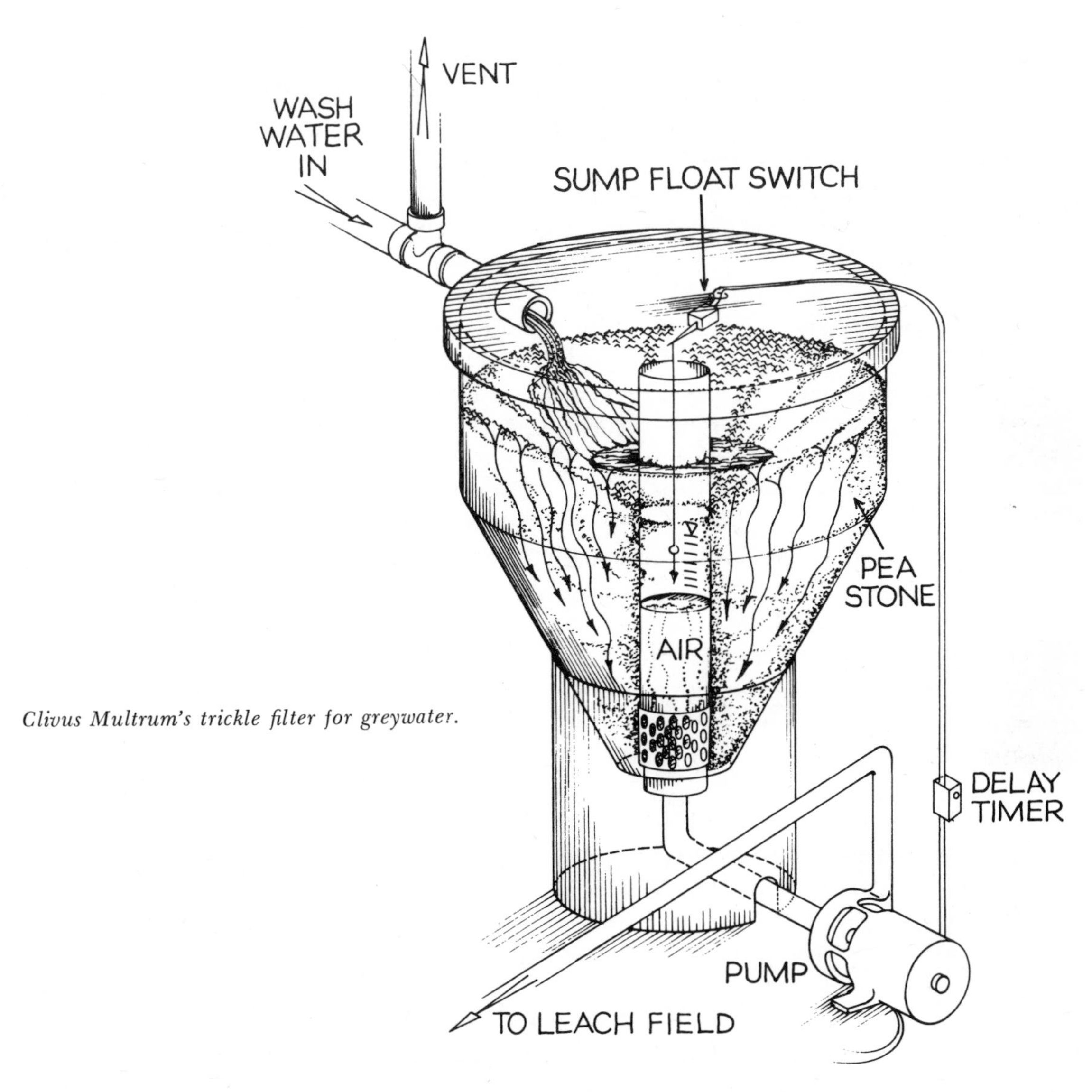

Clivus Multrum's trickle filter for greywater.

simulation studies, a typical greywater mix flows into a septic tank and then is distributed to experimental filters to test the effectiveness of various filter media. They feel sand filters followed by properly designed soil absorption fields offer a cost effective and reliable treatment method for individual homes and possibly even small cluster housing. Sand is used because it is cheap, readily available, and works well. A submerged concrete tank contains the filter, underlined with pea gravel and stone. They use a siphon with a small submersible pump which doses water on a perforated plate on the filter top for uniform distribution. Greywater comes through a septic tank which is necessary to protect the filter and to accommodate surge flows. A crust builds up on the top half to one inch of sand as it is used. First it is aerobic, but it shortly goes anaerobic as the slime builds up. A simple raking operation every four months is needed to restore it to aerobic conditions. When an aerobic effluent from an aerated tank is used, this raking is needed only once every 300 days.*

Slow sand filters for greywater are becoming more popular among professional sanitary engineers. Both J. T. Winneberger in California and Rein Laak in Connecticut have arrived at the conclusion independently that slow sand filters, when used with a septic tank or greywater pretreatment tank, can give significant BOD_5 and suspended solids reduction. Any coarse sand such as cement sand will do. Sand also traps pathogenic organisms, lowers the temperature, and traps grease. Water is applied at very slow rates, approximately one to two gallons per square foot per day, which means that one square foot of sand filter surface is needed to handle each gallon or two of greywater. This could be quite large and possibly also costly.

The *trickle filter* is another design being tested now in Clivus Multrum's wastewater treatment system. It is housed in a fiberglass cone four feet high and four feet wide at the top. It uses one inch of crushed rock. It basically strains and aerates the greywater, enriching it with aerobic microorganisms which are passed to the leach field. It also equalizes the temperature of the greywater, which is a great benefit when reusing the waters for irrigation. It sells for about $450, not including the drain field, and is designed for a family of four. It has already been approved of for use in Kentucky, and some other states are issuing permits for its use on an experimental basis. In Sweden, it has been in use for two years, and a soil scientist has recently found it is operating so well that earthworms are actually living in it!

Methods of Treating and Disposing of Greywater

Subsurface Disposal

This is most commonly used for treating and disposing of household wastewater in rural areas, because surface disposal is not allowed in most parts of the country. There are basically six ways to treat and dispose of greywater below the soil's surface: seepage pits, absorption trenches, seepage beds, evapo-transpiration beds,

* For more information on the SSWMP, write to Small Scale Waste Management Project, 1 Agriculture Hall, University of Wisconsin, Madison, WI 53706.

mounds, and leaching chambers. None of these methods was developed to treat greywater specifically; in fact, they are normally used for combined wastewater. There are a number of excellent references which describe in detail the design and construction of these systems in conjunction with site, soil, and climate.* Here brief descriptions are offered to help you to recognize your options.

Whichever soil absorption system you should choose, the reduction of wastewater flow and nutrient loading due to the waterless toilet will allow for a corresponding reduction in size and cost for greywater treatment. Because aerobic conditions will be easier to promote and maintain using greywater, the greywater system's effectiveness and service life probably exceed that of a conventional combined unit.

Seepage pits, also known as soakaways or drywells, are simply excavated pits three to four feet wide and six to seven feet deep located about 100 feet from the house. They are either left empty and reinforced with open-jointed masonry or backfilled with large rocks, 20 inches or more. Wastewater percolates out of the pit through the bottom and sidewalls. Seepage pits should be constructed only in permeable soils, and because of the danger of ground water contamination there should be at least four feet between the bottom of the pit and the water table. As with any soil absorption system, care must be taken to insure that there is no chance that effluent will seep into the home's water supply. Locating the pit about 100 feet downhill from the well should provide an adequate margin of safety.

In ideal soils and when the influent is free of grease and fat, seepage pits give good service and need little maintenance. When they are used without a pretreatment septic tank they are called cesspools. Both cesspools and seepage pits can be two pits combined in a series with the distance between them at least three times the diameter of the largest pit.

* For more information on septic tanks and leach fields, see *The Manual of Septic Tank Practice* (U.S. Public Health Service, 1972) and the *Manual of Grey Water Practice, Parts I and II,* edited by Winneberger. These excellent beginnings of research on greywater describe "mini" septic tank systems and greywater quality data. Another fine book, sympathetic to on-site disposal and ecologically sound treatment of wastewaters, is Peter Warshall's *Septic Tank Practices: A Primer in the Conservation and Re-use of Household Wastewater,* revised edition, 1976. Available from Peter Warshall, Box 42, Elm Road, Bolinas, CA 94924, for $3.

For more information about evapo-transpiration beds, see Alfred P. Bernhart, *Treatment and Disposal of Waste Water from Homes by Soil Infiltration and Evapo-transpiration,* University of Toronto Press, 1973; available from Dr. Bernhart at 23 Cheritan Ave., Toronto, Ontario M4R 1S3, for $16. See also C. B. Tanner and J. Bouma, "Evapotranspiration as a Means of Domestic Liquid Waste Disposal in Wisconsin," February 1975, from the Small Scale Waste Management Project.

For information about mounds, see Converse et al., "Design and Construction Procedures for Mounds in Slowly Permeable Soils with or without Seasonally High Water Tables," March 1976, from the Small Scale Waste Management Project, and Manitoba Department of Public Health, Winnipeg Canada, 1962. For information on above ground filters, see Saleato, J. A. *Environmental Engineering & Sanitation,* Wiley Interscience 1972 second edition.

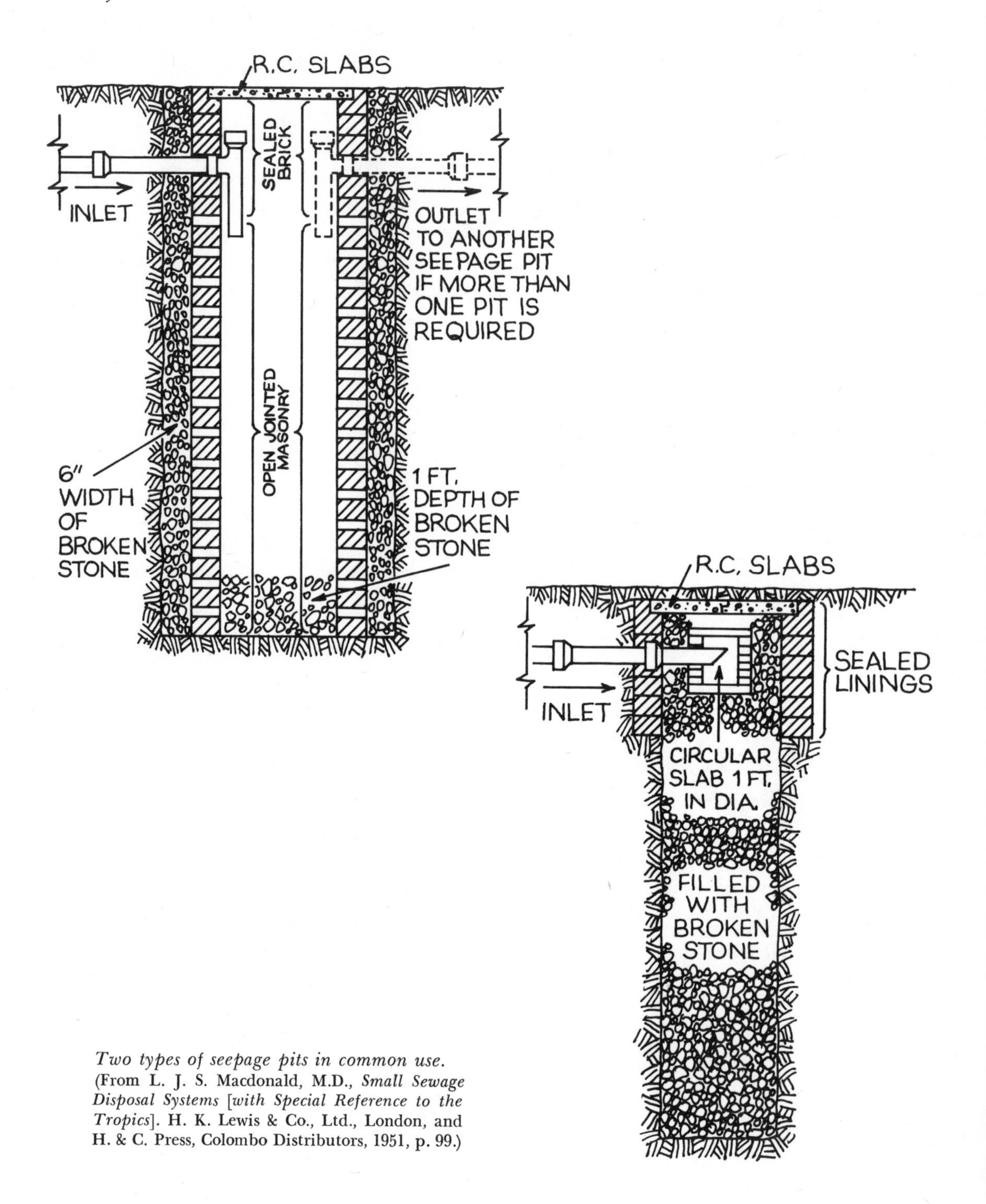

Two types of seepage pits in common use. (From L. J. S. Macdonald, M.D., *Small Sewage Disposal Systems [with Special Reference to the Tropics]*. H. K. Lewis & Co., Ltd., London, and H. & C. Press, Colombo Distributors, 1951, p. 99.)

Absorption trenches (also called leach fields and drain fields) are the usual second component of septic systems. Wastewater from a septic or aerobic tank is distributed by perforated pipes or tiles into a gravel bed, where it leaches out into the surrounding soils where bacterial populations break the pollutants down into inorganic compounds. Seepage beds, evapo-transpiration beds, and mounds are just fancier versions of this basic absorption trench design: Seepage beds recycle to the ground water; evapo-beds utilize water evaporation from the soil and plants; and in mounds, the water and nutrients are transformed by plants into biomass.

Absorption fields are constructed of trenches at least two feet below the surface which radiate from the pretreatment tank and distribution box. Generally the trenches are 18 to 36 inches wide and two to four feet deep. The sides are scratched

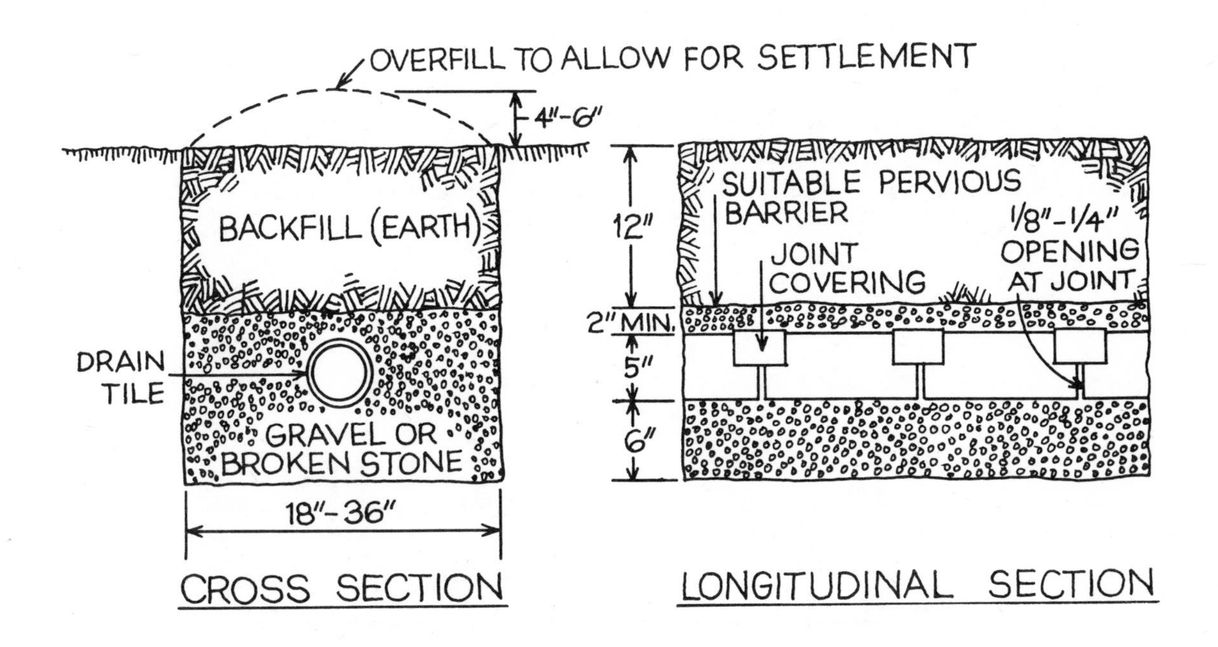

Section of absorption trench.
(From *A Septic Tank System for Sewage Disposal,* Agricultural Engineering Extension Bulletin #386. Department of Agricultural Engineering, New York State College of Agriculture, Cornell University, Ithaca, N.Y.)

or roughed to keep the soil pores open for maximum infiltration. The bottom is filled with six to 12 inches of crushed stone ½ to 2½ inches in size. Above this the distribution pipe is placed. The trench is then carefully backfilled by several more inches of crushed stone, topped with untreated building paper or straw, and covered with soil. Utmost care must be taken when backfilling the absorption trenches, since smearing or compacting the soil at this interface usually leads to clogging and ineffective treatment.

The length of trenches required to treat different volumes of greywater depends mainly on the permeability of the soil. Very clayey soils may need as much as four square feet of bottom and sidewall per gallon per day, while a sandy soil might be able to handle the same load in one square foot. Some think that aerobic conditions can also make a difference. Dr. Alfred P. Bernhart of the University of Toronto has suggested that the aerobic microorganisms keep the trench warmer and thus increase the evaporation rate, but Dr. Rein Laak of the University of Connecticut has pointed out that whether the soil is anaerobic or aerobic has more to do with the loading rates than the microorganisms. Both anaerobic and aerobic systems can result in adequate treatment. However, if greywater is to be recycled into a garden site or any area where plants are set to use the water, aerobic conditions are favored.

One way to create aerobic conditions is to use an aerobic pretreatment tank as described earlier. Another is to vent the distribution pipe to the outside and bring air into the line. Alternatively, a tipping

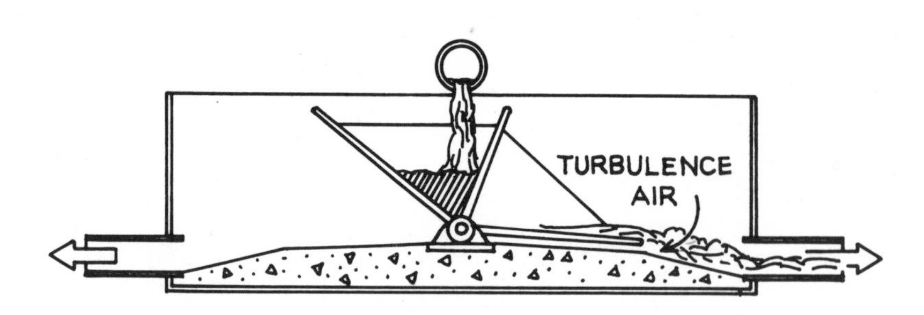

Tipping alternator. One-half of the tipping device fills with wastewater and the shifting center of gravity makes the device tip over and discharge liquid into one seepage bed. Now, the second half of the tipper fills with wastewater and subsequently discharges into the other seepage bed. Some aeration occurs through turbulence at discharge points.
(From Bernhart, *Treatment and Disposal of Waste Water,* p. 24.)

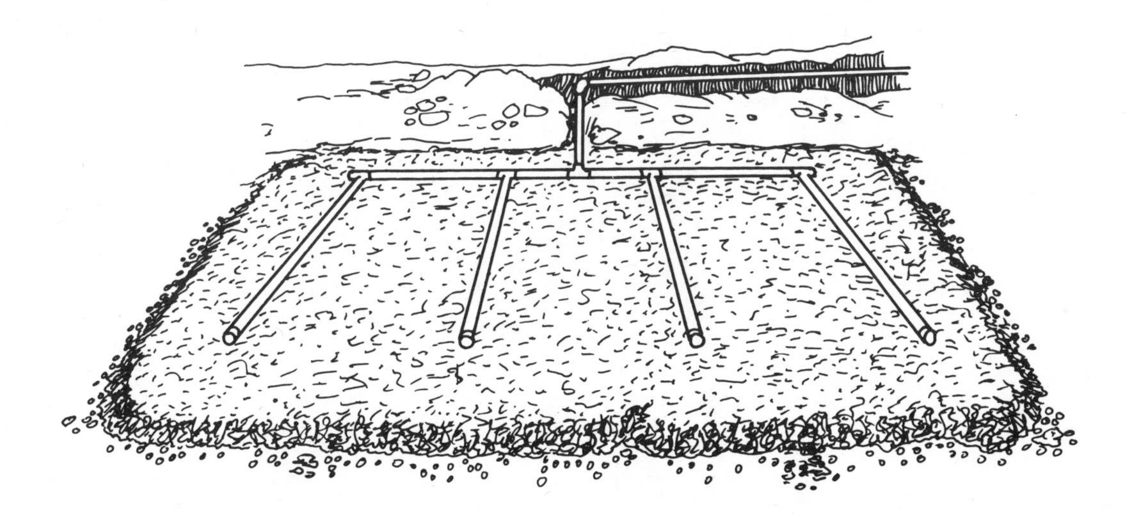

Seepage bed. To prepare it correctly, lay the distribution pipe on the stones, taking care to lay the laterals level. Slope the manifold slightly toward the inlet pipe. Remove dips and rises in laterals, then cover distribution system with 2 inches of stone. Take care not to disturb stone after it is laid, and do not drive on top of the distribution system.
(From J. C. Converse, R. J. Otis, and J. Bouma, "Design and Construction Procedures for Fill Systems in Permeable Soils with High Water Tables," revised March 1976. SSWMP, p. 17.)

device that alternates the greywater flow from one trench to another will partially aerate the water as it splashes into the drain pipes. A practice which is also becoming more widespread involves the use of two separate absorption trench systems, one always being given a chance to "rest and breathe."

Although ordinary absorption trenches do not discharge water near enough to the surface for the use of many plants, they can be used to irrigate orchards, berry patches, and hedges. Planning a leach field with such uses in mind will minimize the danger of nitrate pollution of ground water.

Seepage beds are simply compact leach fields where the bottom of the bed is the primary soil interface. Instead of many trenches wandering over the yard with spaces in between, there is one large bed with distribution lines spaced every five to six feet.

Evapo-transpiration beds evolved from seepage beds. They are shallow beds, sometimes lined with an impermeable mem-

(continued on page 182)

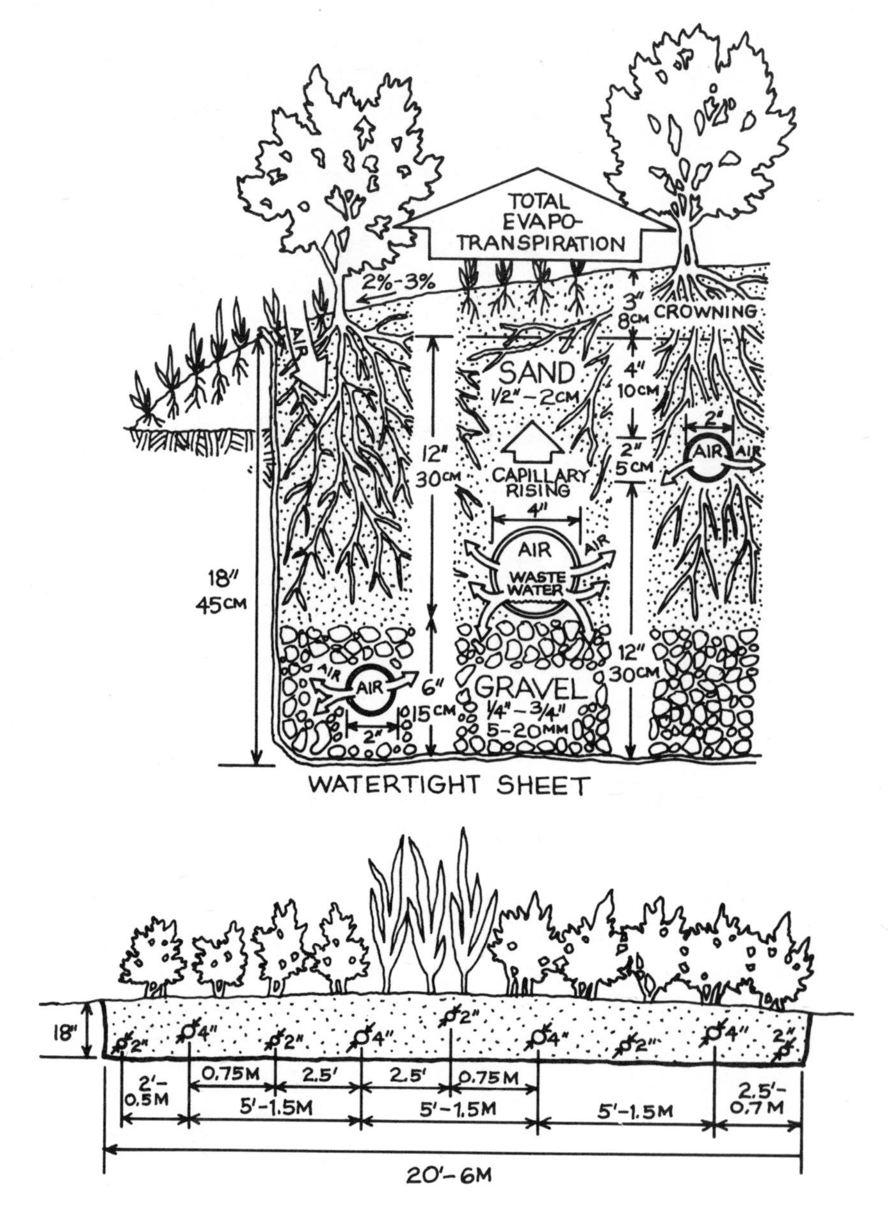

Evapo-transpiration bed. All wastewater evaporates, since soil infiltration is prevented by a watertight plastic sheet. Aerobic conditions are maintained by air circulation through pipes and from surface. Plants grow on the mounded area.
(From Bernhart, *Treatment and Disposal of Waste Water*, p. 49.)

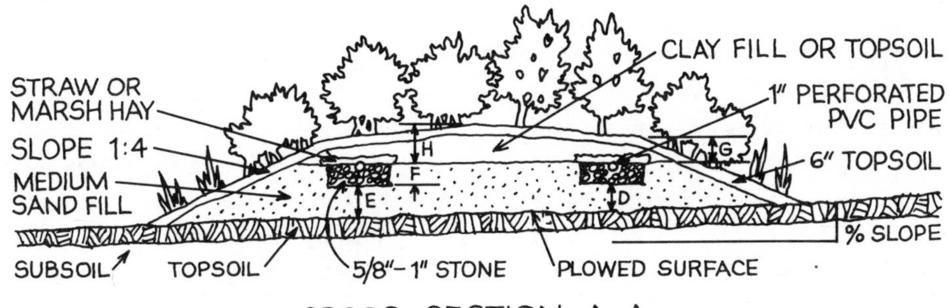

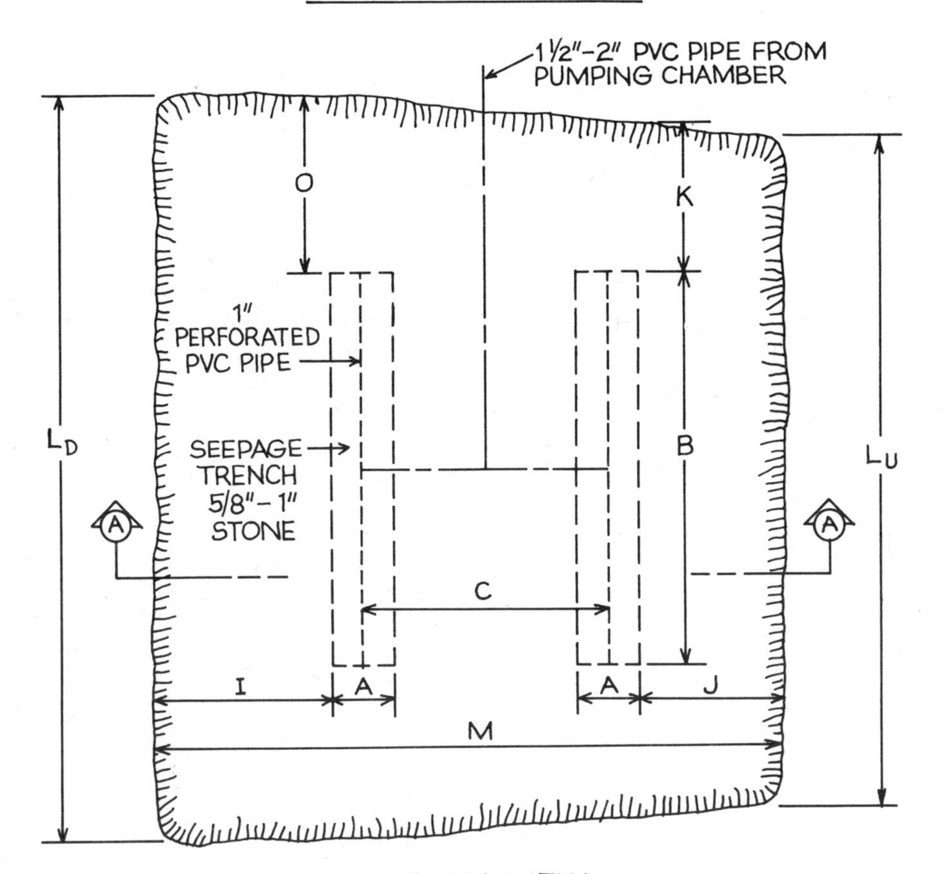

Mound system for a two- or three-bedroom home on level or sloping site.

(From Converse et al., "Design and Construction Procedures for Mounds and Slowly Permeable Soils with or without High Water Tables," revised March 1976. SSWMP, p. A–13.)

brane in order to protect ground water. Gravel is laid at the bottom to provide storage of effluent for eventual evaporation, and plants are arranged on the soil surface. These plants utilize the nutrient-laden water and evaporate the water through its leaves in a process known as transpiration. The interior of the bed is vented by special aeration ducts.

Where feasible, evapo-transpiration is an important aid in the treatment and disposal of greywater. However, climate is critical. Evapo-transpiration is most successful in areas with low humidity, high temperatures, and good air turbulence. The length of day and amount of sunshine are also influencing factors. Thus, evapo-beds should be most suitable to southern regions where the growing season is long. Tests in Wisconsin have shown that climate conditions were unsuitable for adequate transpiration during the winter months. These results suggest that evapo-beds are not feasible where there is extended snow cover and/or long periods of below freezing weather.* They may, however, be useful for summer homes in the North where lawn irrigation is required.

Mounds, first developed in northern Manitoba more than twenty years ago, also have much in common with seepage beds. The principle of mounds is to bring additional soil materials, usually a medium like sand, and mound it over the area to be used for soil absorption. The mound is constructed a few feet above the normal grade, and the wastewater is pumped up to the raised absorption field designed with sufficient capacity to treat the water. The trapezoidal cross section of the mound enables water to infiltrate the upper layer of the surrounding soils without surfacing. This is a great advantage in situations where shallow soils cover creviced bedrock and where there is a high water table. Like seepage beds and evapo-beds, mounds are planted with permanent greenery such as grass, shrubs, berry bushes, or evergreens to stabilize the soil and increase the evapo-transpiration rate. Mounds have been used successfully under severe winter conditions, and they are gaining wider acceptance in the field. Maine has endorsed their use throughout the state. Specialized variations are being examined, and preliminary tests show they might be able to be used in an even wider range of climates and soil conditions than originally considered.

Leaching chambers are precast concrete structures used in place of absorption trenches. They have open bottoms and open sides, usually about four feet wide, one foot high, and up to eight feet long. The water from a septic tank will fill up the hollow space under the leaching chamber and gets distributed evenly across the soil beneath the chamber. This allows for aerobic conditions and 100 percent effective use of the leaching bottom area. They can be laid in a series, interlocked end to end for trenches or side by side for leaching beds of any size. Each chamber has a manhole to allow the user to enter easily for raking and scraping beneath the chamber, thus giving many years of added life to the system. They provide a cheap, pre-

* Tanner and Bouma, "Evapotranspiration in Wisconsin."

fabricated leach field, because the owner can dig his or her own field, have a truck bring in the chambers, and maintain it him or herself, all without the expense of hired labor. Because they can be maintained easily, they are thought to save money in the long run.

At the present time, for people with thin or impervious soils or with high water tables, mounds, evapo-beds, and leaching chambers are the only on-site, subsurface treatment and disposal schemes possible.

Another creative idea (not tested) would be to build a simple greenhouse over an evapo-bed or part of a mound on which a forage crop such as comfrey is grown. The transpired water (essentially distilled) would condense and run down the greenhouse sides where it could be collected for reuse. The crop could be harvested as animal fodder or mulch for the garden. Having the greenhouse as a lean-to

Clivus Multrum's first method of greywater disposal consisted of nothing more than a perforated plastic pipe laid over a 3-foot-high pile of leaves. This setup, the company claims, can handle about 60 gallons of greywater a day. To prevent possible freeze-ups, the pipe is insulated in winter with a 1-foot-deep covering of leaves.

south wall of a barn or house could make the investment pay double dividends as a solar heater.

The Clivus Multrum people now have two and a half years experience using filtered greywater as irrigation water. The first method of treating and disposing of the greywater was to hose it out to the garden into a three-foot-high, 30-foot-long windrow of leaves. They used a 1½-inch pipe with ¼-inch holes drilled every foot, which ran the length of the windrow. The pile needed no turning nor aerating, and the one pile, made in the fall, was sufficient to handle about 60 gallons per day throughout a year. It was covered with one foot of leaves for insulation, but this was not essential because the greywater is 60° to 70°F (15.5° to 21°C). They used this method for two years, but now have switched to a solar-heated greenhouse. The composted leaves from two years' greywater treatment were used as soil, in which salad greens and root crops are grown. Greywater is now piped into the greenhouse, where both the vegetables and earthworms are reported to be thriving.

Surface Disposal

A commonly used method of surface disposal is a *lagoon* or *stabilization pond.*

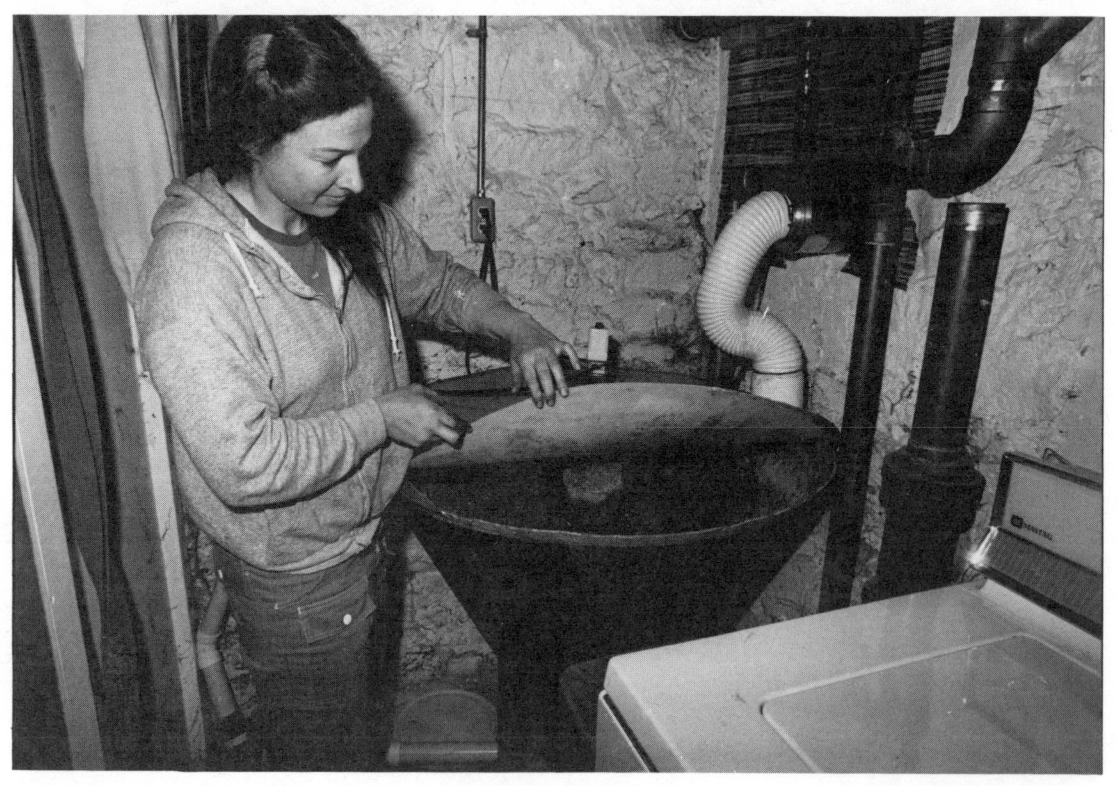

Abby Rockefeller lifting up the cover on the trickle filter, the first stage in her greywater treatment and disposal setup. At right, note how water pipe from sinks and laundry was diverted from sewer hookup to trickle filter.

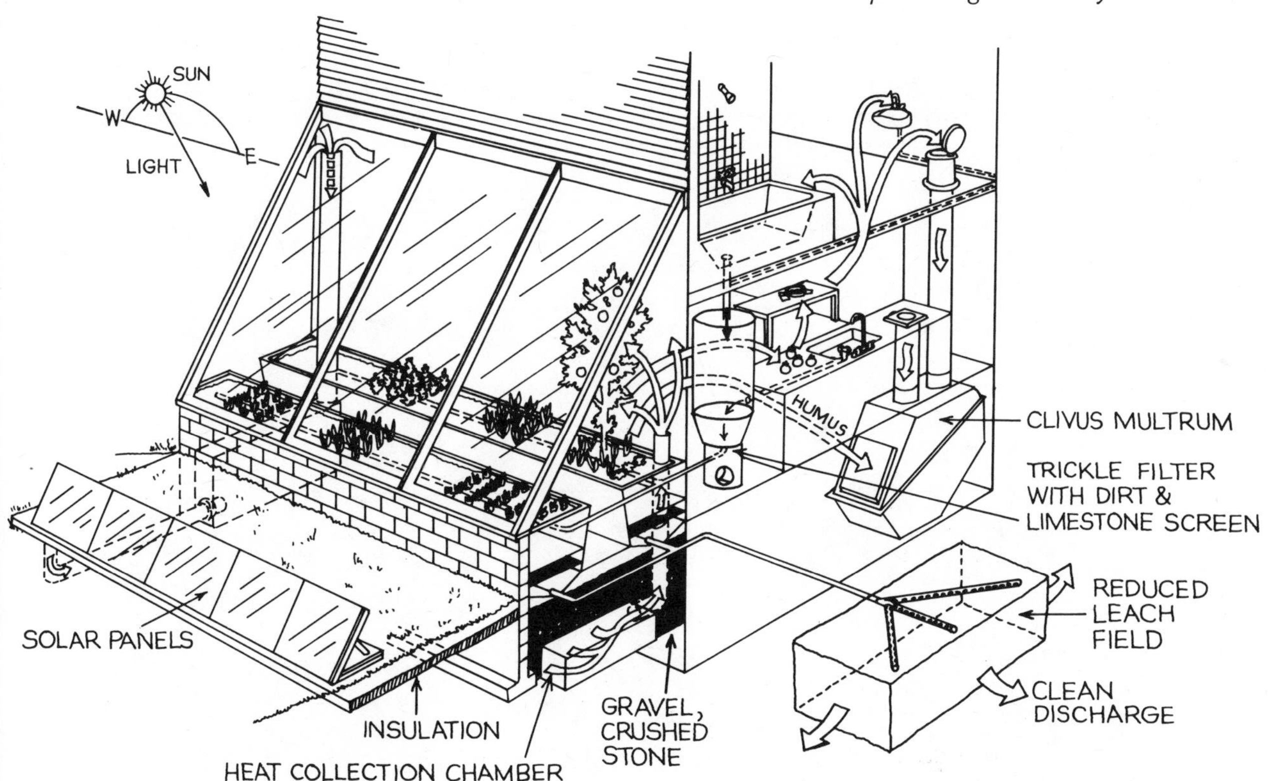

Clivus Multrum's experimental total household wastewater system. All water from bathroom, kitchen, and laundry gets primary treatment in the trickle filter. It then goes into the solar greenhouse planting beds where it irrigates and fertilizes plants as it filters through the soil and drainage material. After leaving the greenhouse it is discharged into a small leach field.

A lagoon is a large open holding pond, usually shallow, in which bacteria degrade organic matter into carbon dioxide (or methane in the anaerobic stabilization pond) and water. Algae use the carbon dioxide, the nutrients in the water, and the energy of sunlight to grow more algae—thus an algae farm, which can be cropped for feeding fish in a secondary pond. Alternatively, ducks and other water fowl could feed on the lagoon and both keep it clean and provide a food source. If the algae are not physically removed, they might become so dense on the surface as to block the light needed by the lower-lying photosynthetic bacteria. If this condition continues, the BOD may actually increase, and the amount of dissolved oxygen will decrease. Anaerobic or septic conditions will follow and the pond will give off odors. Thus physical or biological removal of the algae is critical. There are

Inside the Clivus Multrum experimental solar greenhouse. Note the deep planting beds through which the greywater is filtered.

also a number of other plants which can purify water by converting the nutrients into bio-mass, including water hyacinths and duckweed. The use of water hyacinths to cleanse sewage has been quite successful in Florida, and experimentation is needed in the purification of greywater.

Ponds often are constructed with watertight linings. Impervious soil is adequate but leakage cannot be ruled out absolutely. Consequently, concrete, stone, and plastic are commonly used liners. The New Alchemy Institute has reported use of a "biological sealer" * that may be practical.

To prevent offensive conditions from occurring, aerobic ponds should be at least half filled with water before they receive greywater. If not overloaded, they could actually handle raw sewage, but some primary settling in a grease trap or filter is helpful in preventing the pond from going septic. It is a good practice to build two ponds for assurance of high treatment.

* William O. McLarney and J. Robert Hunter, "A New Low-Cost Method of Sealing Fish Pond Bottoms," *The Journal of the New Alchemists* 3 (1976).

The second pond could be used for fish feeding on algae. The fertile effluent of fish ponds has been shown to be excellent for irrigation or garden vegetables.*

Important advantages of a lagoon are: it is cheap to construct, easy to maintain, requires no power or mechanical equipment, gives good retention, can provide up to 95 percent BOD removal (including pathogens), produces valuable algae, and handles surge loadings easily. When properly designed, they are the cheapest of all methods. Moreover, it is a holding basin for fire protection, irrigation, ice skating, and possible recycling systems.

Disadvantages are: up to 50 percent water loss through evaporation, a tendency to go septic and cause odors, a tendency towards incomplete virus removal, requirements for adequate space, and climatic limitations. If the waters are to be reused in a house, e.g., recycled into the bath or laundry, they should be disinfected first.

Irrigation is another method of surface disposal for reuse. The greywater is usually given some kind of pretreatment where large food particles and suspended solids are strained out so as to keep the distribution lines free from clogging and to keep putrescent food particles and grease off the garden.

Irrigation systems vary widely. In Muskegon County, Michigan, for instance, secondarily treated and chlorinated effluent from the municipal sewage lagoons is sprayed into corn and wheat fields as an effective tertiary treatment method.

Greywater irrigation systems are also being tested by the Farallones Institute at their Integral Urban House in Berkeley and their rural center in Occidental. In Berkeley, the daily production of 75 gallons of greywater is mixed with 1½ gal-

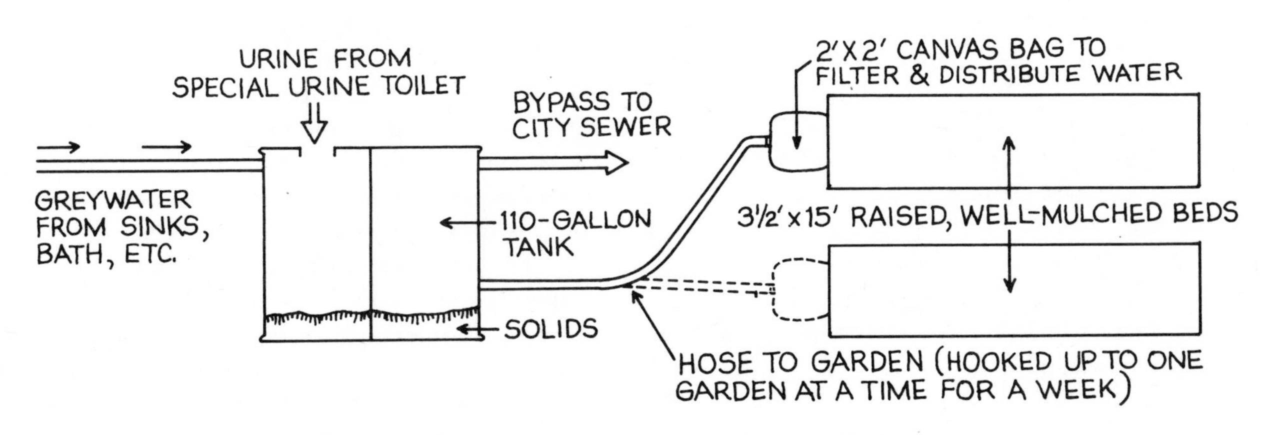

Integral Urban House's greywater irrigation system. Each raised garden bed is irrigated for a week at a time before the hose is moved to the alternate bed.

* McLarney, "Further Experiments in the Irrigation of Garden Vegetables with Fertile Fish Pond Water," *The Journal of the New Alchemists* 3 (1976).

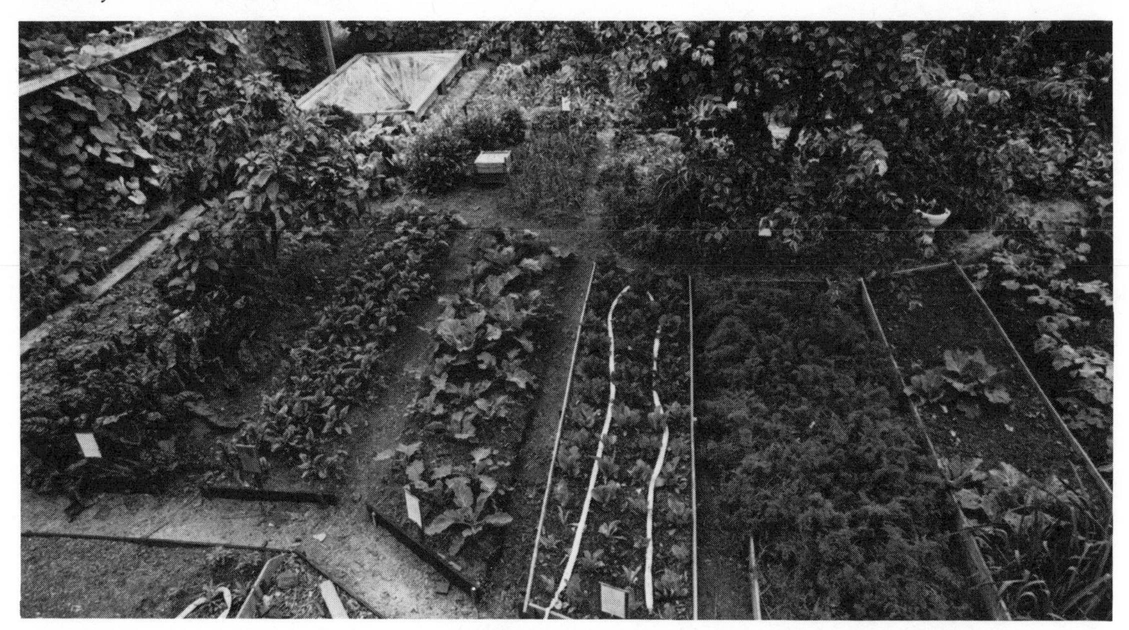

A bird's-eye view of the Integral Urban House's backyard garden beds. The hoses, irrigating the bed in the center, are hooked up to a 110-gallon wine water settling tank.

lons of raw urine which enters the pipes from their special urine toilet. This water is held in a 110-gallon tank (two 55-gallon drums welded together) and daily hosed out to their vegetable gardens. The tank acts as a settling tank and has a bypass to the city sewer in case its capacity is exceeded. Each day the hose is moved to another garden bed so that each bed receives the greywater-urine mix every two weeks. In the week between, the bed is doused with tap water to flush any noxious materials out. These beds are about three to four feet wide and 12 to 18 feet long, and they are well mulched, following Ruth Stout's methods.*

For simple filters, they use wire basket strainers on each drain and attach a canvas bag at the end of the hose. This bag is a two feet by two feet square piece that balloons outward when it fills up with water. It acts both as a filter to keep putrescent material off the garden and as a distributor for uniform watering. When the bags get clogged every three to four days they are removed and cleaned. To clean the hose, they simply connect it to the city water tap so the pressure forces out any accumulated scum. Similarly, the settling tank is backwashed as scum builds up.

For the first year of operation, this system has worked well, although they have not done any extensive testing on the soil. They have not tested for pathogens, but they explain that water-borne viruses and bacteria from urine do not do well in the soil. The vicious competition from the bacteria in the rich organic soil virtually

* For an explanation of Ruth Stout's permanent mulch system see *The Ruth Stout No-Work Garden Book* (Emmaus, PA: Rodale Press, 1971), one of several books she's written on her gardening methods.

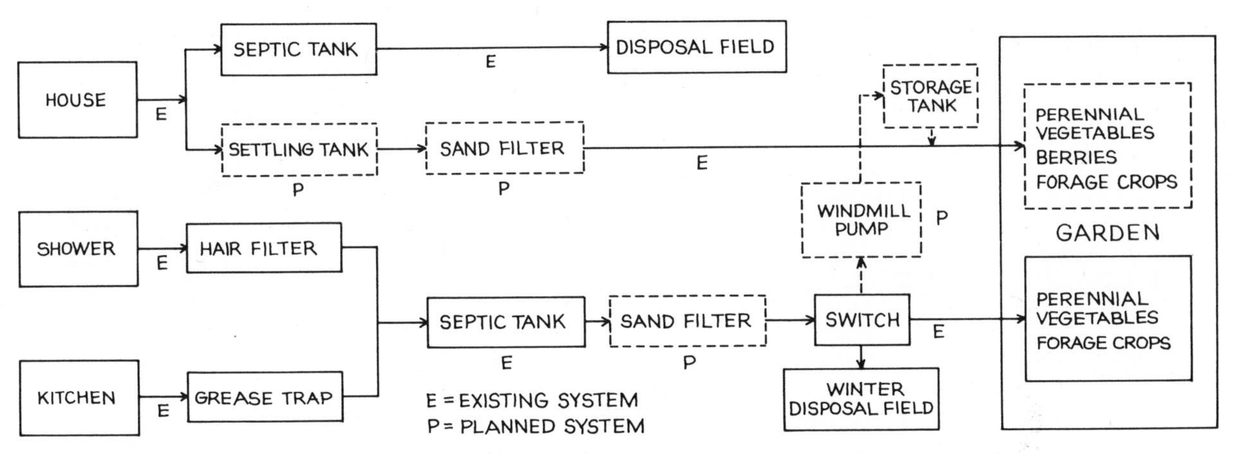

Farallones Institute's Rural Center greywater irrigation system, that combines septic tank and sand filter with garden irrigation and subsurface disposal field.

eliminates any pathogenic organisms from urine. Moreover, they explain, plants' roots cannot take up viruses and bacteria.

The Farallones Institute's Rural Center in Occidental, California, is putting in a subsurface leach field under their berry patch for use during the six-month rainy winter season. They use a 1,200-gallon septic tank which gives them three to four days retention time for good settling of solids, which should minimize clogging problems. For the growing season, they send greywaters out one-inch perforated pipe into their French-intensive garden beds where the pipes are covered with six inches of straw mulch. The mulch aids in water retention and keeps the septic waters below the surface. Eventually they want to lay down permanent lines in permanent beds and rig up a switch to control the flow to alternate beds. These permanent beds will be planted in water-demanding crops such as artichokes, asparagus, comfrey, and alfalfa.

Neither the Integral Urban House's nor the Rural Center's method has been tested over the long term, and it is unclear what effect the greywater, as a very unbalanced nutrient source, could have on the soil. The effect of soaps and detergents on the ionic balance and salt content of the soils could be detrimental. Sodium salts from soaps adhere to clay particles and could build to levels harmful for plants and elderly people who eat them. In Berkeley, they use a side dressing in their gardens of gypsum every two months so that the calcium in the gypsum will theoretically exchange places with the sodium in the soil, changing the sodium into a form that readily leaches out of the soil in the next rain. However, flushing the soils with clean water may cause just as much leaching and will not introduce any new ions.

Greywater Reuse in the Urban Garden

Here are answers to three of the most commonly asked questions concerning the use of household wastewater for garden irrigation:

How much greywater can be used in an urban garden?

Use only as much greywater in your garden as is required for reasonable irrigation; scale your wastewater recycling effort to suit your garden water requirement. A good, safe rule is that a square foot of loamy garden soil, rich in organic matter, is capable of handling one-half gallon of greywater per week. Sandy, well-drained soils will accommodate more water; clayey, poorly drained soils, less. If your garden area suitable for greywater application is 500 square feet, then up to 250 gallons of wastewater may be discharged each week. This rate might be greater during the summer months when surface evaporation and plant transpiration are considerable and less during the winter when evapo-transpiration is minimal. Frequently check soil moisture to determine precise application rates for your garden.

Follow these suggestions for sound wastewater application:

1) Apply the greywater to flat garden areas; avoid slopes where runoff might be a problem.

2) Avoid using wastewater on root crops which are eaten uncooked, such as carrots and radishes, or leafy salad vegetables such as lettuce.

3) Use the wastewater on mature vegetation or well-established vegetable plants, not on young plants and seedlings.

4) Minimize wastewater application to acid-loving plants such as rhododendrons and citrus, since wastewater is alkaline.

5) To the extent possible, disperse the greywater application over a large garden area.

6) When available, use fresh water for garden irrigation on a rotating basis with greywater to help cleanse the soil of sodium salts.

7) Apply thick compost mulches to areas receiving greywater to improve natural decomposition of waste residues.

What about soaps and detergents? Are they harmful to the soil and plants?

As a general rule, soaps are less harmful than detergents, but either presents potential problems over periods of sustained use of greywater containing them. The common problem of soaps and detergents is that they both contain sodium, an element which in excessive amounts is harmful to soils (destroys soil aggregation) as well as to plants (induces tissue burn). The best strategy is to minimize the use of cleaning materials, and wherever possible choose soaps rather than detergents. Gentle soaps, such as soap flakes, are preferred to those heavily laden with lanolin, perfumes, and other chemicals. Where detergents must be used, select those which do not advertise their "softening powers" (softeners are rich in sodium-based compounds). If you plan on reusing washing machine water, bleach should be minimized or eliminated, and boron-based (Borax) detergents absolutely avoided. Phosphates in detergents are not as great a problem in soil application as they are in sewage discharge into water bodies; nevertheless, low-phosphate detergents are preferable because they generally contain less sodium. Ammonia is preferred as a cleaning and deodorizing agent and washing soda may be used as a scouring powder.

How should the greywater be applied to the garden?

Apply the wastewater directly to the soil; *do not* overhead sprinkle or allow the wastewater to contact the above-ground portion of food plants. Wastewater is best conveyed from the output buffer tank to the garden by a standard ¾-inch garden hose. A central hose may feed several lateral short hoses by way of a "Y" junction such that the wastewater is distributed evenly over larger areas of soil. The lateral arms should be rotated around the garden frequently to reduce the possibility of localized flooding or excessive residue buildup.

At the end of each hose lateral, attach (by hose clamp) a cloth bag (cotton or canvas) to intercept particulates and soap residues conveyed in the greywater. The bag will allow for dispersed water outflow while trapping undesirable materials. The bag should be removed periodically, washed, turned inside out and allowed to sun-dry, and reused.

Researched and prepared by Tom Javits
with the assistance of the Farallones
Institute's Integral Urban House
research staff.

Putting It All Together

Now that you are aware of the basic alternatives for greywater treatment and disposal or reuse, it is time to figure out your own needs and soil conditions in order to choose the best alternative for you. To review briefly, the illustration here shows the various options available in the proper sequence.

Another couple of years of experimentation will be needed to refine greywater management methods and to explore the effects on soils and vegetation. This chapter is meant to encourage you to try your hand at small-scale greywater treatment fit to your own needs.

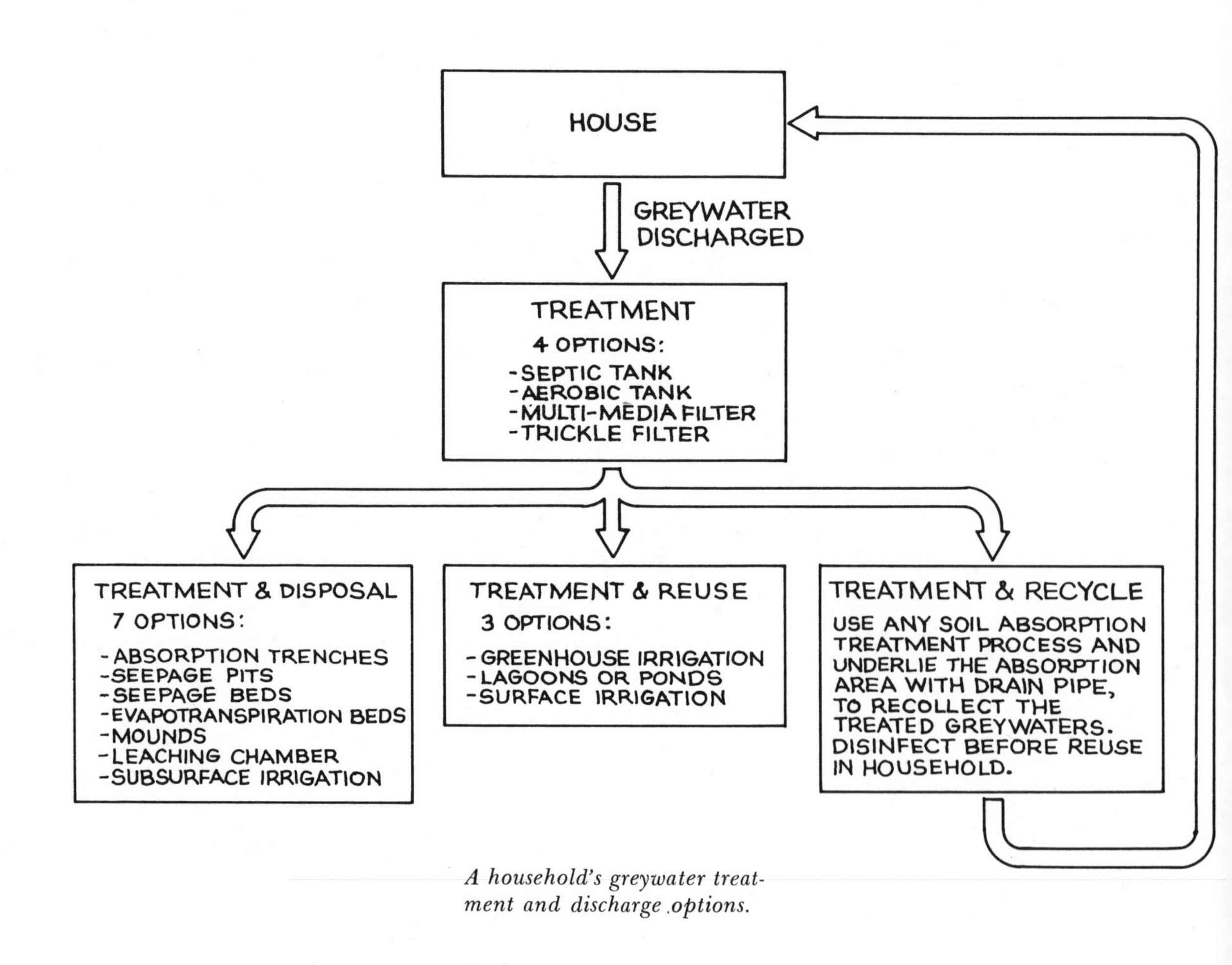

A household's greywater treatment and discharge options.

General Rules for Laying Out Any Soil Absorption System

1. A crested area or a level area are preferred over a sloping site. If a slope is used, up to 12 percent slope is tolerable if the percolation rate is between three and 29 minutes/inch. For less than six percent slope, a percolation rate of 30 to 60 minutes/inch is allowable. In general the greater the slope is, the larger the system will be.

2. The design, size, and orientation of the absorption system are dependent on the house, amount of water use, and topography. The absorption field should be laid perpendicular to the ground water floor.

3. For distribution pipes use, schedule 40 or 80 PVC perforated pipe of the diameter required by the design. Holes are made in a straight line along the length of the pipe, using a ¼-inch drill, every six to 12 inches. However, it is easier to buy it with the holes already bored.

4. If a backhoe is used to dig absorption field, dig it "only when soil moisture is low to avoid compaction and puddling. If a fragment of soil occurring approximately nine inches below the surface can be easily rolled into a wire, the soil should not be plowed, since the moisture content is too high. If, on the other hand, the soil is friable or dry and falls apart when rolling into a wire, the soil can be plowed because the danger of causing compaction and puddling is minimal. Once compaction and puddling occur, they reduce the ability of the soil to accept water, thus increasing the chance of failure. *Once plowing is completed, keep all vehicular traffic off the plowed area.* Try to avoid the occurrence of a long period between plowing and construction. If it rains after plowing is completed, wait until the soil dries out before the start of construction. Immediate construction after plowing is highly preferable." * However, if you want to dig it by hand, it would be far easier to dig when the ground is wet, and then let it dry out before putting in the lines.

5. Do not smear or compact the sides and bottoms of the trenches. Take a rake to them if they become smeared at all. This step is critical for proper drainage, for premature clogging will occur if the subsoil is compacted.

6. Lay the distribution pipe *level* with the holes downward. Use a shovel to cover the pipe with about two inches of stones. Be careful here, because PVC pipe has a tendency to bounce around. It must be level.

7. Place three to four inches (uncompacted) of straw or marsh hay over the top of the trenches.

8. Cover the trenches with soil in mounds, fill systems and evapo-transpiration beds. This should include six inches of good topsoil over the entire system.

9. Do not drive on top of the absorption field, *ever*.

10. Landscape soil absorption systems on the surface by planting grasses, except for evapo-transpiration beds. The mixture of 90 percent bird's-foot trefoil and 10 percent timothy has been suggested if it is not going to be cut. If a manicured vegetation is desired, a mix of 60 percent bluegrass, 30 percent creeping red fescue, and 10 percent annual rye is suggested.†

* Converse et al., "Design and Construction Procedures for Mounds," p. 12.

† Ibid., p. 13.

An Owner-Built Greywater System

Peter Franchino built himself a house in Michigan and incorporated into it one of the most ingenious greywater systems we have yet to come across. His filter—earth bed/hotbed setup—strains, degreases, and partially cleans the water from his two sinks and then discharges them, effectively watering, tempering, and possibly fertilizing vegetable growing areas. Peter's new proposed system would also have increased filtration and absorption capacity (to handle laundry and/or bath water) and would make room for a solar hot water heater.

Like most others we've come across, this greywater system is new and has not been in operation long enough to disprove or prove its feasibility. And at present it handles a very small volume of water a day. We do think, though, that it has merit, and we let Peter explain his system here as an example of the kind of greywater experimentation in progress.

Existing System

The scraps on our plates are food for cats, chickens, compost, and worms. Dishes are rinsed in plain water. This water is spread onto the earth bed where vegetables and worms are growing. Dishes are either stored in the large sink tank or washed.

Sink waters flow into the filter via 1½-inch plastic pipe. Bath water can also be included, but in my household scheme we have a sauna and river for bathing.

The filtered water flows into the lower 1½-inch pipe and down into the perforated pipe that lies along the bottom of the earth bed.

The water level is maintained at the stone and sand depth where it is absorbed by the sawdust. Any excess passes through the four-inch drain to the outdoor hotbed. The absorbed moisture rises by capillary wick action and evaporization to the compost and dirt where vegetables and worms are growing.

The sink waters enter the *filter* and initially pass through a modified five-gallon bucket where grease and food chunks are trapped. The strained water is gravity-forced through the sand, charcoal, and pea stone and out the holes in the bottom and side of the 20-gallon can.* Then it flows into the 32-gallon can and up and out the lower pipe to the earth bed.

The five-gallon bucket is suspended on the 20-gallon can so that the outflow holes in the bucket are below the minimum water level. This position traps the floating grease in the bucket.

Cleaning the bucket is recommended every couple of months, although six months elapsed without doing so. To empty the bucket, remove the elbow and nipple by releasing the short hook while

Peter Franchino has helped to design and build a number of houses. He's an organic gardener with a few chickens and lots of earthworms.

* I haven't determined the flow rate through the can with sand, charcoal, and stone. There is about a seven-gallon surge capacity in the 20-gallon can before water flows over the can and bypasses the sand, charcoal, and stone filtration.

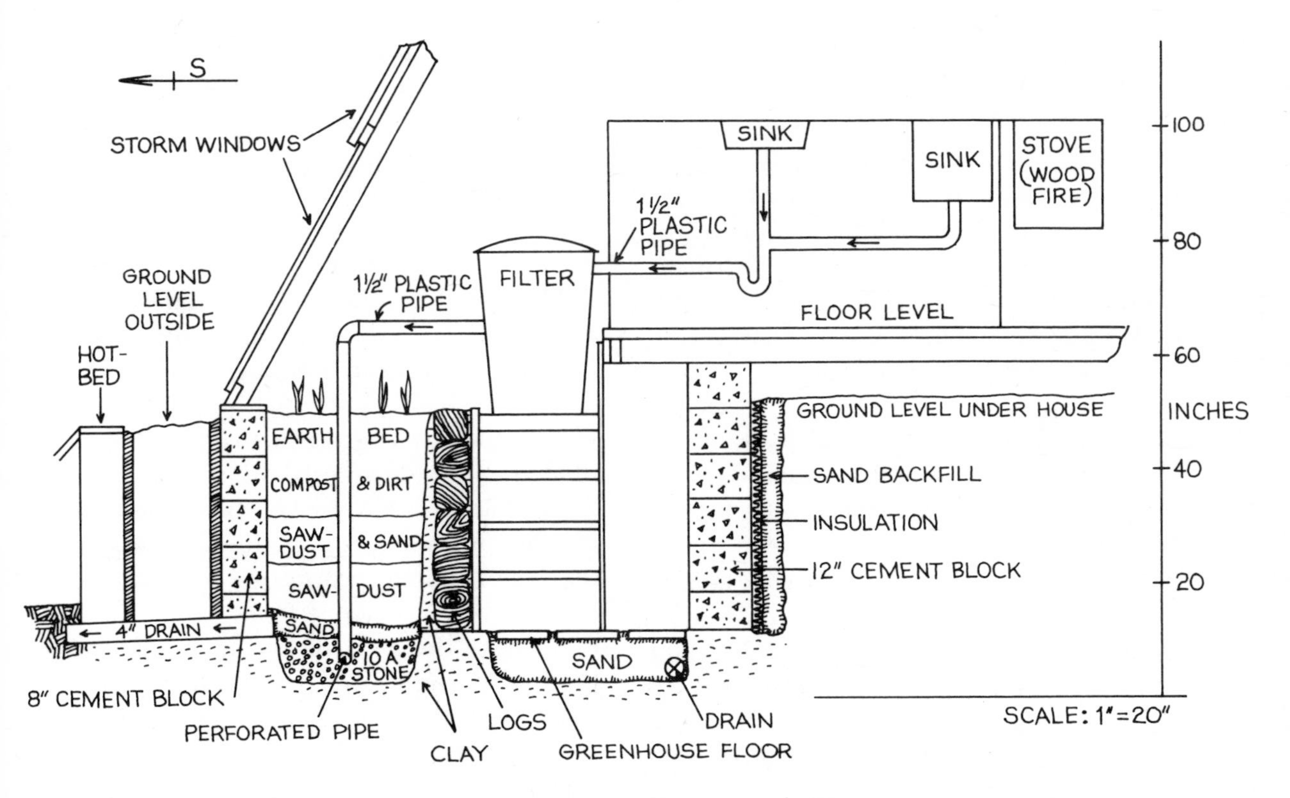

Franchino's existing greywater system. Water from the sink travels first through the bucket filter, then through the in-house earthbed, through the outdoor hotbed, and finally to the outdoors.

moving the bucket, elbow, and nipple away from the coupling. Put a plastic 1½-inch cap on the nipple, invert the bucket while holding the elbow in place. Remove the bucket, take off the rubber strap and screen, bury the contents, and clean with hot soapy water.

Clean the entire filter every six months. To do so, remove the five-gallon bucket and coupling. Dip out the water and remove the 20-gallon can. Release the unions and remove the 32-gallon can. Clean all with hot soapy water and put in new sand, charcoal, and stone.

The in-house earth bed was constructed with a minimum of manmade materials. Its foundation is in clay. Cement blocks stacked without mortar are the outside walls which face south, east, and west. The inside wall is made of cedar logs and lined with wet clay. Straw was put in the blocks and around the outside walls for insulation. The overall dimensions are four feet wide, 12 feet long, and four feet deep.

The interior of the bed, in ascending order, consists of eight inches 10A stone, two inches sand, 12 inches sawdust, eight inches sawdust and sand mixed, and 18 inches compost and black dirt.

The 1½-inch perforated plastic pipe (two rows of ⅜-inch-diameter holes drilled in underside) lies on top of a layer of 10A stone. The pipe is parallel to the drop in the bed's floor, one inch over 12 feet.

The earth bed with the worms also acts as a garbage disposal. Food scraps are buried in trenches between the growing vegetables.

A layer of hay mulch around the vegetables serves to retain moisture for the plants and worms and also makes the worms feel at home so they don't leave.

There have been no odor or water leakage problems even with this primitive construction as yet.

The outdoor hotbed is for late fall and early spring planting plus wintering-over worms. It is 40 inches wide by 12 feet long.

The floor of the bed is lower than the in-house bed to promote a one-way flow of the greywater.

The walls are eight-inch cement blocks stacked without mortar and lined on the inside with ferrocement. The top of the south wall is lower than the north wall. One-inch Styrofoam encases the outside perimeter.

The interior of the bed is the same as the in-house bed except above the sand and sawdust level. A layer of "hot" compost is added in the fall to provide warmth for the vegetables and worms.

There also is no perforated pipe. The drain from the in-house bed puts water at the higher end of the hotbed's floor. The water flows through the stone and sand to the lower end where a drain passes any excess to an outside perforated pipe covered with stone and earth.

This system handles a household size of three. The in-house earth bed alone absorbs all our wastewater. To increase the capacity to accommodate a significantly larger household and/or a bath would necessitate increasing the surge capacity and/or flow rate of the filter. This increase is dealt with in the proposed system section.

In the existing system, the top of the earth bed is lower than the wood-burning stove. Cold air settles in the greenhouse and restricts wintertime growth. A hot water radiator is needed in the greenhouse to maintain enough heat for prolific win-

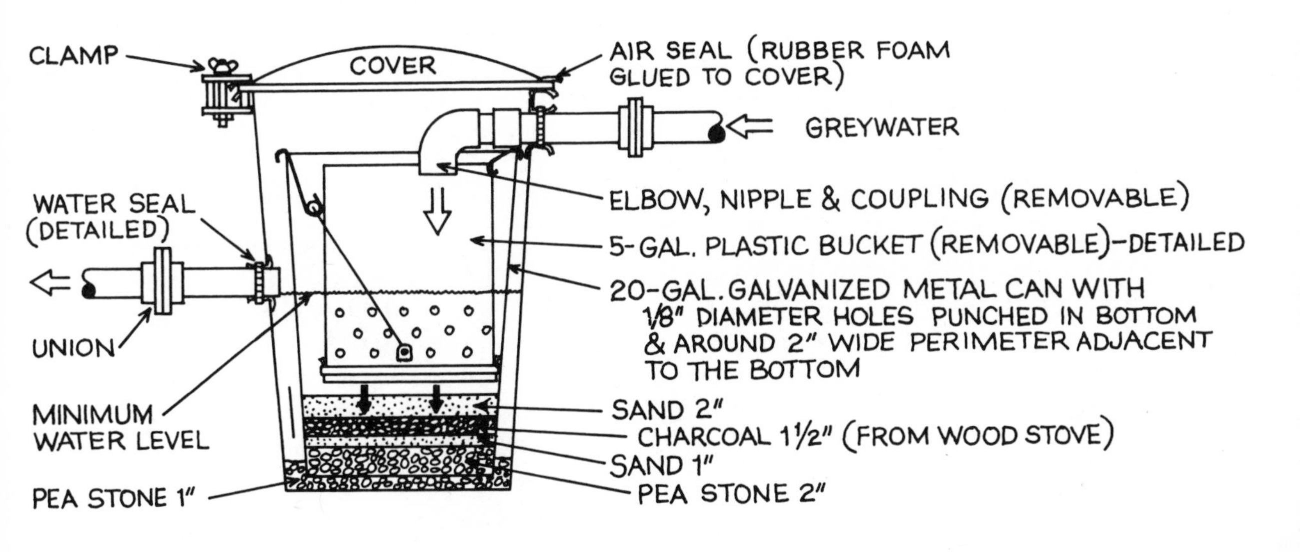

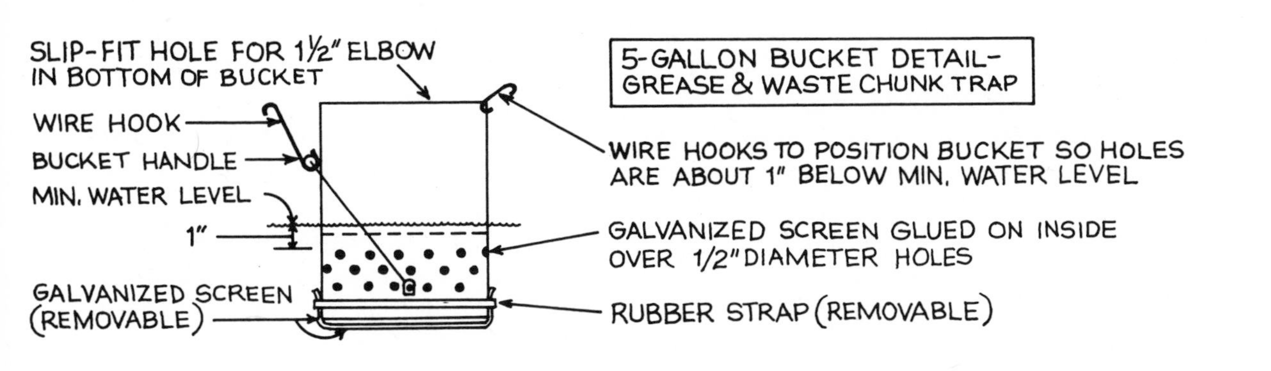

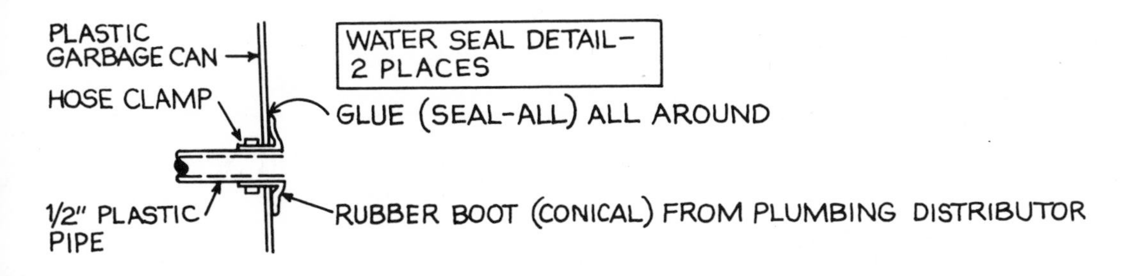

Details of the bucket filter system.

tertime growing.

The amounts of sand, stone, and charcoal used were somewhat arbitrary. The charcoal can be increased (for greater absorption) and the stone diminished.

Inspection while cleaning showed a whitish paste which ringed the 32-gallon can at the level of the exit pipe. It is probably a soapy residue. A soapy residue would tend to clog the stone and sand in the earth bed. This is a prime consideration since cleaning the bed is an undesirable (but not impossible) operation.

The Proposed System

The proposed system incorporates the same processes of the existing system. The greywater is strained, degreased, and cleansed in the filter. The filtered water passes to the bottom of the earth bed where it is absorbed by the sawdust. The moisture rises by capillary wick action and evaporization to the vegetables and worms growing in the black dirt. Any excess water passes to an outside hotbed to be absorbed and utilized.

The changes in the system are increased filter capacity, increased absorption capacity, improved placement of vegetables relative to heat source, improved construction techniques, and a possible solar heating system.

The filter's capacity is increased to accommodate a larger household and bath water. Inlet and outlet piping is three inches in diameter. The exterior container is made of wood and can be as long as the filter pit to accommodate the needed number of 20-gallon cans (other sizes can be used).

The cans function in the same way as in the existing filter. Greywater is strained and floating grease is trapped by the five-gallon bucket. The strained water flows down into the 20-gallon cans and is gravity-forced through the sand, charcoal, and stone and out the can's holes into the exterior container. The outlet carries the water to the earth bed.

A drain and valve, to remove the water from the filter, connect to the earth bed at a height above the drain to the hotbed.

The five-gallon bucket and 20-gallon cans are easily removable for cleaning. The procedure is the same as for the existing system. If the exterior container is small, it also can be removed if unions are properly incorporated.

The top of the earth bed is now above the floor level where the wood-burning stove is. Heat from the stove reaches the vegetables. The depth of the bed is likewise increased and the absorbing capacity is expanded.

Standard construction materials and techniques used in making a cement block septic tank are utilized in making the earth bed.

The addition of a solar heating unit appears compatible with the greywater use system. The earth bed is used as a heat storage medium.

Heated water from the solar collec-

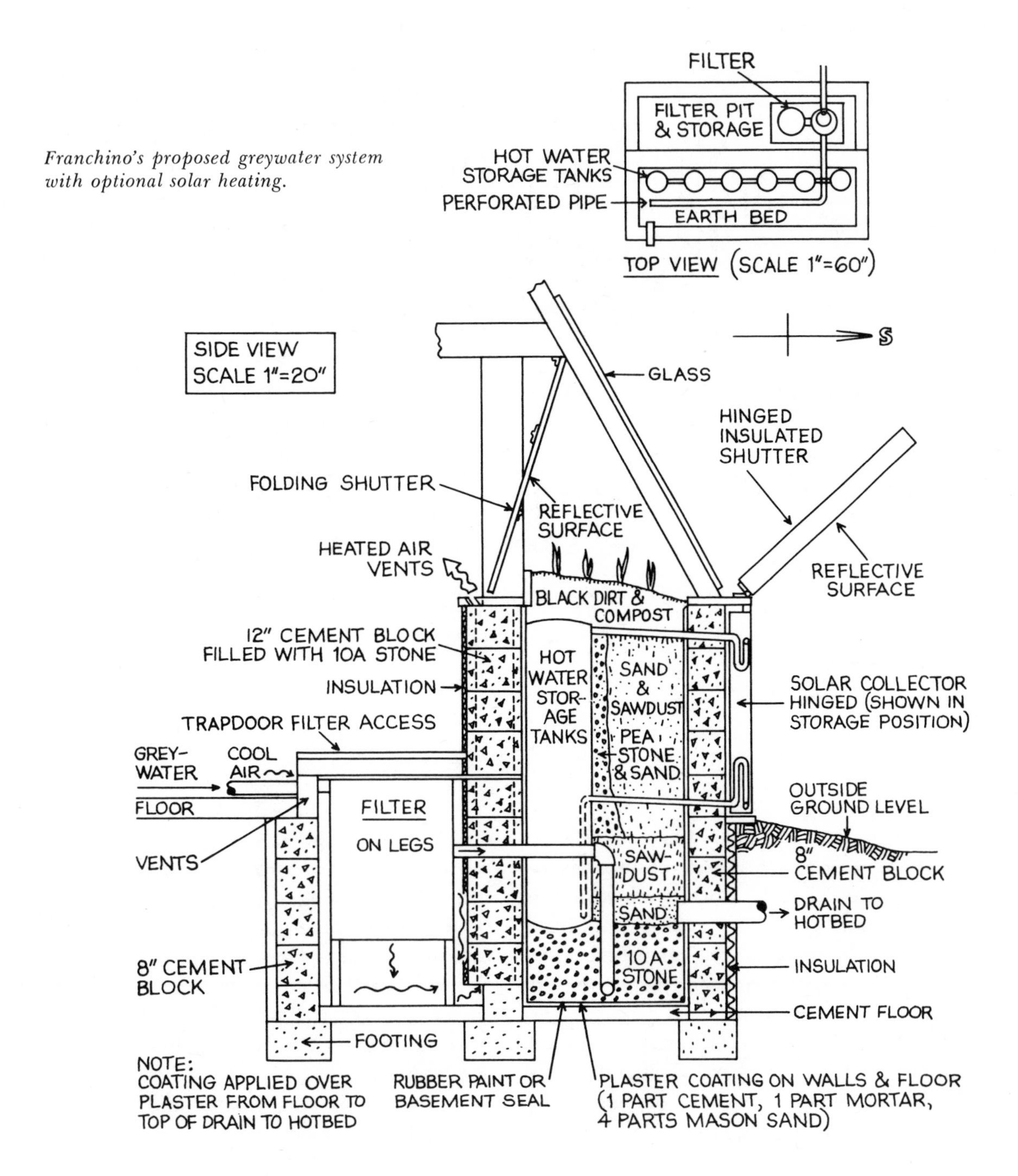

Franchino's proposed greywater system with optional solar heating.

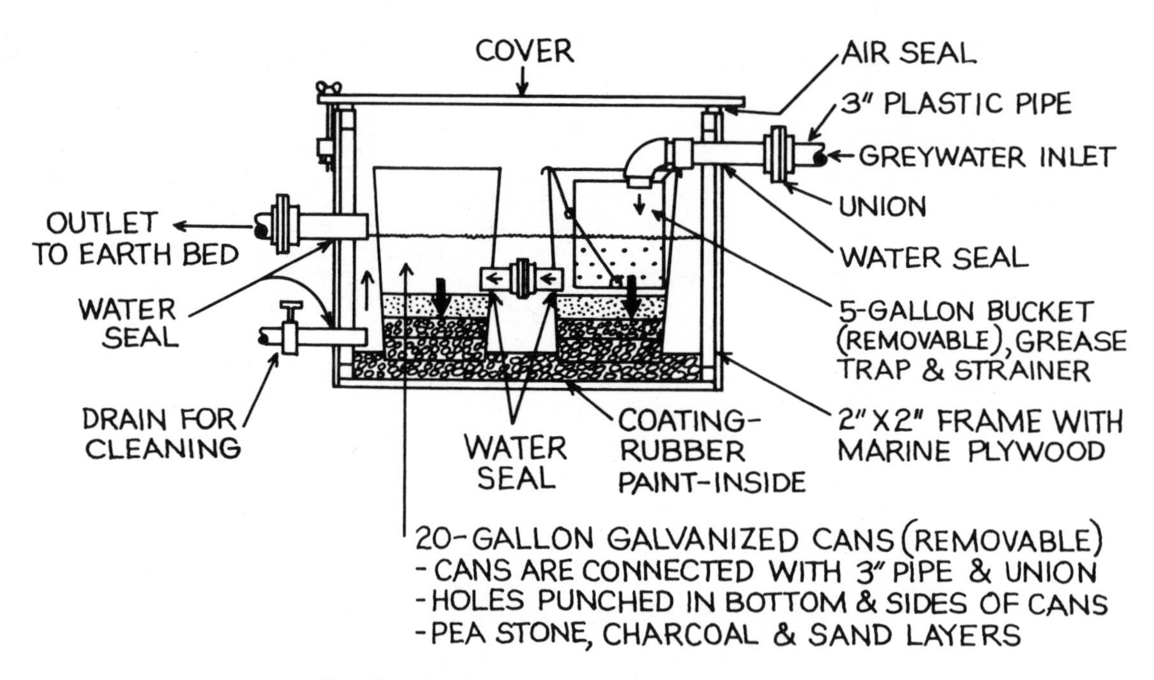

Detail of the larger filter in the proposed system.

tors moves into the hot water storage tank by thermosiphoning. The tanks are placed next to the 12-inch block wall which is filled with 10A stone. Pea stone and sand surround the tanks in the earth bed. The block wall and the sand and pea stone absorb heat from the tanks.

When heated air is needed, vents at the top of the 12-inch block wall and on the floor near the filter pit are opened. The cool floor air drops into the pit and the rising hot air in the blocks draws in this cool air to be warmed and circulated.

When laying the 12-inch block wall, care must be taken to keep the vertical air columns free of excess mortar. Stone larger than 10A might be needed to permit a greater air flow.

A layer of sawdust might be needed between the hot water tanks and the black dirt to insulate the worms and vegetable roots from excessive heat.

After reading Peter's description of his greywater system, a few questions came to us and we asked Peter to answer them.

Us: How long have you been using this system?

Peter: Since May 10, 1976.

Us: What volume of wastewater is it presently handling?

Peter: About seven gallons a day.

Us: What is the mean low temperature during the winter? What effect does this have on water evaporation in your house earth bed and the hotbed?

Peter: The normal low temperature for the month of January is 10.4°F (–12°C). I would assume the evaporization of the outside hotbed in the winter is very little; the indoor bed should not be affected greatly.

Us: Is the present system dependent on any heat source other than solar and the heat carried by the greywater itself? What would happen if you should be away for an extended period during the winter?

Peter: My house is heated by a wood-burning stove. If I left for several days without putting insulation around the filter, it would freeze. If I left for a longer period of time, cleaning the filter without refilling with water would be necessary. In the proposed system, the filter could be drained easily. The water in the earth bed itself is below frost level and is insulated enough from the inside of the house so it won't freeze.

Us: You mentioned that the rinse water from dirty dishes is spread onto the earth bed. Are you actually putting food particles on the bed? Doesn't it smell?

Peter: The food particles put on top of the bed are small and are soon eaten by the worms. An occasional turning of the hay mulch conceals the food's unsightliness. There is no odor.

Us: Do you have a backup system? When you remove the parts of the filter for cleaning, what do you do with the hot soapy water?

Peter: There is no backup system. I clean the filter outside. The hot soapy water used for cleaning can be buried, put in a compost pile, or thrown to the wind. A way to put this water into the pipe going to the earth bed could easily be done. Of course it wouldn't be filtered.

For most applications a backup system would be advantageous. One way to accomplish this is to "Y" the piping from the filter to the bed, with one branch going to the bed and the other going directly to the hotbed. Incorporate valves to direct the greywater to either the inhouse bed or the hotbed. If the system has no hotbed, the inhouse bed can be divided in half, each portion being independent, and appropriate plumbing installed to direct the water to either bed. Plumbing to bypass the filter could also be installed.

Us: Is there any impermeable liner for the earth bed? If not, what kind of leaching or drainage do you get from it? How does this affect the foundation of the house?

Peter: The impermeable liner in the present system is clay. I possibly should have lined the block walls with clay also, but the water bulk is retained in a clay bowl. The block walls are adjacent to the moisture absorbed by the sawdust. I don't believe any leaching would be significant to cause any harm.

The proposed system has an impermeable liner on all walls and floor of the bed.

Us: Why do you use sawdust in the earth bed? From what kind of wood?

Peter: Sawdust (for water absorption) is used because it is free, available in most areas, and natural.

The type I use is from deciduous trees. The worms prefer it. Sawdust from decay-resistant trees (cedar, redwood) may be preferable in other applications.

Us: How much evaporation do you get? Does this have any adverse consequences for your house? What about in the summer?

Peter: I don't know the amount of evaporation. It depends on the temperature of the soil, the air, the humidity, the amount the plants assimilate, and who knows what all.

In the winter the vapor is welcomed. In the summer, it is just another drop in the bucket.

Us: Have you done any testing on the soils? Do you consider the greywater simply as irrigation waters, or are you depending on its nutrient content for soil fertility? What vegetables do you grow?

Are there any you have decided not to grow for any possible reason related to the greywater quality?

Peter: Because I used compost, black dirt, and worms, I'm not greatly concerned with the nutritional value of the greywater. However, it was the intent of the system to utilize the nutrients, but lacking information as to what the plants could expect, I decided to be safe and not experiment.

I grow Swiss chard, lettuce, endive, kale, collards, parsley.

I think all plants (root crops also) could be grown without concern except for the chemicals used in the soap.

Chapter 7

What's in Store for Flush Toilet and Greywater Alternatives?

To most Americans, even the mention of a composting toilet brings turned-up noses in utter disgust. The thought of living in a house with one's own excrement nearby is anathema to a Puritan-influenced culture that has gotten quite used to modern sanitation. High technology sanitation has encouraged most of us to regard human wastes and nearly everything else related to them as only worthy of being disposed of in the quickest and most convenient way possible. With every flush of the toilet and release of the drain plug, the responsibility for waste treatment is gladly passed on to the government. Similarly, outside the city, where drains are typically connected to a septic tank and leach field, the soils are expected to treat and dispose of inordinately large volumes of sewage with little or no help from the user. In both cases, rinsing the wastes "out of sight, out of mind" has long reinforced the myth of abundant clean water in a throw-away society.

Owner Acceptance

In our research, we found that composting toilet owners are, for the most part, a different breed of people from the "flush and forget" kind. In developing this book we interviewed about 125 users of compost toilets across the country, asking them why they had installed the units and what were their experiences with them. These interviews showed most favorable reactions to toilets, and most everyone who replied indicated they would buy (or build) another composting unit if the need should ever arise.

We learned that the major market is in vacation homes, where intermittent use of the home does not justify the installation of a large and costly leach field for treatment and disposal of large volumes of wastewaters. In some such homes, the actual building of a home may be dependent on separation of black- and greywaters via a waterless toilet, or a discharge permit will not be granted. People on vacations don't seem to mind the so-called "inconvenience" of a flushless toilet, and many enjoy "roughing it." The toilet is seen as a pleasant variation from their regular routine.

The second large market is the permanent homes of homesteaders and owner/builders. According to Zandy Clark, Mullbank (Ecolet) dealer in Maine, these homesteaders "almost exclusively have a philosophical desire to get rid of the flush toilet." Many environmentally minded people are so committed to the waterless toilet concept that they have actually designed their home around the Clivus Multrum. Greg and Jennifer McKee of Lopez Island, Washington, told us that their new "house will utilize many energy-efficient features of which we feel an 'alternative toilet' is one of the most important. Flush toilets are very inefficient 'waste disposal' systems since they create more sewage than they were designed to dispose of in the first place. We feel that body wastes are only wastes to the body and can be very useful elsewhere."

Many owners believe the main advantage of a composting toilet is that it does not waste water, and actually expressed delight about not having to contaminate drinking water. One user went so far as to say that she thought it is criminal to waste water flushing the john. This is particularly true where there is not enough water in the first place, but it also applies in areas where the soils are so thin that they cannot accommodate the disposal of the large volumes of wastewater implied by flush toilets.

An approximate 10,000 gallons per person per year is needed to flush toilets beyond the normal household demand for washing, bathing, and food preparation. On Maine's rocky coasts, it could cost up to $8,000 to construct a soil absorption field to dispose of this additional wastewater. Geological conditions in many regions are not suited to handling large volumes of water, thus creating very unpleasant situations where septic tanks and leach fields fail. One composting toilet owner we spoke with in upstate New York remarked that "many homes in the neighborhood with septic systems have continuous odors and are health hazards because of sewage-saturated soils. We like the Clivus because it eliminates this problem."

For this reason, many composting toilet owners are proud to make a contribution to solving the water pollution problems of their communities. Jim and Mary Dietrich of Bondville, Kentucky, commented, "Our county has poor drainage and there have been four or five cases of typhoid fever in different areas of the county, said to have been caused by faulty septic systems and water systems. We may be the first in our area to use the Ecolet and we are proud to show it off to others. We have also been able to solve what could be a bad situation with a reasonable alternative."

But there are other than just environmental considerations. Toa Throne users

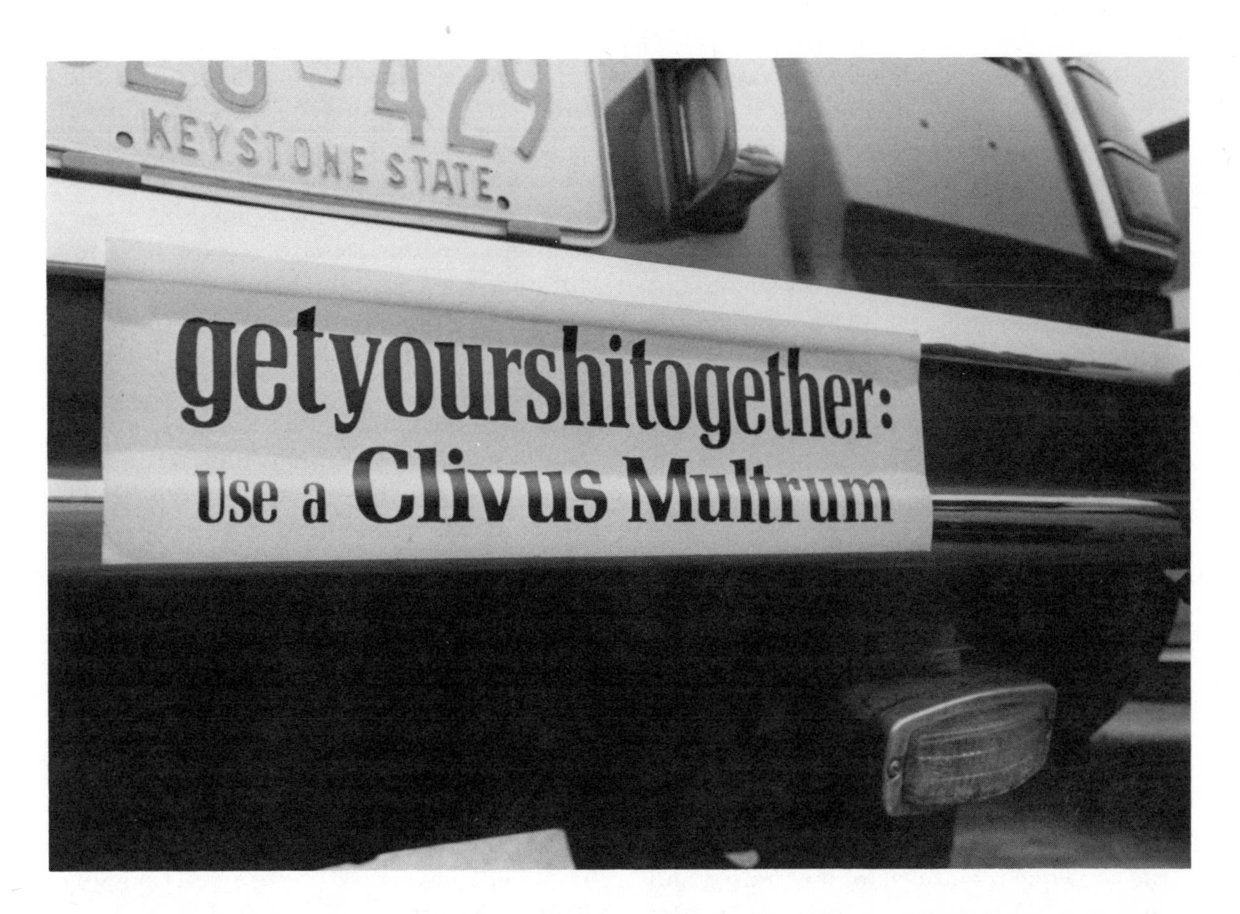

in particular report that, to their surprise, no odors linger in the bathroom after they use the toilet. One owner told us, "It is funny what you get accustomed to—now when I go back and use a flush toilet, I am amazed at how odors linger. I never smell my feces using the Toa. There is no lighting of matches, no air fresheners, and no need for a fan in the bathroom."

Of course there are negative reactions to composting toilets, though surprisingly enough, nearly everyone we interviewed was incredibly tolerant and would do it over again if the situation presented itself. The high cost of the Clivus and its installation is perhaps the most frequently heard complaint, though even the CM owners note that they are less expensive than the alternatives in the long run. One person said he was tired of explaining it, and guests would occasionally suggest, "don't you think you ought to flush it?" Commonly people find that their guests tend not to use it as frequently as they would if it were a conventional toilet. "It gets treated like an outhouse, and friends who grew up with an outhouse are least favorable—they are *tired* of 'hearing it drop'."

Other negative responses to composting toilets can be viewed as adjustment problems that most everyone goes through when they switch from the use of a conventional flush toilet. The frequent, but hardly serious problem, is what do you do with the old habit of reaching back to flush the toilet? A few people jokingly suggested to us that a flush lever and a sound track are needed to simulate the noise regularly associated with using a toilet.

A second, more uncomfortable phase of adjustment is due to the stabilization process of the composting activity itself. As the populations of decomposer organisms are introduced to their new home in the composting tank, there is often a temporary imbalance in the numbers of predators and prey. These imbalances are normal to the establishment of any ecological system, but the consequent outbreak of flies, mites, and spiders is not a normal occurrence in most people's houses. The succession of these three critters early in the life of the Clivus Multrum has come to be viewed as the triumvirage of successful stabilization, indicating the toilet is off to a good start. Nevertheless, when they enter the house they become pests, and remedial action will often be required to keep them in check. Even so, proper maintenance can minimize, if not avoid, the annoyance. More than one person told us that it is not a serious problem except for the most fastidious.

Another concern about the composting toilet often expressed by parents is how their children will react to it. Here again the concerns seem to be more about comfortable adjustment than anything else, because once the concept is explained to children, they usually become more fascinated than frightened. When children see the relationship between composting toilets, gardens, and composting, they come to care for the life inside the tank. One family reported near devotion, claiming their "Clivus is like a pet—we even 'feed' it." The 24 children in a preschool in Aspen, Colorado, are very proud of their Clivus. "In fact," says the school's founder, "the kids love it!" This can be taken as good advice to parents who are worried that their children might go exploring it and fall in. More likely, children will be so good to it that they begin to "feed" it their toys —and beware, retrieval is a bit awkward.

All in all, the people who use alternative toilets have very positive reactions to them. One woman commented that since she believed in the principle, she is willing to give it a break: "Doing anything new takes more time, money, and effort before it works well." It would be a mistake to deny there are problems with such systems, but as Russ Lanoie, composting toilet dealer in Conway, New Hampshire, has said, "I find it hard to conceive of the transition from flush to composting toilet as being any more of a hassle than the switch from 'clean and safe' electric heat to 'clumsy, dirty, dangerous, and unpredictable' wood heat which so many of us have made in recent years. Just as people have relearned the art of heating with wood, so can people learn the art of composting, as natural a process, and just as much in harmony with nature."

Government Acceptance

As was discussed in chapter 2, public health officials and the sanitary engineering profession have leaned heavily on sewers and centralized sewage treatment plants for management of sanitary sewage in urban areas. In less populated areas, septic tank-leach field systems have served the same purpose as sewers and central treatment plants—they both take away the wastes, never to be seen nor smelled again. But each of these methods of treatment and disposal fall under different categories for regulation.

Like the public sewer, regulations for on-site treatment grew out of the concern to protect the public health and the desire to accommodate modern home conveniences. Improved methods of treatment and disposal, a major focus of public health agencies, have attempted to *control* the spread of water-borne pathogens, rather than *prevent* the entry of the pathogens in the drinking water in the first place. Unfortunately, this concern has been translated into a wide spectrum of rules and regulations, often varying widely from state to state, that are sometimes not even related to protection of the public health. For instance, how does a four-inch pipe versus a three-inch pipe protect health? These regulations might even be used for political objectives, and in many rural communities, the septic permit is the only mechanism available for controlling land development.

In addition to such abuses, septic tanks and their component leach fields do not always insure the protection of the public health. When these systems fail, they may cause ground water pollution, contamination of drinking water supplies, and puddling of raw sewage. In many situations they are inappropriate because of poor soils, exposed bedrock, high water tables, lack of space, or prohibitively high costs.

Although failing septic systems can be corrected by reconstruction of the leach field or use of innovative designs for the field (such as described in chapter 6), both of which are less expensive alternatives to centralized treatment, neither the federal government nor the engineering profession favors this option. On-site methods of sewage disposal are used with the understanding that the method is interim, in spite of the fact that properly maintained systems are far more reliable protectors of the public health for at least as long as central sewers and treatment facilities.*

This centralist approach, deeply ingrained in the thinking of sanitary engineers, public health officials, and federal environmental policymakers, is one of the main biases facing those who are advocating the legalization of composting toilets and on-site greywater management options.

The effort by government officials and sanitary engineers, started on the premise of preventing disease, has evolved into a very powerful preference that today excludes many innovative technologies, regardless of the merits of the alternatives in disease prevention. Let us look more closely at how these preferences are manifested today, and also at the related institutional impediments to the acceptance of composting systems:

1. The federal government supports 75 percent of the cost of the expansion and upgrading of publicly owned sewers and treatment plants. State and local governments pay the remaining 25 percent. There is no similar grant program for the repair, installation, and management of on-site sewage systems unless the systems are publicly owned. Technical advice on low-cost alternatives for small towns has only been available since the early part of 1977, and this information is also biased toward centralized, albeit smaller-scale, treatment options. Very little, if any, information is available about water conservation, waterless toilets, and greywater management.

2. On the local level, county sanitarians are reluctant to give approval for alternative systems because they are unfamiliar with anything other than the standard septic tank. They most likely will refuse to make a decision, regardless of how much information is presented to them, because they do not want to take the responsibility for possible failure. When approached about the use of a composting toilet and greywater reuse scheme, they may indicate that these would be a return to the more "primitive" pit privy, and thus may be a step backwards in terms of public health protection.

3. Public health officials also are reluctant to approve of any on-site system that requires knowledge or responsible action by the user. There can be no assurance that individual users will always maintain the toilet properly. Because the art of composting and an understanding of composting toilets are somewhat alien to the way sanitary engineers and public health people think, they apparently figure that "ordinary" people cannot be trusted to understand.

* Gary Plews, Manager of the On-Site Sewage Disposal Program in Washington State, stated at National Sanitation Foundation's Third National Conference on Individual On-Site Waste Water Systems (Ann Arbor, MI, November 17, 1976) that because of "the philosophy that public sewers are the only permanent way to handle the problem, . . . on-site sewage disposal continues to be viewed . . . as interim or temporary."

4. Both the federal government and the National Sanitation Foundation have been reluctant to test the health and safety aspects of composting systems. While several tests have been done in the Scandinavian countries, only spot checks of various composting toilet samples have been done in this country. This alone is not enough information to approve or disapprove of composting toilets. In areas where there are no other compelling reasons to switch to their use, the health profession's conservatism wins out. (See the related discussion in the next section on safety.)

5. Public health departments do not always have the money or the manpower to monitor alternative systems. Even Maine, which has given broad approval to composting toilets, has only 24 staff people to cover the whole state. Because the technology of composting toilets is so young, the concern of health departments to oversee their operation and maintenance is understandable. When they allow several kinds of waterless toilets, in addition to the various new leach field designs, they incur greater costs in personnel training. For these reasons, it is cheaper and safer for states to give blanket approval to only those systems which their engineers learned in school—those being the conventional septic systems.

6. Land use patterns will likely be altered by approval of alternative on-site methods, if strict land use planning is not enforced. A major point of institutional resistance relates to the scattered development of rural areas. Generally, a septic permit requires adequate soils with suitable drainage, and consequently residential development is most often on the best agricultural lands. If waterless toilets were suddenly found acceptable, people could settle virtually anywhere they wanted. Planners are distressed at the thought that people could live at distances from fire, police, transportation, electric, and school services. The cost of providing these services to such scattered development would be considerable.

These are the major substantive concerns of states, local county health departments, and planning district commissions that have prevented the widespread approval of both composting toilets and alternative forms of leach fields. In addition, other concerns are based on misunderstandings. Comments such as "the Ecolet and similar units (are) an inadequate answer for the total household sewerage flow" * show a common misunderstanding of composting toilets. Of course they will not handle the *total* household sewerage flow, because they are not supposed to! The state has a rule that composting toilets are

* J. Howard Duncan, Kansas Bureau of Environmental Sanitation, 1976: personal communication.

approved for use as long as no garbage is deposited in them. This, of course, makes no sense at all because proper composting of human wastes needs garbage to raise the C/N ratio. Many states hinge their skepticism toward composting toilets on the basis of the added cost which is seen as duplicative of the costs of septic systems. Since greywater has to be treated and disposed of anyway, the reasoning continues, the required septic system would greatly reduce the need for a waterless toilet. The Commonwealth of Virginia requires that one solution be found to the "total problem," and the design standards are set so that any dwelling with pressurized water must have a full-sized leach field, regardless of whether it generates all those wastewaters.

In the midst of these negative reasons explaining why governments on all levels have been slow to approve use of composting toilets and greywater schemes, there are equally, if not more, compelling reasons why counties and states have had to take notice and approve of experimental units. Poor soils, drought, expense, land use, and environmental protection have all been sufficient reasons for acceptance in many states.

Donald Hoxie, Director of Health Engineering in the Maine Department of Human Services, explains that the decision to recognize "separated systems" in the new Maine Plumbing Code was the direct result of "recognizing the limitations of our soils and realizing that the basic problem of sewage disposal is designing a system capable of handling 200 to 300 gallons, day in and day out." It became very apparent that anything the Department could do to encourage a reduction in water usage and nutrient loading would simplify the engineering aspects of treating and satisfactorily disposing of the sewage.*

In the West Coast states, the large numbers of people who have built low-cost, ecologically attuned homes and have experimented with alternative sanitation systems, have, as California state architect Sim Van der Ryn explains, "found themselves at odds with many features of the building codes designed for the urban dweller and purchaser of tract homes." For these people, the composting toilet and greywater reuse are essential components of low-cost housing. In California particularly, but also in Washington, Oregon, and Maine, owner/builders have forced their states to allow for dry or composting toilets in alternative homes as opposed to the flush toilet. But the battle for approval of composting

---

* Steve Smyser, "New Visibility for On-Site Waste Treatment Systems," *Compost Science* 17, no. 5 (Winter 1976): 14.

According to Eugene Moreau, Chief of Waste Water and Plumbing Control in Maine's Department of Human Services, Maine presently allows a 40 percent reduction in the size of drain fields in separated systems and may be further reducing the size shortly, depending on the water conservation measures practiced by the particular household. (Paper presented at the 7th Annual Composting and Recycling Conference, University of Massachusetts, Amherst, MA, May 4–6, 1977, sponsored by Rodale Press.)

It may be surprising to many that Maine also encourages privies in many instances, especially in rural areas of the state. They have been recognized and allowed in Maine since the inception of their first plumbing code in 1933.

toilets is only half the job—the traditional septic tank-soil absorption field is basically the only approved method of greywater disposal, and this alone makes many of the low-cost homes not very low cost.

The Oregon Department of Environmental Quality administrator Ken Spies indicated that there is an "urgent need to develop an acceptable alternative to on-site sewage disposal methods so that as many property owners as possible can . . . develop their parcels of land even though they are not accessible to public sewers and cannot be served by conventional subsurface system." In Oregon, like many other well-populated states, the need to conserve prime agricultural land for production of food, fiber, and timber is finally being acknowledged, and efforts to prevent its conversion to housing and commercial developments are accelerating. Some alternative to the septic system and the flush toilet is desperately needed, however, because these less agriculturally valuable lands are entirely unsuitable for large volumes of wastewater disposal in perhaps as many as half the permit application sites.

Such situations, along with the unprecedented 1976–1977 drought in California, have hastened the wheels of change. The rationing of drinking water in California to one-third the regular use can most easily be obtained through simple water conservation measures and elimination of the flush toilets. Many Western states have begun to issue experimental permits for both commercial and homemade composting toilets. Greywater irrigation is being openly advocated in drought-stricken areas, although only as a short-term measure until the drought ends. Water conservation also strengthens the argument for the reduced leach field size. Some states do this already: Minnesota, Connecticut, Maine, and Washington allow for substantial reduction. Massachusetts will also allow reduction in the leach field, but the space for a full-sized field must be available just in case the home ever gets sold to someone who may want to use a flush toilet. It is likely that California will permit reduction of the field in the near future.

Most states have reluctantly accepted the notion of a composting toilet, but are satisfied that their use will generally be limited in rural areas to replace the privy or in recreational developments. In general there is a requirement that if a municipal sewerage disposal facility is nearby, then alternatives are not allowed, except in Washington and New Mexico. In those states, discharge of greywater into an existing sewer is allowed even in cities, as long as the human wastes are disposed of in a safe manner.

Aside from the situations where states have allowed waterless toilets and experimental permits to individuals, the more persuasive argument for the widespread use of composting toilets is the cost of sewers and central treatment facilities in small towns. Preliminary estimates show that the reduction of drinking water demand, wastewater volume, and pollution load in the sewage due to the use of waterless toilets could save between 50 and 80 percent of the cost of construction, operation, and maintenance of centralized treatment, depending on the particular situations. Compared to the tremendous economic burden accompanying "sewering-up," it is no wonder that several small towns have had to face the threat of bankruptcy just because of their sewage works.

Walton, New York, is only one of the many small towns where vast sums on the order of $9 million are being spent to sewer and treat the wastes of three to four thousand people. All this when a combination repair of septic systems and transition to waterless toilets would have fewer destabilizing effects on the community and much less money would be spent.

The state of Washington has taken the lead among states which have approved the use of waterless toilets and experimental greywater systems, and other states can be expected to model their regulations after Washington's. In contrast, Maine has given approval to specific individual commercial and homemade units, while Washington has adopted various performance standards to which each alternative toilet applicant must conform. The minimum performance criteria allow for a variety of commercial units and owner-built designs, as long as they can insure good structural integrity of the units, safe and reliable performance, proper ventilation, and compliance with sludge disposal guidelines. Washington has also developed a model for decentralized management of privately owned units by a public on-site management district. Washington's administrative rules and guidelines were established to transfer the responsibility for performance of privately owned systems to public management and should answer many questions raised about how to guarantee proper maintenance and safety of alternative on-site methods. California also has considerable experience with on-site management districts.

A similar experiment in public management systems is being tested out by the Appalachian Regional Commission (ARC) in Boyd County, Kentucky. The ARC's program replaced chronically failing septic tanks with aerobic tanks at government expense in several homes. Although the units are in operation at individual homesites, they are owned, monitored, and maintained by the local health department. This demonstration project has shown that the concept of a public management system does work when applied to on-site treatment and disposal of wastes. Other states can be expected to follow suit as the pressure to use on-site systems continues.

The actual stance that a particular state might take when a permit is requested is difficult to anticipate. The smaller composting toilets have been accepted in as many as 25 states, but only with the provision that the local county officials give their approval. In general, uniform codes apply throughout the state, but the actual determination is on the basis of the specific case. Most states have a policy where "each situation is reviewed on its own merits." A flush toilet will generally be required if a house has pressurized water, unless soil conditions require alternate sewage handling; and if there is no standard plumbing, the regulations say pit privies or sealed vault privies (which most manufactured composting toilets are) are allowed. Only Maine, New Hampshire, and Kentucky give blanket approval to one particular toilet—the Clivus Multrum—without provisions for local authorization.

Most states and local health departments do not even acknowledge the existence of composting toilets, which leaves the owner/user an out. This silence could be interpreted as meaning they are either disallowed or allowed, and in neither case could damage be brought without first getting something on the books with which to prosecute. In Vermont, the code appears to exempt composting toilets from health regulations because "it does not connect to the water supply or waste piping in the building." In Kansas "there is no need for specific product or installation approval," but the final compost product does fall under sludge disposal regulations.

Getting Permission for an Alternative System

If you want to approach your state for a permit for a composting toilet and greywater system, you ought to be prepared to make many such attempts unless you live in one of the progressive states which allow them. The power to approve or reject a permit, including experimental permits, usually resides in the local county health department, but frequently permission may be required at the state level. Start at the bottom with your local sanitarian, and don't let him intimidate you with the myriad rules, regulations, and guidelines. Arm yourself with plenty of information on the relative merits of a composting toilet, a comparison between composting toilets and septic systems, cost data, and leach field designs signed by a registered engineer. Make sure you have these designs in hand, even if you don't plan on implementing them. You probably will have to go over his head, maybe several times, and each time there could be more examination and interrogation. Be persistent with your pursuit and don't give in! You might even be referred to the top of the health department eventually, but if you still have no luck, go on to the governor. Ron Davis in Oregon ended up calling the governor every day, person to person collect, for a few weeks. Finally an aide answered the call, and two weeks later the governor's office granted him an experimental permit. And this happened in a state that had gone on record in favor of experimental units, though they obviously were reluctant to implement their test program!

Trying to get permits, even experimental ones, may be a tremendously difficult and draining process. The situation in California illustrates the dilemma many people have faced:

In trying to get permits for composting toilets and the like, many people run into a "Catch-22" situation. Health departments say they can't approve any system that doesn't use the flush toilet because it would violate the Building Code, and building departments say they can't consider alternatives until the health department approves them. Health departments want data before approval, but there is no way to get data without first setting the system in place. In some localities alternative systems have been approved as long as the home owner agrees to provide a standard type system as a "back-up." The problem, besides the high additional cost ($1,500–$2,500 for septic tank and leaching lines), is that the amount of leaching areas required, based on "average" water consumption of more than 100 gal/person/day, restricts applications to generally flat areas with soils that percolate well.*

Building Codes and the Dry Toilet

In other cases, ridiculous rulings which are meant as concessions make it even more difficult to put a composting toilet and greywater system in. As Alex Wade reports in *30 Energy-Efficient Houses You Can Build*,† the health department allowed a reduction in the required size of the leach field by 10 percent in one house he designed:

> Since the standard septic tank required for this house is 750 gallons, and the next smaller one is 500 gallons and smaller than the 10 percent allowed, you can see where that leaves us. Flush toilets use a minimum of 40 percent of the water in a house. If we were allowed to reduce the septic tank accordingly, we could have used that commonly accepted figure, the 500-gallon system. As it is, we would have had to stick with the 750-gallon tank which would have cost $3,000, and then add a $1,500 Clivus onto that. Needless to say, my clients couldn't afford that, and the health department with its unreasonable regulations effectively killed the Clivus.

These experiences are enough to test anyone's patience with governmental processes, but there is really little else an owner can do legally. Of course, concerted group action speaks louder than the voice of one, so if you are inclined to organize, do it for faster, more widely applicable results.

It is likely that as the applications for permits increase and the costs of flush toilets become more apparent, there will be pressure to relax the restrictions on the use of compost toilets. The situation seems to

* Sim Van der Ryn, "Building Codes and the Dry Toilet," *Environmental Action Bulletin* 7, no. 12 (June 12, 1976): 7.

† Rodale Press, 1977.

be changing constantly, and even if official sanction is not given to these units, we can be sure that the numbers of illegal installations are increasing.

Public Health and Safety

One of the most serious questions raised about composting privies and toilets is about their safety in terms of the potential spread of disease. Sewerage has been known to cause major epidemics throughout the history of humankind, and use of composted human wastes and partially treated greywater has raised many an eyebrow for this reason. There is a possibility that infectious organisms are in the finished compost and the liquid buildup from compost toilets. Even though these may be handled in a hygienic manner, final disposal of incompletely composted matter onto the garden as the manufacturers have promised may actually be a way of spreading disease. Alternatively, disease could possibly be spread by flies, mosquitoes, and rodents, should any of them gain access to the composting tank.

How great a health risk are composting toilets? How remote is the possibility for the spread of disease, and is it a matter of great concern? These questions cannot be answered definitively at this point because the necessary long-term testing has not been done. The best we can do today is to address these questions from two standpoints: first, what we know now about the finished compost product; and second, what are the relative risks of using composting toilets and greywater recycle systems as compared to the risks of conventional treatment methods.

The finished compost from both commercial and homemade composters has not been tested in a wide enough sampling to conclude anything about its safety. The humus from the Clivus Multrum has been studied more than any other unit's end product, primarily because of the continuous testing program that its manufacturer has initiated. Consequently, much of what we know about the finished humus is based on the decomposition process inside this large tank. It is likely that the finished product from the smaller units is not comparable biologically for reasons discussed below.

To begin, it would be worthwhile to look at the actual biological processes that go on inside the composting tank in order to estimate what kind of pathogen kill we can expect. As was explained in chapter 3, in aerobic decomposition the microorganisms already present in the wastes and soil starter utilize air, nutrients, and water to convert the organic matter into humus. As they do this they give off a tremendous amount of heat, enough to heat up the pile to 140° to 160°F (60° to 71°C). This heat, if sustained, will kill pathogens. Note that the high temperature is obtained only in aerobic decomposition, which is why a continuous supply of air is critical for proper pathogen destruction. The heat given off in anaerobic conditions is only 1/40th of that from aerobic composting!

The important aspects in a composting toilet then are how hot it gets, how uniform the heat is, how long it stays hot, how well air flows through it, and what happens if it does not get hot enough.

This temperature data has not been monitored during the process of stabilization, but spot checks in the Clivus Multrum show 150° to 160°F (65° to 71°C) internally. The Mullbank's (Ecolet's) temperature has been measured at 68° to 103°F (20° to 39°C), which is clearly not enough to kill off human parasites and viruses. Although the Clivus Multrum runs hotter, the fringes of the pile do not always get this hot, and thus the humus can never be assumed to be 100 percent pathogen-free. Test data on the other units are not available. For an added measure of safety, the new Bio Loo has added a sterilization chamber that submits the finished product to temperatures of 180°F (82°C) for three hours before discharge.

These comments are in stark contrast to the statements made by the researchers testing for pathogens and the promoters of composting toilets. Their conclusions have all been based on the bacterial coliform group, which is an indication of the possible presence of other pathogens. Coliform are undoubtedly degraded successfully at the cooler temperatures at which the small units operate. The bacterial analysis of the finished product of the Clivus does show a well-stabilized bacterial community resembling that usually found in soil.* Similar analysis done on the Mullbank (Ecolet) final humus also suggests that the bacterial decomposition within the tank has produced a humus-resembling soil. However, whether these products are safe for use as a fertilizer and soil amendment cannot be concluded from these studies.

Dr. Daniel Dindal of SUNY-Syracuse has examined the Clivus humus to learn more about the role of soil invertebrate organisms in the degradation of human wastes. Nineteen different kinds of invertebrates are commonly associated with the decomposer community, which included the flies, mites, and spiders often found near the Clivus. Dr. Dindal explained to us that many of the invertebrates found are predaceous, and they are thought to eat pathogenic microbes in the wastes. "Though we have not substantiated this notion, our studies suggest that the environment in the composting tank, created in part by the invertebrates, is sufficiently hostile to result in high rates of pathogen kill. The complexity of the microbial community caused by the naturally occurring high heat and the aerobic conditions also contributes significantly to pathogen kill." †

However, these microbial and inverte-

* H. Wayne Nichols, "Analysis of Bacterial Populations in the Final Product of the Clivus Multrum," Center for the Biology of Natural Systems, Washington University, St. Louis, MO, December 1976.

† Private communication, March 11, 1977.

brate analyses do not give positive indication of whether or not intestinal parasites and/or viruses are inactivated in the composting process. Dr. Clarence Golueke has commented that *satisfactory* study has not yet been made on pathogen, parasite, and virus survival in *any* of the on-site disposal systems presently on the market.*

Safety requires that we remember that human pathogens will undoubtedly be present in feces and urine at some time. Even if a person is not sick, he or she may be a carrier. Tests of a sample from a homemade composting privy in California have shown the presence of hookworm, whipworm, and other pathogens. The sample was thought to be fully composted and had even sat undisturbed for six months prior to examination. These results underscore the need for cautious handling of the finished compost. They certainly suggest that a six-month period is not sufficient to kill off remaining parasites, and that extra precautions must be taken in the final disposal of the compost.

One frequently recommended method is burial in a trench with at least 12 inches of soil cover in some permanently planted area, such as a berry patch or orchard.† Dr. Golueke commented to us that "time, inter-microbial competition, and unfavorable environmental conditions eventually take their toll on all but the most resistant disease-causing organisms. The problem is to be sure that a sufficient time intervenes and that no short-circuiting takes place." **

This position on the safety of the finished product is the very conservative, cautious one that most health departments can be expected to take. It is not, however, an indictment against the use of composting toilets; instead, it is meant to underscore the need for careful handling of the finished compost. Normal hygienic measures should be used when the humus is removed from the tank, being careful not to spill the contents, cleaning the implements thoroughly afterwards, and washing one's hands. Certainly these simple rules are not hard to follow and the task seems less demanding than the more frequent cleaning of the porcelain toilet bowl to which we are accustomed now.

The relative risks of composting toilets appear to be less than those associated with either the conventional septic system or the sewers and central treatment plant. In each method, pathogens will be present, and care is needed. The difference is whether they are discharged at the outlet of a sewage plant following disinfection, disposed of underground in a leach field, or degraded in a closed container nearby. In all cases there is the potential for adverse environ-

* Private communication, February 7, 1977.

† This follows the World Health Organization's regulations given by E. G. Wagner and J. N. Lanoix in *Excreta Disposal for Rural Areas and Small Communities,* World Health Organization Monograph Series No. 39, 1958.

** The effect of time on the die-off of pathogens cannot be underestimated. While no specific data are available, Dr. Golueke and others have suggested that retention of the material for at least nine months to a year would be sufficient assurance against contamination. Burial would provide the same. It is also thought that the three to four year retention time in the large composting toilets also provides this margin of safety.

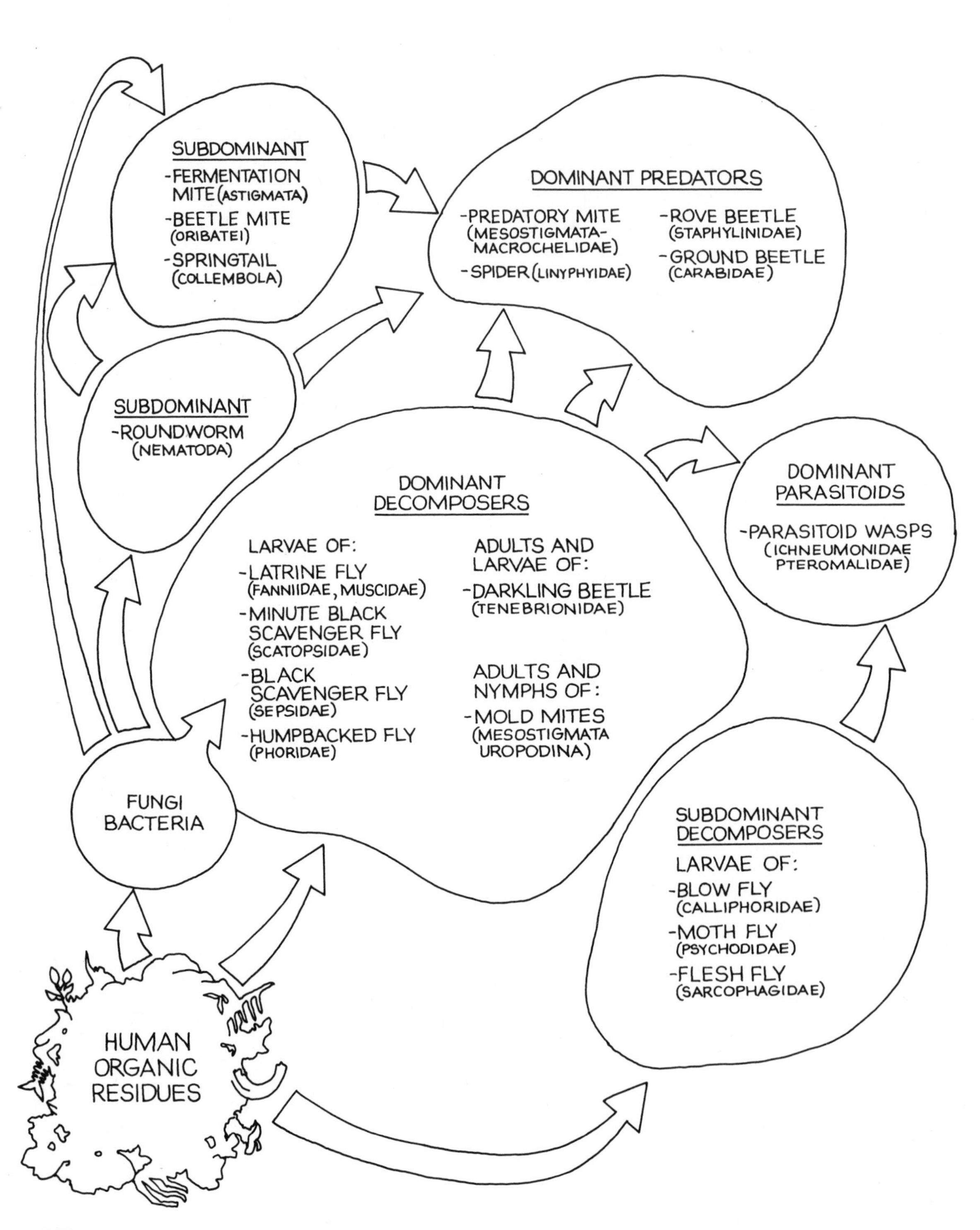
SUBDOMINANT
-FERMENTATION MITE (ASTIGMATA)
-BEETLE MITE (ORIBATEI)
-SPRINGTAIL (COLLEMBOLA)
DOMINANT PREDATORS
-PREDATORY MITE (MESOSTIGMATA-MACROCHELIDAE)
-SPIDER (LINYPHYIDAE)
-ROVE BEETLE (STAPHYLINIDAE)
-GROUND BEETLE (CARABIDAE)
SUBDOMINANT
-ROUNDWORM (NEMATODA)
DOMINANT DECOMPOSERS
LARVAE OF:
-LATRINE FLY (FANNIIDAE, MUSCIDAE)
-MINUTE BLACK SCAVENGER FLY (SCATOPSIDAE)
-BLACK SCAVENGER FLY (SEPSIDAE)
-HUMPBACKED FLY (PHORIDAE)
ADULTS AND LARVAE OF:
-DARKLING BEETLE (TENEBRIONIDAE)
ADULTS AND NYMPHS OF:
-MOLD MITES (MESOSTIGMATA UROPODINA)
DOMINANT PARASITOIDS
-PARASITOID WASPS (ICHNEUMONIDAE PTEROMALIDAE)
FUNGI BACTERIA
SUBDOMINANT DECOMPOSERS
LARVAE OF:
-BLOW FLY (CALLIPHORIDAE)
-MOTH FLY (PSYCHODIDAE)
-FLESH FLY (SARCOPHAGIDAE)
HUMAN ORGANIC RESIDUES

(opposite page)
Proposed food web of a composting toilet microcosm.
(Modified after D. L. Dindal, "The Soil Invertebrate Community of Composting Toilet Systems," in U. Lohm and T. Persson, eds., *Soil Organisms as Components of Ecosystems.* Proceedings of VI International Soil Zoology Colloquium. Ecol. Bull. [Stockholm] vol. 25 [in press].)

Preliminary study has shown that the biological decomposition in a composting toilet is the work of nineteen different kinds of invertebrate organisms that inhabit the ecosystem inside the toilet. As in any ecosystem, these organisms are related to each other in what ecologists call a food web. The proposed food web of a composting toilet ecosystem shows that wastes are first digested by various fungi and bacteria, most of whom are already present in the wastes. These become a food source for many of the mites, maggots (fly larvae), adult flies, and nematodes which comprise the first level in the food chain. In turn, the predators of the second level feed upon the first level organisms. For instance, the beetles will prey on larvae which have eaten the bacteria and fungi, which grew up on a diet of human and other organic wastes. The parasitoid wasps (second level) lay their eggs in the fly maggots (first level), and these wasps will eat the maggots when they hatch. This is an example of how one organism can control the population size of another.
As the ecosystem within the composting toilet stabilizes during the early period of use, outbreaks of a particular organism may occur. This means that the predator populations have not grown sufficiently in size to eat enough members of the other organism's population. If the ventilation system of the toilet is not working properly, then these excess organisms may leave the toilet, enter your house, and become pests. Use of a pesticide to control the outbreak will impede the growth of the predator organisms' populations, and will push the development of all the populations in the composting tank back to where it was before the population boom occurred. But because these population booms are inherent in the early stages of the stabilization process, the pesticide use will only insure the regular recurrence of outbreaks of the same organisms. As organic materials accumulate with the continued use of composting toilets, a more stable ecosystem develops within the tank and outbreaks of pests become rare. Therefore, Dr. Dindal advises that patience will be necessary for both you and your composting toilet.

mental impact, although the scope of the impact is much greater with the treatment plant, and the septic field can also cause pollution. Only the composting toilet will contain, and thus control and prevent, the spread of water-borne pathogens. With respect to the question of viruses and parasites, the proper testing has not yet been done on any option.

Thus the question of relative safety of composting toilets hinges on how large an area negative impacts can be felt and how well these impacts can be controlled. And in this regard, the composting toilet is beyond comparison.

As discussed in chapter 6, the use of greywater as a nutrient-laden irrigation water in gardens has not been examined closely for its long-term effects on either the soil or the plants. However, the level of pretreatment, the degree of original contamination, the general health of the

CONTENTS OF DIFFERENT SUBSTANCES IN THE
IN COMPARISON WITH DIGESTED

Substance	Toilet number 1	2	3	4
Dry matter (dm)	88.8	69.9	79.0	84.9
pH	6.6	8.0	7.0	6.9
Kj.-N in % of dm	3.16	2.76	3.06	2.13
NH_4-N in % of dm	1.70	1.44	1.80	0.33
NO_3-N in % of dm	0.11	0.33	0.09	0.06
Tot.-N in % of dm	3.27	3.09	3.15	2.19
Org.-N in % of dm	1.46	1.32	1.26	1.80
Org. C in % of dm	16.2	24.7	23.5	28.5
Org. matter in % of dm	44.2	51.9	51.2	47.8
Phosphorus (P) in % of dm	2.88	0.66	2.19	1.33
Potassium (K) in % of dm	3.86	1.90	6.80	2.17
Calcium (Ca) in % of dm	0.46	0.79	0.55	2.53
Magnesium (Mg) in % of dm	0.03	0.09	0.04	0.33
Zinc (Zn) mg/kg dm	51	37	18	154
Lead (Pb) mg/kg dm	56	14	14	53
Cadmium (Cd) mg/kg dm	1.1	0.7	0.8	0.6
Mercury (Hg) mg/kg dm	0.12	0.15	0.07	0.35
C/N ratio (org. N)	11.1	18.7	18.7	15.8

* mean value of 25 Swedish sewage plants
** mean value of 90 Swedish sewage plants, in cooperation with S. Oden
Kj.-N = Kjeldahl-nitrogen

FINAL PRODUCT OF DIFFERENT MULLBANKS, SLUDGE AND FARMYARD MANURE

5	6	7	8	Mean	Digested sludge	Farmyard manure
86.6	91.0	94.1	95.2			
6.9	6.6	6.7	6.6	6.9	6.9	8.5
1.70	3.21	2.22	1.93	2.5	4.9	2.2
0.35	0.73	0.26	0.42	0.9	1.0	0.3
0.05	0.08	0.11	0.10	0.1	0.1	0.1
1.75	3.29	2.33	2.03	2.6	5.0	2.3
1.35	2.48	1.96	1.51	1.6	3.9	1.9
20.7	28.7	21.0	18.1	22.7	34.2	44.1
40.2	57.9	45.2	41.9	47.5	59.0	76.0
1.00	1.26	1.49	1.12	1.5	1.5	0.6
2.00	4.40	1.95	1.88	3.1	0.4	2.1
1.36	1.24	3.19	2.95	1.6	2.5	1.5
0.22	0.12	0.64	0.59	0.3	0.4*	0.1
60	96	247	184	106	1700**	80
37	15	30	22	30	180**	10
0.4	0.8	0.9	0.8	0.8	6.0**	0.2
0.20	0.31	0.50	0.41	0.3	6.0**	0.04
15.3	11.6	10.7	12.0	14.2	8.8	23.2

Source: Kalgu Valdmaa, "Function of the Ecolet Biological Compost Toilet" (Uppsala, Sweden: The Royal Agricultural College Sweden, 1975).

people, the application rate, and the manner of food preparation are all important factors in how safe it is. It is thought best not to use greywater if someone is sick in the house, and diapers should be rinsed separately if irrigation is intended. It is generally suggested that root crops should not receive greywater, but it is unclear if this is because root crops are more likely to be eaten raw or because the vegetables themselves have direct contact with the water.

To summarize, the necessary testing on composting toilets and greywater irrigation systems has not been carried out. No definitive conclusion can be made, although there also have not been any serious public health hazards reported due to the use of these alternatives. For final disposal of the contents of a composting toilet, burial in an undisturbed place is recommended for at least two years. Because composting toilets and greywater reuse offer so many advantages, particularly in comparison to the standard flush toilet and centralized treatment, more experimentation should be encouraged in a multiplicity of situations. The technology is new, and we need to learn its strengths and limits. It may be that their use will be restricted eventually (just as conventional septic systems are today), but then we will at least have another ecologically sane option open to us.

Does Toilet Compost Differ from Other Sources of Organic Waste?

This is an interpretation of the tabulation of comparative data from eight Mullbank toilets, digested sludge from 25 to 90 Swedish sewage plants, and from barnyard manure.

pH

The hydrogen ion activities (pH) of compost from the Mullbank toilet and municipal sewage sludge are identical. They are lower, that is, more acid, than the more alkaline pH of the barnyard manure. Bacteria and fungi, the primary decomposer organisms, have optimum pH ranges. Within these optima the majority of the species grow best; they are more abundant under these ideal conditions, and therefore, they are most efficient in decomposing waste. Most bacteria live

1. COMPARISON OF ORGANIC MATTER AND ITS COMPONENTS FROM VARIOUS SOURCES OF ORGANIC WASTES

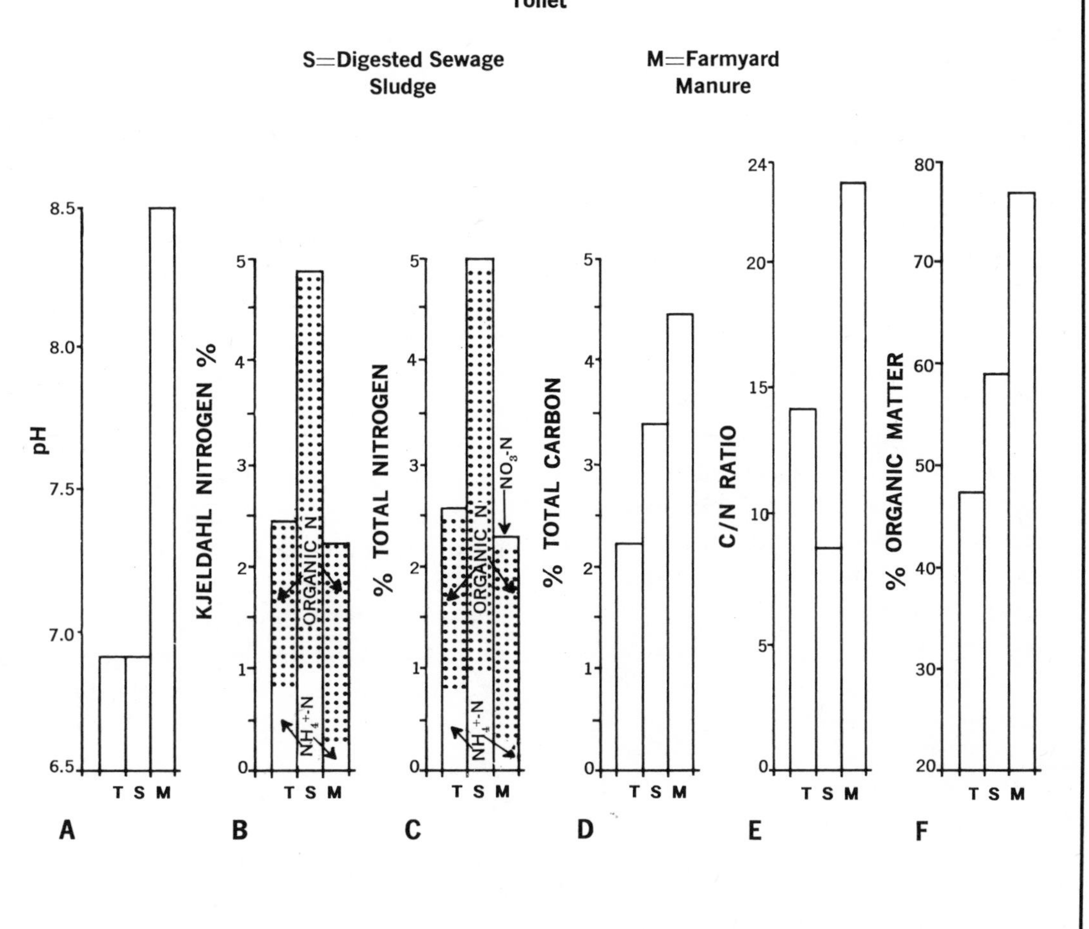

best at pH 6 to 8 and fungi at pH 4 to 6. The pH of the average Mullbank toilet contents (figure 1A) is well within the range, providing an ideal microenvironment for bacterial growth. If a "bloom" of fungi occurs, it indicates that the pH is too low for the most rapid decay, and some source of calcium should be added such as ground eggshells, old ashes from a barbecue grill or fireplace, or crushed clamshells.

Nitrogen (N)

Kjeldahl nitrogen gets its name from a Danish chemist who developed the complex chemical process to assay certain nitrogen sources in organic matter and soil. This process measures both organic nitrogen and that nitrogen incorporated in the ammonium ion (NH_4^+) in natural materials. The amount of nitrate (NO_3^-) is not measured by this chemical technique. Organic N is that which is combined very tenaciously within molecules that make up living tissue. When organic material is used as a soil amendment, organic N will be released very slowly through microbial activity. It then becomes available for the beneficial soil bacteria and plants growing in that soil. The tenacity of organic N is illustrated in figure 1B by the very high concentration of it present in sewage sludge that has already passed through a digestion treatment.

As can be seen from figures 1B and 1C, total N differs from Kjeldahl N in that the former included NO_3^- concentrations in addition to organic and NH_4^+ nitrogen. In general, the amount of NO_3^- present is the same for toilet compost, sludge, and manure. Nitrate leaches out very readily from these organic materials, or it may be immediately available to plants and microbes once the material is used for soil amendments.

Carbon, Nitrogen Relationship, and Organic Matter (figure 1D–F)

A very important requirement for the most efficient and complete decomposition in any composting situation is the proper carbon to nitrogen (C/N) ratio within the organic matter. Microorganisms such as bacteria and fungi must have enough N to build their own cells as their populations grow and increase while decomposing the masses of available C. The optimal C/N ratio for the start of the decomposition process is about 25:1, but as the process reaches stabilization, a more ideal ratio is around 15:1 and ranging upward to 30:1. Nitrogen becomes scarce at ratios above the 30:1 level and C is digested more slowly or not at all as the ratio approaches 100:1. Comparatively, the Mullbank contents possess the most ideal ratio (figure 1E). This shows that microbial breakdown of organic debris is complete.

Nutrient Elements

Compared to sludge and manure, the Mullbank compost has relatively high

2. COMPARISON OF NUTRIENT ELEMENT CONTENT FROM VARIOUS SOURCES OF ORGANIC WASTES

T=Mullbank Composting Toilet

S=Digested Sewage Sludge

M=Farmyard Manure

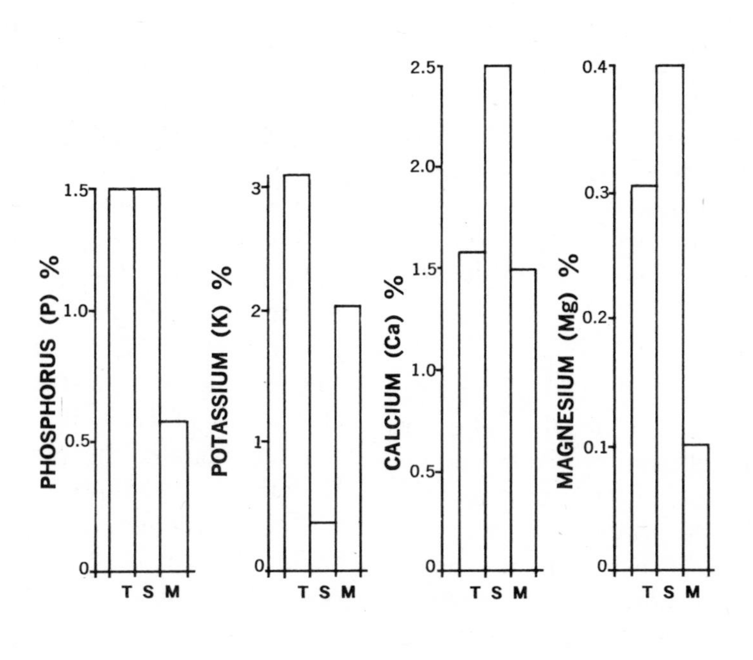

levels of the important nutrients phosphorus (P), potassium (K), calcium (Ca), and magnesium (Mg), as shown in figure 2.

When any of the organic wastes are applied to soil, the nutrient elements play a variety of important roles in the ecosystem. One role of P in all cells of plants, microbes, and animals is the production of chemical compounds that hold and exchange energy when it's needed in any one of the many cell activities. Potassium aids in movement of food, fluids, and waste materials into and out of plant and animal cells. Calcium is a building block of cell walls in plants, giving them durability and structure. Of course, Ca is necessary for shell, bone, and other skeletal structures in various animals. In green plants Mg is most important as a necessary component in chlorophyll, and it serves as a trace element that is required for many chemical reactions in a healthy cell.

Heavy Metals

In general, heavy metals cause the disruption of cellular proteins; in the presence of heavy metals proteins are coagulated like the white of an egg that has been heated. Since proteins are so very important for the normal functioning of cells of all organisms, any disfiguration of the protein causes irreversible detrimental results.

Concentrations of heavy metals within organic wastes are usually determined by the history of the waste material. This is best illustrated in figure 3 where sludge consistently contains the highest level of zinc (Zn), cadmium (Cd), lead (Pb), and mercury (Hg). High Zn levels in domestic sewage materials are usually due to the input of the metal from zinc galvanized plumbing. Cadmium and mercury inputs are generally related to industrial wastes. Pb can concentrate from plumbing of old lead pipes, but is more frequently added through paint wastes and industrial byproducts. In any case, the heavy metal levels from Mullbank compost are all relatively low as would be expected. Also predictable are the extremely low levels of metals in barnyard manure; livestock do not concentrate these metal substances.

The final graph in figure 3 displays the Cd/Zn ratio comparing organic wastes. Engineers and environmentalists have found that a Cd/Zn ratio of less than 1.00 percent is a good measure of acceptable levels of Cd and Zn.* These levels can be tolerated by plants and animals. Cadmium can, however, accumulate in soil to hazardous levels when waste matter having a ratio of 0.5 percent or greater is constantly applied to the same site over long periods.† Since human feces contain only minute traces of heavy metals, the

* Chaney, R. L.; White, M. C.; and Simon, P. W. 1975. Plant uptake of heavy metals from sewage sludge applied to land. Proc. National Conf. Municipal Sludge Management and Disposal, pp. 169–78. USEPA, Environm. Qual. Systems, Inc. and Inform. Transfer Publ. Washington, D.C.

† Page, A. L. 1974. Fate and effects of trace elements in sewage sludge when applied to agricultural lands, a literature study. (EPA–670/2–74–005) Nat. Tech. Inform. Serv., Springfield, Va. (NTIS No. PB–231–171) 96 pp.

3. COMPARISON OF HEAVY METAL CONTENT FROM VARIOUS SOURCES OF ORGANIC WASTES

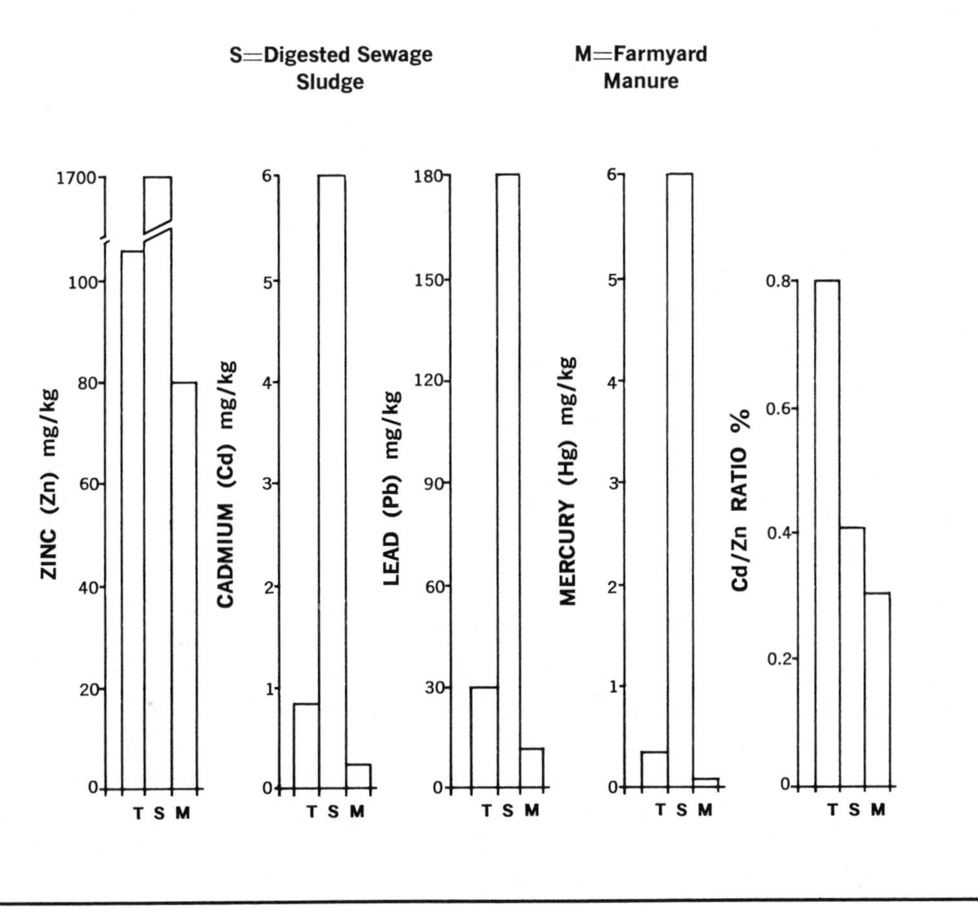

source of metals in the Mullbank must be from other materials introduced into the chambers, such as printer's inks on paper. The incorporation of these items could easily be controlled.

Summary

Considering all the chemical components of Mullbank toilet compost, as reported from the Swedish study, this decomposed waste matter can be used with little difficulty as a soil amendment. Although the total N is about equal to that in manure and half as much as sewage sludge, the C/N ratio is ideal for microbial activity. Also, the pH is optimal for efficient organic matter breakdown by bacteria. Nutrient elements are present in levels comparable to either sludge (P and Mg) or manure (Ca), but K in toilet compost is more abundant than amounts in either sludge or manure. Heavy metal concentrations are much lower in toilet compost than in sewage sludge, but not as low as natural barnyard manure. Cadmium/zinc ratios in Mullbank compost are well within the safe level. Because the compost will be used only once or twice per year when emptying the Mullbank, there appears to be no need to fear a hazardous buildup of Cd or Zn. Heavy metal concentrations in toilet contents can be kept low by either not adding or adding very few waste items containing metals such as printed papers.

Daniel L. Dindal

Costs of Composting Toilets

The cost of composting toilets is one of the major impediments to their general use at this time. The prices range from between $736 for a small unit to $1,685 for a large tank toilet. The greywater handling system is purchased separately, and its price is totally dependent on which method of treatment and disposal you choose. A septic tank with subsurface disposal via a leach field is often the only method approved for handling wastewater for individual on-site systems. These leach field costs are primarily dependent on the kind of soils present and the percolation rate, and consequently these costs will vary tremendously from one location to another.

Typically it will cost $350 for the tank installed and between $625 and $1,250 for a fully sized leach field, for a total of $975 to $1,600. A smaller field reduced by 50 percent for greywater disposal might lower this latter price to between $312 and $625 to total $662 to $975.* Greywater will have to be disposed of anyway, however, so the real cost difference is seen when one compares the $700 to $1,700 range for compost toilets to the $250 for the standard flush toilet.

Even with this big difference in costs, the composting toilet is more economical than the flush toilet where the costs of the water supply and the greywater system are greater than the difference between the composting toilet and the standard toilet. This occurs when there is no pressurized water, no existing septic system, no sewer hookup, or possibly no adequate soil to handle large volumes of wastewater. The waterless toilet also provides savings in water bills and sewerage charges.

In some places an outhouse is just not acceptable, or may even not be legal. Composting toilets have been found quite appropriate in vacation homes, garages, warehouses, and backcountry areas, because in these situations a self-contained unit is the fastest and cheapest toilet to install.

In permanent homes the situation is somewhat different. Running water is usually considered a necessity, and consequently the above mentioned situations do not apply. In these circumstances, the only way the composting toilet will be the most cost-effective option is if an adjustment in the sizing of the leach field is made for greywater, because the added initial cost of the toilet is about the same as the savings from the reduced leach field. This of course is a generalization, as there are many situations where the soils may be so poor, or even nonexistent, that the cost of any leach field, no matter what size, is economically prohibitive. This is why it is necessary that health departments make allowances in the size of the leach field for greywater.

Another initial cost is the cost of installation, and this again is highly variable. The large units require construction of a stand for support and installation of the toilet chute from the bathroom through the floor down to the tank. The Clivus Multrum also needs a garbage chute, preferably in the kitchen, and this can add considerably to the total costs. All units need vent pipes which should be insulated. Installation costs are difficult to predict. Our questionnaire showed that Clivus owners generally spent $150 to $250 retrofitting homes, and new homes built with the Clivus in the original design have no additional costs except the vent pipe and insulation. Mullbank (Ecolet) owners have spent about $50 on the average, and this is primarily for the vent. Data on the average costs of installing the other units are not available.

Many of the commercial models depend on electricity to run the exhaust fan and the heating coil. Although the Clivus and Toa Throne have an optional fan, most users find that the fan improves the

* Cost data from James Kreissl, "Small Flows Program," Municipal Environmental Research Laboratory, Wastewater Research Division, U.S. Environmental Protection Agency, Cincinnati, OH, 1977.

toilets' operation, and you should figure on getting one eventually. The electricity will add an operational cost which will depend on the model used and on your local electric rates. This amount can be figured out by multiplying the watts used by the time (in hours) used, dividing by 1,000, and then multiplying this by the cost of electricity.

Homemade units can cost much less than the commercial models, and some are as reliable if not more so than the commercial ones. This mostly depends on how much ingenuity and effort one puts into it. It is possible to build a unit for as little as $65, though $125 to $150 is more often the amount reported.

According to David del Porto, President of ECOS Inc. in Boston, the high costs of the toilets can be expected to drop by as much as 30 to 40 percent in the next few years. Because ECOS is the major distributor of all composting toilets on the East Coast, David has an unique perspective on the composting toilet industry. He explains that the individual units are too expensive today, which effectively makes them available only to people with moderate to high incomes. Costs are high because the market is so small and because most units are manufactured overseas, which means that import taxes can drive up the costs. Del Porto expects that as the technology itself develops, as governmental acceptance becomes more widespread, and as the sales volume increases, manufacturing will be begun in the United States. Already there are small firms involved in making the toilets in the United States on a custom-ordered basis. Ron Davis of Cottage Grove, Oregon, makes and sells individually made Clivuses for $850, and when shipping and the basic accessories are added on, the package sells for about $1,010. More localized manufacturing and distribution efforts can be expected in the near future, resulting in considerably lower prices.

Most people knowledgeable about composting toilets believe that the market will pick up and prices will drop in a few years not only because the need is growing but also because the federal government has begun to widen its focus to include alternatives to the large centralized treatment plants and alternatives to the standard leach field design. Senator Jennings Randolph of West Virginia has pushed the Environmental Protection Agency to consider the potentials of on-site treatment alternatives as a viable, cost-effective alternative to sewering-up for small towns. The implementation of comprehensive land use planning under Section 208 of the Water Quality Amendments of 1972 makes the opening for full examination of the alternatives on a community basis. Whether or not the notion of segregated wastes is given the attention it deserves is still to be seen, but the economics of composting systems on a community level are far superior to the present methods of handling wastes. And if the federal money that supports centralized treatment facilities were ever made available for on-site treatment, composting toilets could become much more widely used.

As the problems with the present systems increase, the pressure will mount on all levels of government to find a long-range solution that does not depend on costly chemicals, fuel oils, and electricity. Water shortages, lack of good soils, pressures to develop marginal lands, and conservation of resources will all support the growing acceptance of composting toilets. And as these factors grow in importance, the kinks will be worked out of the operation of composting toilets, so that they will also grow in importance as an alternative to the flush toilet.

Good Garden Dirt

When building your castle, consider your throne.
While you're meditating you can also atone,
For keeping all that compost away from the earth,
It's time we quit wasting that good garden dirt.

Chorus: I'm gonna dig up my yard, I'm gonna plant
Some tomatoes, and cucumbers and carrots and beans.
Empty the Clivus, give it back to the earth.
It's time we quit wasting that good garden dirt.

I'm gonna throw away my plunger, get rid of my snake,
Save the water for the garden, give the vegetables a break.
Who needs all that plumbing, taking up the room,
When the copper and plastic can't make my garden bloom.

Chorus

If the health inspector comes, to give me some grief,
He'll end up inspecting a big green lettuce leaf,
He'll say "Seeger, Seeger, how does your garden grow?"
Mr. Inspector, all you have to know,

Just mix yesterday's dinner, with a little bit of air,
Make friends with some microbes, and what you got there,
But a solid foundation for a tomato or two,
Goodbye Mr. Krapper, see what my Clivus can do.

Chorus

Words by Joel Bernstein and Nick Seeger
Music by Nick Seeger

Chapter 8

Every Little Bit Counts—Saving Water

Whether or not you feel a flushless toilet is for you, conserving water can be a reality, and it can be very practical for just about everyone. The difference between using 50 gallons of water a day as opposed to 80 gallons a day may only be a simple and inexpensive flow-restricting device in your shower head and a brick or water-filled plastic bottle to displace water in your toilet tank. Tests with such devices have proven successful in saving water and have presented no inconvenience or significant adjustments for the people using them. Being conscious of your own use of water and making small changes in your personal habits—like taking a shorter shower and no longer letting the hot water run the whole time you wash dishes—can result in even greater water savings.

Cutting down on the amount of water you use can have significant impacts. Obviously, it will affect your water bills or that part of the electric bill that goes for pumping water from a spring or well. Since much of the water you'll be saving is hot water, it will also mean smaller water

In San Francisco they have a large expensive sewer in place already, and they're talking about a billion and a half dollars or so to improve the treatment process. If I were doing a study of San Francisco, the first thing I would say is, "Hey, if we can reduce the flow of sewage by 50 percent then we can reduce the cost of the plant significantly." But I don't hear anybody saying that. For the installed cost of four dollars for every house in San Francisco, or about $2,000,000, we could reduce the flow into this new plant by 30–40 percent by installing a water dam in every tank-type toilet. There's no question that would be cost-effective. It would probably reduce the cost of the sewage plant by hundreds of millions of dollars.

Sim Van der Ryn, California State Architect and head of the Office of Appropriate Technology heating bills. Less water consumption can also mean a smaller and longer-lasting septic tank or other on-site wastewater treatment and discharge system. And if you have an alternative greywater system to go along with a flushless toilet, conserving water may be a necessity.

But the impact of water conservation is more far-reaching. On a municipal level it can affect the volumes of water a city needs, and it can reduce the chances of wastewater runover into ocean and other surface waters. It will affect the energy needed for transportation, the energy and chemicals required to treat the wastewater, and the quality and quantity of wastes discharged into waterways. The extent of this impact was studied by engineers who analyzed the water consumption and treatment of two fictitious housing divisions in the Washington, D.C., area.*

The two new housing divisions were the same in every way, except that one had conventional plumbing and a conventional wastewater treatment system, and the other had water-saving equipment installed in each household and used an advanced wastewater treatment system that is only practical with lower than average levels of wastewater. The engineers found that the water-conserving community:

1. reduced the quantity of water used in households by 68 percent.

2. reduced the wastewater treatment system energy requirements by 56 percent.

3. reduced the wastewater treatment system cost to each household by 33 percent.

4. achieved a virtual "zero discharge" of pollutants to receiving waterways.

Except for residents of the arid West and Southwest, Americans have always taken clean water for granted. Each generation has found new ways to use water faster, until the average daily use per person has risen to approximately 65 gallons. At a time when 70 percent of the human race has no piped water at all, the annual water consumption of the average North American family is 88,000 gallons. With new households hooking on to our already overtaxed sewer mains at a rate of between two and three million a year, Americans will need 800 billion gallons of water a day for domestic consumption by 1980—twice the amount used in 1974. Municipal sewage loads are expected to quadruple in the next 50 years.

Now that the true energy and environmental costs of overloaded sewage treatment plants and of securing fresh water are beginning to be understood, what are our options at the individual household level? Clearly, it's going to be more economical to cut water consumption than to develop new sources, but just how much of an impact can we really hope to have by adopting water-conserving habits and hardware around the home?

* Larry K. Baker, Harold E. Bailey, and Raymond A. Sierka, "Household Water Conservation Effects on Water, Energy, and Wastewater Management: Proceedings of the Conference on Water Conservation and Sewage Flow Reduction with Water-Saving Devices" (The Pennsylvania State University, University Park, PA, April 8–10, 1975).

The Flush Toilet

The chief villain in this depressing scenario, of course, is that paragon of civilized society, the flush toilet. About 41 percent of all water piped into homes is used to flush toilets. At an average of slightly over five gallons per flush, that means the typical user of a flush toilet contaminates 13,000 gallons of pure water a year to carry away 165 gallons of body waste.

To the family that routinely composts its kitchen wastes, recycles its paper, glass, and cans, keeps the thermostat in the 60s, and generally takes pride in maintaining a no-waste organic household, the inconsistency of having to flush away all that clean water can be downright painful—a glaring gap in an otherwise closed loop.

Savings by Water Displacement

One of the simplest and most effective methods of saving water in flush toilets is to place an object into the tank to displace a volume of water equal to its cubic measurement. The most common object is an ordinary building brick. It will displace a little over a quart of water and saves that much every time the toilet is flushed. Most people lay the brick horizontally in the bottom of the tank. But since few tanks empty completely when flushed, a brick in this position doesn't do much good. Standing it up vertically in the tank is much better, but there's a good chance the brick would fall over and possibly crack the tank. Another problem with common bricks is that they slowly disintegrate in water, and eventually clay bits wind up in the flush water and possibly in the valves, affecting the proper functioning of the toilet. If you want to use a brick, get a more expensive ceramic brick that won't dissolve in water.

But better yet, use a set of plastic bottles instead. One-quart liquid soap or bleach bottles will work nicely. Take two and put some pebbles in the bottom of each to weight them down and fill them with water. Stand them upright in opposite sides of the tank.

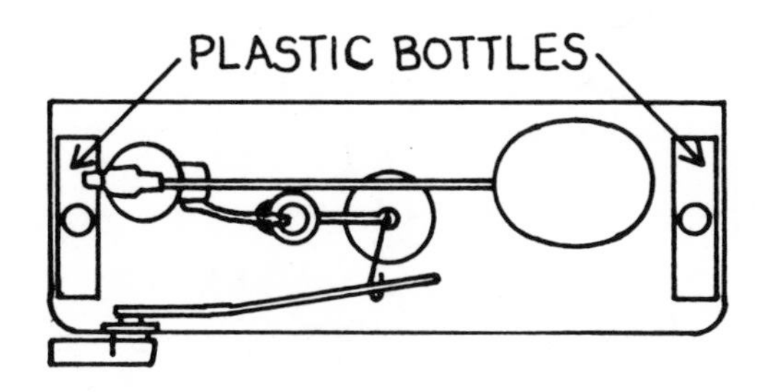

Water-filled plastic bottles, weighted down with pebbles and placed in opposite ends of toilet tank, will displace water quite effectively and cut down on toilet water usage.

Water Volume Reducers

There are some commercial devices that can also reduce the amount of water used each time you flush. There is a three-inch ring or sleeve that goes around the flushing valve to prevent the last few inches of tank water from flushing into the bowl. There are also two pieces of flexible plastic that are wedged into the tank on either side of the flush valve and act like dams. And then there are new float assemblies that can be adjusted to maintain a lower water level in the tank without any reduction in flushing efficiency. A 1972–1975 study in the suburban Washington, D.C., area indicated that the use of any one of these water-saving devices produced a savings of up to 26 percent in household water use, while other similar studies have shown a 15 percent savings.

By bending the toilet float rod downward, you can lower the water level in the toilet tank. Although this reduces the amount of water used, it does not maintain the same hydraulic advantages fostered by one of the water displacement devices or the commercial tank-volume reducers. A reduction in the water level, of course, slightly reduces the velocity of the water being released to the bowl for flushing. Also, in making this adjustment, it is important for you to make sure that you continue to get a good, leak-free cutoff at the valve (after refilling) with the float set at the new level.

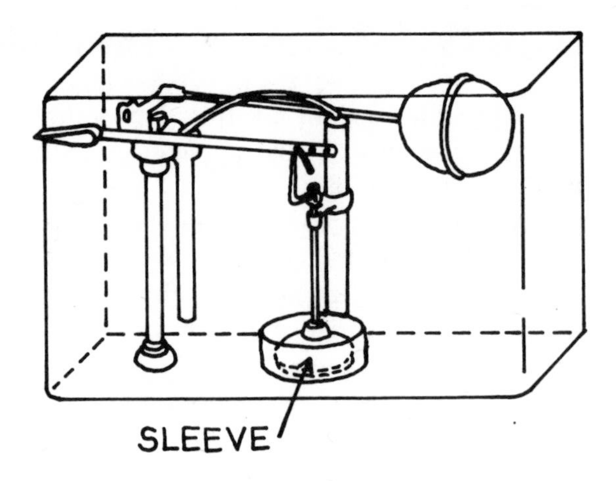

Toilet tank flushing valve sleeve.

Water dam for toilet tank creates a reservoir inside the tank that saves about 2 gallons of water per flush.

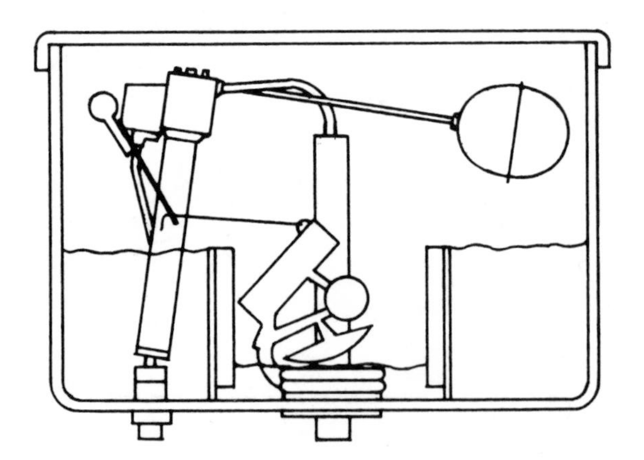

Low-Flush Toilets

If you're remodeling or building a new bathroom, you can put a real dent in your wastewater output by installing one of the new shallow-trap toilets. Such toilets are in common use in both England and Europe, and the average consumption of water needed for flushing is almost half of what it is in this country.

By virtue of its smaller water reservoir and the special design of its bowl, the shallow-trap toilet uses about 3½ gallons per flush as compared to five or six gallons for the standard toilet—a savings of about 30 gallons a day for a family of four. A growing number of local sanitary districts are following the lead of the Washington, D.C., Suburban Sanitary Commission (WSSC) and the state of California in requiring the use of 3½-gallon flush toilets in new housing and other construction. Several of the leading toilet manufacturers we

Water Usage for a Family of Four

	England		United States	
	Percent	Liters/day	Percent	Liters/day
W. C. flushing	35	196	39	380
Personal bathing	35	196	31	300
Laundry	10	56	14	130
Washing up	10	56	3	30
Car washing, garden	6	32	—	—
Drinking, food preparation	4	24	11	100
Utility sink	—	—	2	20
		560		960

Source: Thomas P. Konen, "European Plumbing Practices: Incentives for Change" (Paper delivered at Urban Water Conservation Conference, California Dept. of Water Resources, Los Angeles, January 16–17, 1976), p. 24.

spoke with told us that they are all making 3½-gallon flush toilets now and feel sure the 5-gallon flush will be phased out.

Pressurized Toilet Tanks

Not actually a water volume reducer or a low-flush toilet, the pressurized tank toilet replaces the conventional toilet tank and uses air pressure to assist water in flushing the bowl. The air in the tank that is replaced by the entering water is compressed rather than being allowed to escape, and when the flush button is pushed this compressed air forces the water down into the bowl. Because of its increased pressure the water can completely clean the standard size bowl with only two to 2½ gallons. As Murray Milne is quick to point out,* this is the same amount of water that was used by most older and now unavailable toilet models which had their tanks hung high above them so that the water's velocity increased as it made its way down from the suspended tank.

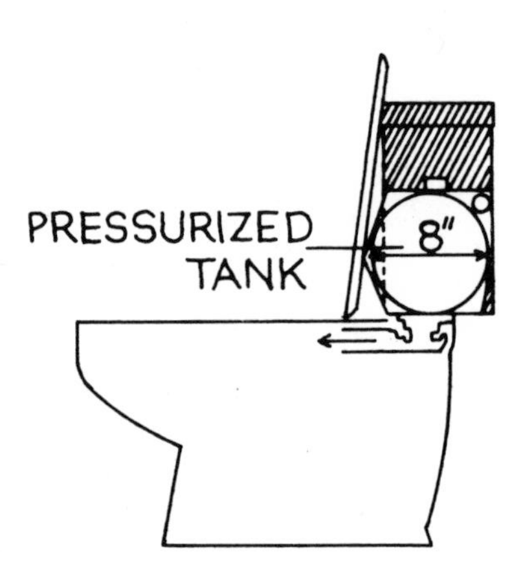

Position and size of pressurized tank that takes the place of the conventional toilet tank.

Dual and Variable Flush Devices

Using five gallons of clean water to dispose of one-half pound of feces is extravagant enough. Using (and paying for) the same amount to flush away one pint of urine is patently absurd. There are now several models of dual-flush toilet devices on the market that reduce the flush cycle to 2½ gallons for solids and 1¼ gallons for liquids. The different cycles are initiated by a short, sharp pull on the flush handle for the smaller amounts of water, and a longer, more persistent pull for the larger amount of water. Another model operates by pulling the handle in one direction for a small flush and in the other for a regular flush. Although similar in effect, the several dual-flush attachments differ substantially in design. Some are easily incorporated inside the conventional toilet tank; others require modification in the toilet.

In addition to dual-flush devices, attachments are also available that can allow a user to adjust the amount of flush water merely by holding down the flush handle for as long as necessary to effectively remove all urine and/or feces from the bowl. What makes this work is a weight attached to the

---

* Murray Milne, *Residential Water Conservation* (University of California/Davis: Water Resources Center, 1976).

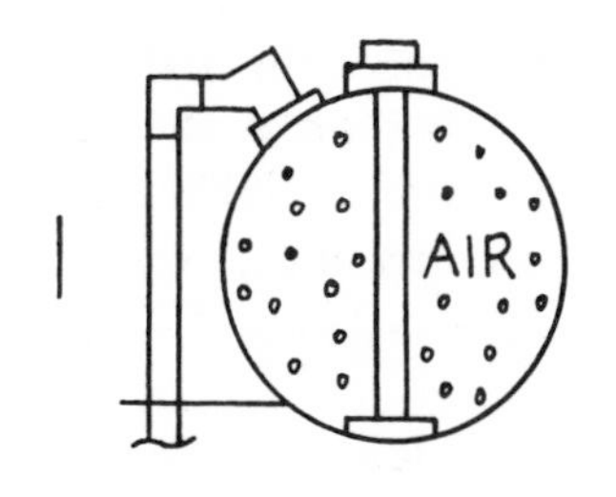

EMPTY TANK CONTAINING AIR.

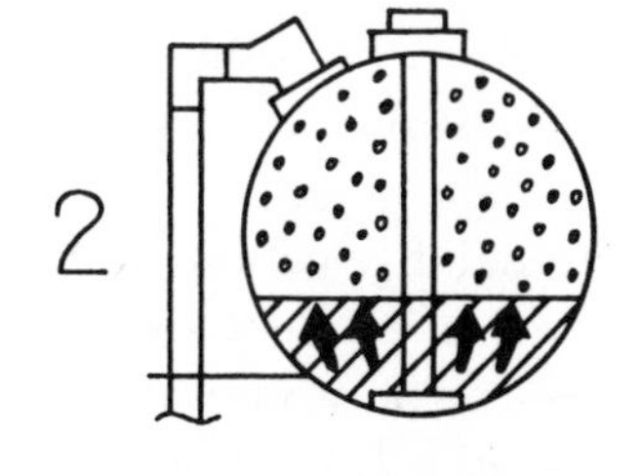

WATER ENTERING TANK COMPRESSES THIS AIR.

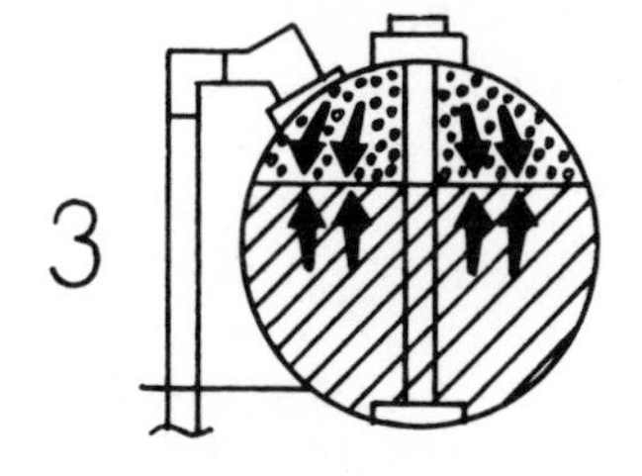

WHEN THE FORCES OF AIR PRESSURE AND WATER PRESSURE ARE EQUAL, WATER FLOW STOPS.

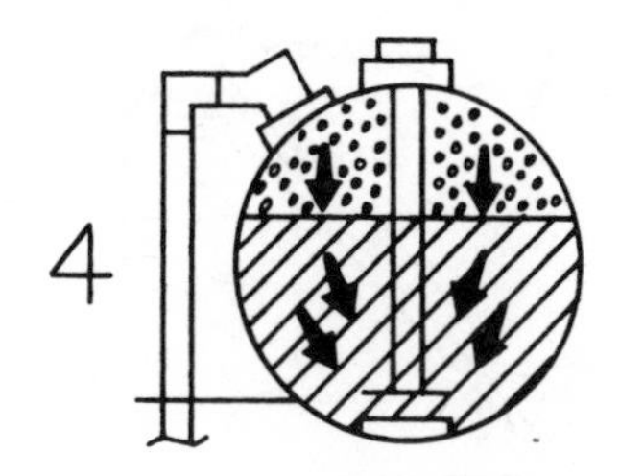

DEPRESSION OF THE PUSH BUTTON LIFTS MAIN VALVE INSIDE TANK. THIS ALLOWS TANK WATER TO "ESCAPE" INTO THE BOWL, BEING PUSHED BY THE FORCE OF COMPRESSED AIR AS WELL AS PULLED BY GRAVITY.

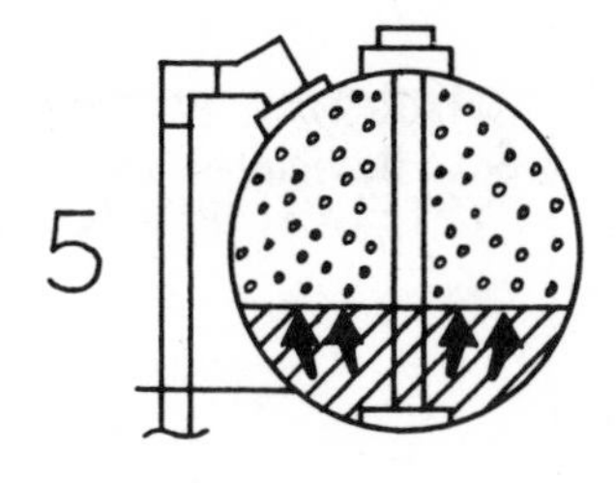

THE DROP IN PRESSURE INSIDE THE TANK AUTOMATICALLY CLOSES THE MAIN VALVE; FLUSHMATE IMMEDIATELY REFILLS FOR NEXT USE IN APPROXIMATELY 60 SECONDS. (REFILL CYCLE OF CONVENTIONAL TANK-TYPE TOILETS IS BETWEEN ONE AND TWO MINUTES.)

How a pressurized toilet tank works.

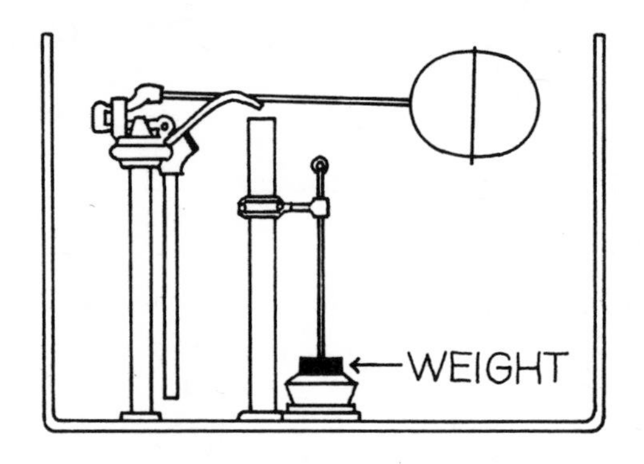

The weighted ball in this toilet tank allows users to adjust the amount of flush water merely by varying the pressure they put on the flush handle.

tank ball that brings the ball back to its resting position as soon as the flush handle is released, instead of having to wait for the end of the entire flush cycle.

Home Wastewater Recycling Systems

If you're putting in a whole new plumbing system or building a house, you may want to consider a wastewater recycling system. Such a system is hooked up to a conventional five-gallon or 3½-gallon flush toilet and effectively treats and reuses blackwater. The wastewater recycling system takes water from the flush toilet, first through an aeration tank and then through a charcoal filter to remove objec-

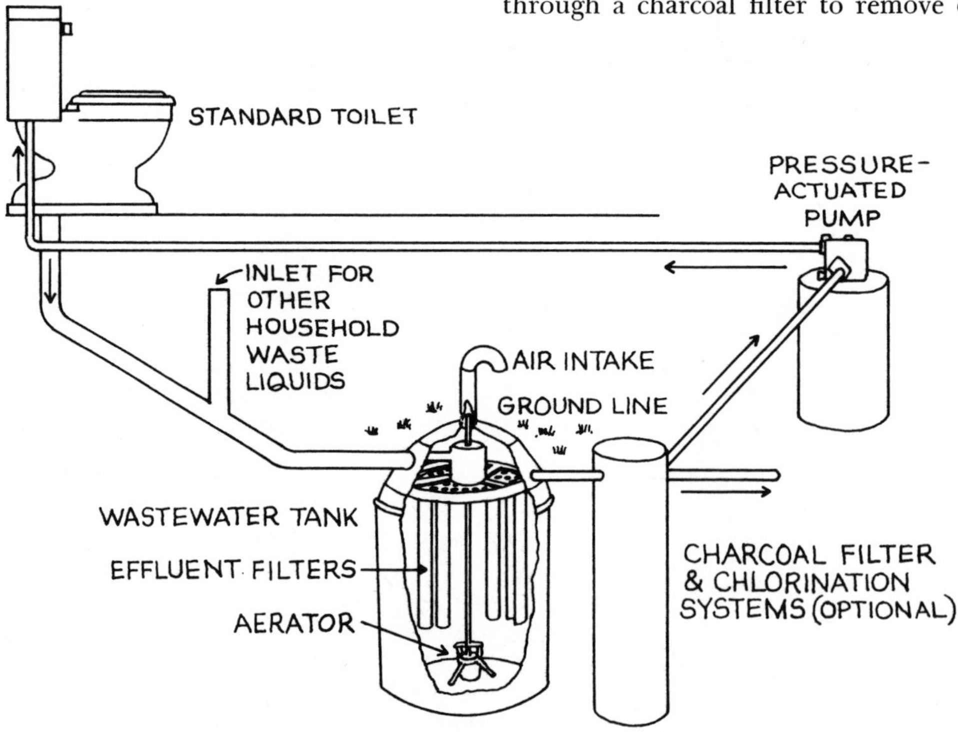

This household recycling system is said to save between 30 and 50 percent of the fresh water used for toilet flushing.

tionable colors. From there it is pumped up to the toilet tank again, and the cycle is completed. The system can be adapted to recycle greywater as well. An extra tank may be added to treat the greywater separately from the blackwater. An overflow drain, which is necessary when greywater is treated but is optional for blackwater alone, is connected to a disposal field or, in some places where it is required by law, to a sewer.

Such a recycling system is rather costly, and at this time is really only practical for areas where water is in short supply. With or without greywater recycling, the system saves between 30 and 50 percent of the fresh water used for toilet flushing. Energy use is minimal, and maintenance consists of occasional removal of the sludge in the aeration tank and disinfectant cleaning of the bowl.

Checking for Leaks

All your good efforts of conserving toilet flush water will do you no good, though, if you have a leaky toilet. A worn or poorly seated tank ball or a defective toilet tank valve can silently leak many hundreds of gallons of water a day.

To prevent water waste you should check for toilet leaks a few times every year. Begin by looking inside the toilet tank. Check to make sure all the mechanisms are working properly.

The water level should be about one-half inch below the overflow pipe. The proper level is marked on the backside of most toilet tanks. If the water level is too high more water than necessary is being used for every flush, or worse still, water is constantly flowing out through the overflow pipe. Lower the water level by bending the float arm down slightly so that the valve shuts off and water stops rising after it has reached the proper level. If the water keeps rising after you've bent the arm, then try another strategy: replace the ballcock valve.

Check also to see that the rod moves freely through the guide and that the arm inside the tank that is attached to the outside handle is aligned properly so that it responds well to a pull on the handle when flushing.

You can spot a toilet leak by adding a few drops of food coloring to the tank water. If there's a leak, colored water will show up shortly in the bowl. Officials in the Washington, D.C., suburbs and in Marin County, California, distributed free water kits to residents that contained a dye tablet just so they could detect toilet leaks this way.*

A new device to help tell users of toilet leaks and control the amount of water coming into the toilet tank is the *leak-signalling ballcock*. Instead of using a conventional float and rod arm to control water flow, this ballcock relies upon water pressure to open and close the valve. The new ballcock doesn't really conserve water itself. Rather, what it does is help people detect leaks quickly enough so that little water runs out

* See Appendix (Toilet Tank Devices) for more about this dye tablet.

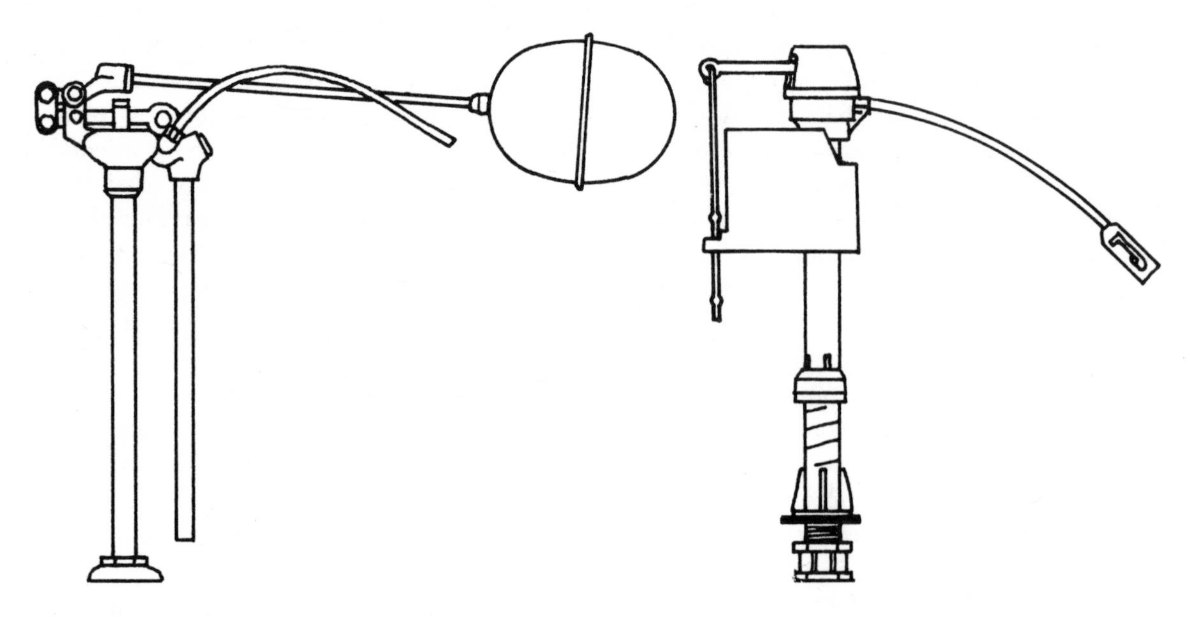

Conventional ballcock (left) and leak-signalling ballcock (right).

of the tank unnoticed. It does this by rapidly opening and closing the valve completely to refill the tank as soon as about a gallon of water has leaked out of it. The sound of this constant on and off of the valve should let anyone but the hard of hearing know that something is wrong. In contrast, a regular toilet's ballcock will open the valve slightly and allow the tank water to refill almost constantly as the tank water leaks out. This slow water flow can very easily go unnoticed.

At 33¢ per 100 cubic feet and a toilet which is constantly leaking, the following are the possible water savings:

Water and Money Saved by Checking Toilet Leaks

Frequency of signal once each	Gallons of water saved/day	$Savings/year
5 mins.	216	$34.78
15 "	72	11.59
30 "	36	5.74
60 "	18	2.90

Source: Milne, *Residential Water Conservation,* p. 222.

Showers and Faucets

After the toilet, the heaviest water user in the house is the shower. Approximately 30 percent of the total household water consumption goes for showering and bathing—roughly 80 gallons a day for a family of four. Of course, the amount of water used per shower varies considerably depending upon showering habits and the amount of water that flows through the shower head each minute. Flow rates vary greatly among different makes of shower heads, from 3 GPM to about 10 GPM.

Flow Restricters

Sizeable water savings can be obtained by installing a flow restricter for shower heads (and also for sink faucets). In 1973–1974 a test was run by the WSSC on 25 single-family homes fitted with 40 restricters which limited shower flow rate to 3.0 GPM. Duration of showers was not controlled. In December the average savings per household was 50 gallons a day, and in January 30 gallons a day. Flow restricters did not disturb the functioning of faucets or shower heads and few, if any, users were ever aware of their presence. It's interesting to note that the progressive WSSC has changed the D.C. area plumbing codes to require that all new buildings install 3.0 GPM shower heads.

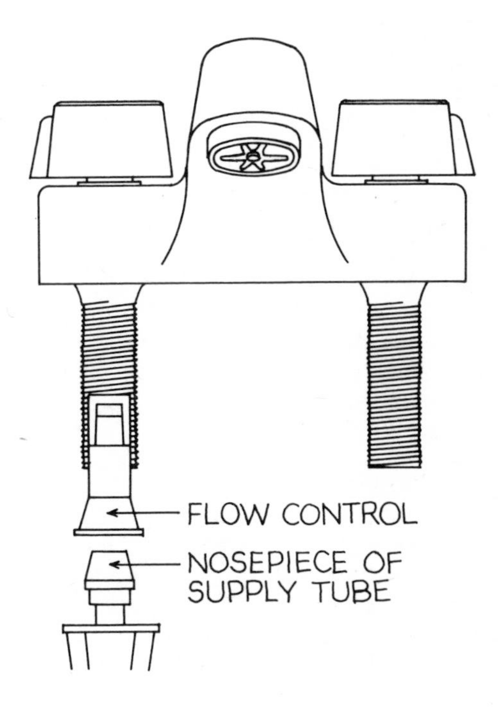

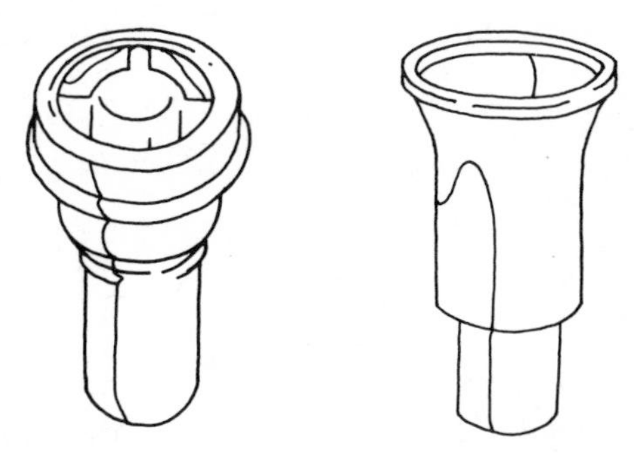

Inexpensive, plastic flow restrictors can be installed quite simply in faucets and shower heads.

Shower Flow Rate at 40 PSIG

Shower Head	Shower Head Sizing Guide	
	Manufacturer	Flow Rate GPM
Act-O-Matic	Sloan Valve	5.5
Boyd	American Std.	9.5
Victor	American Std.	10.0
Anystream No. 1 52240	Speakman	4.5
Anystream No. 2 52240	Speakman	4.5
Anystream No. 1 52250	Speakman	9.0
Anystream No. 2 52250	Speakman	7.0
Anystream No. 3 52250	Speakman	6.0
Brown 620-B	Chicago Faucet	8.0
Bubble Stream	Wrightway Eng.	4.0
Central Brass No. 3033	Central Brass Mfg.	3.0
Chase 188-355	Chase Brass & Copper	10.0
Crane Rainbow No. 8-2556	Crane Co.	5.5
Crane Capri 8-2550	Crane Co.	3.0
Crane Crestmont 8-3592	Crane Co.	3.0
Crane Crestmont 8-3590	Crane Co.	3.0
Crane Criterion 8-1561	Crane Co.	5.5
Dearborn	Dearborn Brass	7.5
Eljer	Eljer	6.0
Federal	Federal Huber	—
Gyro-Manystream	Gyro Brass Mfg.	—
Harcraft B-2	Harvey Machine	8.0
Homart 2091	Sears	—
Homart 2080	Sears	—
Indiana Brass	Indiana Brass Mfg.	—
Kohler K-7325	Kohler	—
Kohler K-7332	Kohler	—
Magic Fountain	Magic Fountain	2.8
Milwaukee "Premier"	Milwaukee Faucet	3.5
Sterling S-194	Sterling Faucet	6.0
Sure Flow 35	Sure Flow Brass Mfg.	3.0
Sure Flow 36	Sure Flow Brass Mfg.	4.5
Symmons Clear Flow	Symmons Eng.	3.5-7.0
Symmons Super Flow	Symmons Eng.	4.0-9.0
Universal/Rundle	Universal/Rundle	—
Wizard	Logan Mfg.	3.5-5.0
PP-15-020	Price Pfister	4.5
Moen 1533	Std. Screw	4.0
Delta	Delta Mfg.	4.0
Gerber	Gerber Brass	4.0
Sayco	Stephen Young	3.0

Source: *Commercial Water Heaters: Electric Engineering Handbook.* State Stove and Manufacturing Co., Inc. Ashland City, TN; Henderson, NV.

There are a variety of flow restricting devices available for use in shower heads. These devices are usually nothing more than valves that fit into the supply lines for showers (and faucets). The most effective kinds are those that have openings that vary in diameter.

Under normal or high water pressure the openings are small so that the water flow rate is reduced, but a decrease in the pressure of the water coming into the house causes the throat of the valve restricter to open and maintain a constant water flow. There are both metal fittings threaded at both ends and molded plastic spoollike devices. The cost of such restricters varies from $1 to $20.

The simplest and cheapest way to restrict water flow is to simply install in the shower head itself or in the pipe leading to it a simple washer with a hole of appropriate size (a ⅜-inch washer will usually fit snugly into the head of a ½-inch fitting). Although the washer is very inexpensive, it's not as effective as the other flow restricters. Because its opening is fixed, the water flow rate is determined by the water pressure, and turning on another faucet in the house may cause the flow of water from the shower head to drop below an acceptable level.

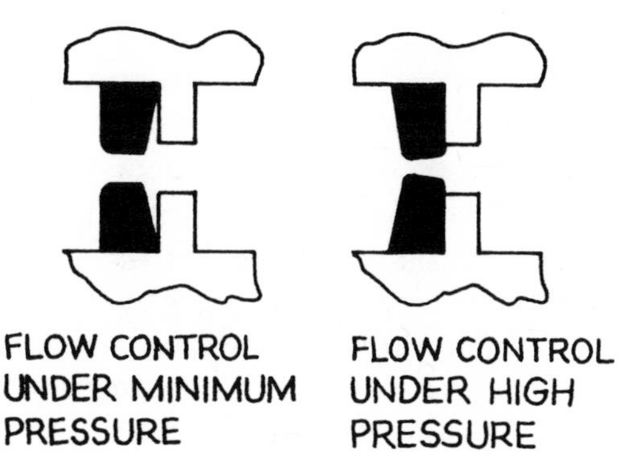

Flow control with flexible opening adjusts automatically to water pressure and maintains a steady flow.

Faucet Aerators

Unlike the conventional faucet, which allows the water to gush out in a single stream, aerators mix air with the water as it leaves the faucet, which gives the illusion of more water flowing from the tap than actually is. Faucet aerators are inexpensive, easy to install, and most use about 50 percent of the water that would normally be used with a regular faucet. New water-saving aerators are available now that can cut water flow to as much as one-tenth of the normal flow.

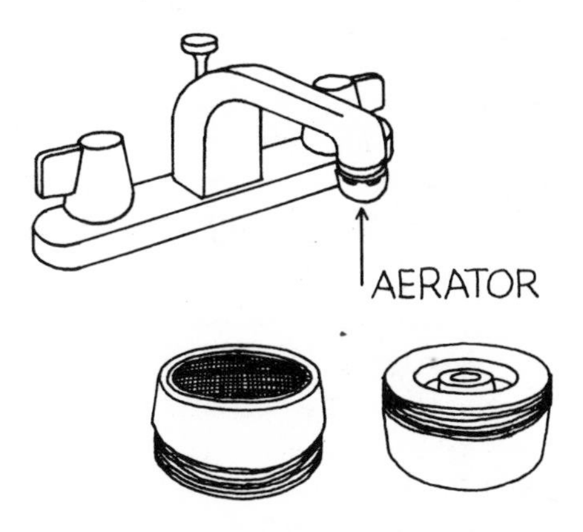

A faucet aerator can reduce water use by as much as one-third by mixing air with the water as it leaves the tap.

Spray Taps

Spray taps don't mix air with the water; they break it up and shoot it out in droplets, very much like a shower head does. Although used all over Europe, they're not commonly used in the United States. Spray taps use about 50 percent of the water normally used, with some claiming that these taps can save more than 90 percent of the water normally used.

Air-Assisted Shower Heads

Air-assisted shower heads act much like faucet aerators. They reduce the amount of water used by mixing it with air. An air compressor is located either above the ceiling or behind the shower wall. It starts up immediately when the shower is turned on and forces air down through the shower head where it mixes with the water to create a spray.

Although these shower heads are expensive (they cost about $275 in early 1976) they can save about 90 percent of the water used for showering. The real advantage, though, is not the savings in water cost, but in the cost of the energy necessary to heat the hot water—the greatest heating expense in the home, after space heating. The manufacturer has estimated that the cost of hot water heating can be cut down one-half to two-thirds over a 20-year period if air-assisted shower heads are used. Here's an analysis of water heating costs for 20 years in the Los Angeles area:

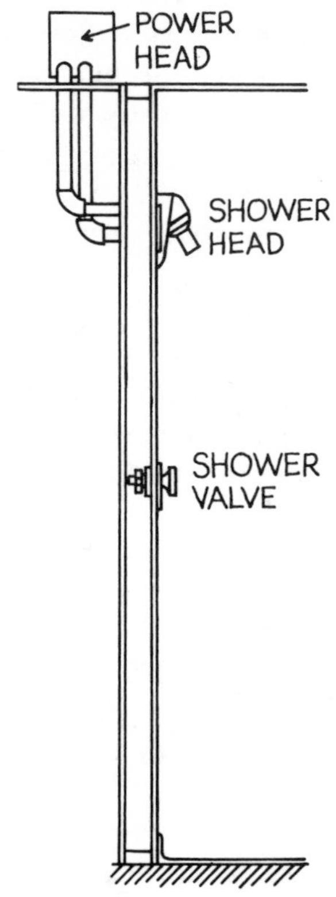

Air-assisted shower head compresses air and forces it down through the head to increase the pressure of the outcoming water.

Annual Water Heating Costs for the L.A. Area

Source of Heat	Conventional	Air-assisted	Savings
Gas (70 therms natural gas/month)	$1,114	$538	$ 576
Electric (500 kwh/month)	2,368	694	1,674

Source: Milne, *Residential Water Conservation,* p. 256.

Reducing House Water Pressure

Normal water pressure runs 40 to 50 pounds per square inch. Anything higher than this is unnecessary and usually wastes water at sinks and showers. The WSSC now requires that all homes in its high pressure areas install water pressure valves that keep the pressure at a maximum of 60 pounds per square inch, and most other municipalities throughout the country prohibit pressures greater than 100 pounds per square inch.

If you have especially high water pressure you can cut down on your water use by having a plumber install a pressure-reducing valve on your house's main water line. It is possible to reduce the pressure without investing in such a valve. Turn on all the faucets in your house and then turn the main water valve down until the flow through the faucets on the top floor of your house is reduced but still at an acceptable level. This adjustment is a fixed one, and if the water pressure is reduced because of a seasonal change later in the year, you'll have to readjust the main valve. You'll also have to make sure that the adjustment in water pressure doesn't disturb the normal functioning of water-using appliances, like washing machines and dishwashers.

Washing Machines

Automatic clothes-washing machines account for about 15 percent of the water consumed in households where they are present. Front-loading models that rotate on a horizontal axis use 22 to 33 gallons per cycle; top-loading machines require 35 to 50 gallons. Although front-loading washers use about one-third less water, they're not readily available anymore in this country because of persistent operating problems with them. Moreover, most front-loading models hold less wash than the top-loaders and this might negate any water savings. Washers with suds savers store the used wash water in an adjacent 20-gallon tank or service sink during the rinse cycle and reuse it during the second wash cycle, thereby saving on both hot water (about 20 percent savings if this water were used once) and detergent. Some new machines have this holding tank built into the bottom of the washer. Suds savers may not be a common feature on many washing machines, but the device can be obtained from some washer manufacturers through dealers.

In many washers the amount of water can be adjusted according to the size of the wash load. This feature, which may save as much as 12 gallons a week (especially if you do several small loads of wash), is certainly worth considering if you're going to buy a new machine.

Check your washer occasionally while it's operating to see if there is any overflow that goes down the drain or through the laundry tub when the washing chamber is filling. Many machines have a water level adjustment to eliminate this kind of waste.

Just about every machine now made has water temperature controls, and there are lots of detergents that can be used effectively with warm or cold water. Washing in warm or cold water is not only gentler on your clothes, but is also easier on your hot water bills.

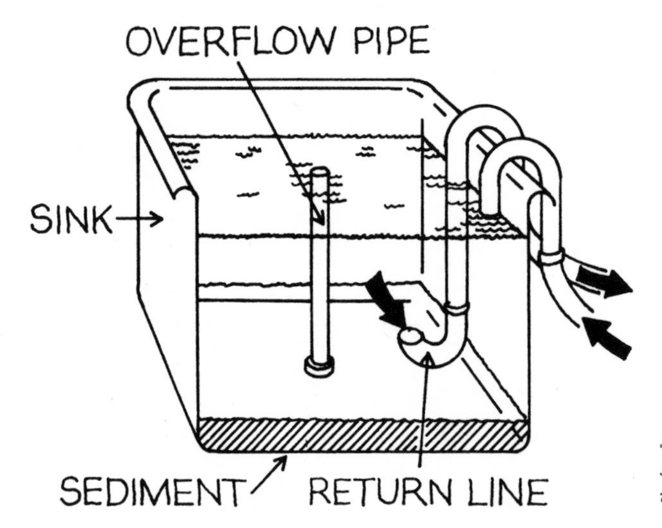

Suds-saver feature makes it possible to use the same wash water for both the first and second wash cycles.

Annual Laundering Cost

	Electric water heater	Gas water heater
Hot wash and warm rinse	$52	$15
Warm wash and warm rinse	35	10
Warm wash and cold rinse	18	5

Source: John George Muller, "The Potential for Energy Savings through Reductions in Hot Water Consumption" (Washington, D.C.: the Federal Energy Administration).

Specifications and Performance of Washing Machines

	Regular Cycle [1]		Durable-Press Cycle [1]		Presettable Combinations
	Water Used (hot wash/cold rinse)		Water Used (warm wash/cold rinse)		Temperatures (wash/rinse)
	Hot [2]	Total [2]	Hot	Total	
Whirlpool LAA5800	24 gal.	51 gal.	15 gal.	71 gal.	H/W, H/C, W/C, C/C
Whirlpool LAA5805	24	51	15	71	H/W, H/C, W/C, C/C
Maytag A207S	16(27)	37	8	49	H/W, W/C, C/C[3]
Hamilton WA373	25	52	9	56	H/W, H/C, W/W, W/C, C/C
Frigidaire WCD3T	23	55	13	63	H/W, H/C, W/W, W/C
Gibson WA83312A	24	48	12	54	H/W, H/C, W/W, W/C, C/C
Sears Kenmore 32611	16(30)	35(39)	10	44	H/W, W/C, C/C
Wizard Citation 5WC2340	25	53	12	53[4]	H/W, H/C, W/W, W/C, C/C
Blackstone BA702	20	46	10	56	H/W, H/C, W/C, C/C
Norge LWA2050	24	53	11	53[4]	H/W, H/C, W/C, C/C
Kelvinator W624G	25	52	9	56	H/C, W/W, W/C, C/C
Westinghouse LA470P	23	60	12	54[5]	H/W, H/C, W/W, W/C, C/C
Wards Signature 6224	23	50	10	50[5]	H/W, H/C, W/C, C/C
General Electric WWA7030P	22(34)	45	11	45[6]	H/W, W/C, C/C
Hotpoint WLW2600	9(20)	36	5	36	H/W, W/C, C/C
Speed Queen DA6123	23	41	8	42[4]	H/W, H/C, W/W, W/C, C/C

[1] Water pressure at 40 psi with 8-pound load and maximum fill. Variations among samples of the same model may cause variations of 1 or 2 gallons.

[2] Figure is for hot wash/cold rinse. Where this combination is available only by resetting temperature control after wash fill, the gallons used if not reset are shown in parentheses.

[3] H/C, W/C, C/C on durable-press cycle.

[4] Regular cycle is used for durable press.

[5] Short cycle is used for durable press.

[6] Using regular cycle, judged preferable to "permanent-press" cycle.

KEY: H-Hot; W-Warm; C-Cold.

Source: *Consumer Reports 1975 Buying Guide,* p. 66.

Dishwashers

Automatic dishwashers use water lavishly, too—from 13 to 25 gallons a day. The best conservation technique is simply to load to capacity for each use.

Washing all dishes and pots and pans in a dishwasher without rinsing them first would save you water, but since most people do rinse dishes or wash pots by hand before they load the washer, the water-savings rationale for having a dishwasher is questionable. It's estimated that one would use 15 gallons of water to hand wash the equivalent of a dishwasher load of dishes if the dishes were washed in a filled sink or dish pan and rinsed under running water. If the dishes were both washed and rinsed under free running water, the consumption would be about 25 gallons. Some dishwasher manufacturers suggest shortening the full machine cycle to accommodate some types of washing loads. A shorter run means some savings of water.

If you're going to buy a dishwasher, look for one that allows you to turn off the drying element and let the dishes dry naturally without the aid of heat; this will obviously cut down on your fuel bill. If this is not possible on your present dishwasher, you can skip the drying cycle by opening the door of the washer after the rinse cycle is completed.

Saving on Hot Water Heating

Of the various water-consuming devices in the household, the dishwasher is the only one that really justifies the 140°F (60°C) water that Americans maintain in their 40- to 50-gallon water heaters 24 hours a day. A recent Stanford Research Institute study concluded that a whopping three percent of all the energy consumed in the United States goes for heating domestic hot-water tanks. Hopefully, the hardware will soon be available in this country to enable us to follow the lead of West Germany and France in installing small "point of use and heat on demand" water heaters above faucets in our kitchens, bathroom sinks, and bathtubs. Although many people might find these "point-of-use" water heaters inconvenient at first, they really make sense for conserving heating fuel and water. Only a small amount of water is heated at a time, and it's not kept hot when it's not needed, like during the sleeping hours. Rather, the heater in each small water tank is activated simply by opening the faucet. Since the units are located where they're used, there is no hot water piping and no heat is lost through transfer.

(Do not confuse these "point-of-use" water heaters with instant hot water taps. These new taps are located on the sink and

Residential Water Heater Tank Insulation: Minimum-Cost Thickness,[a] Economic Benefit, and Energy Benefit with Present Range of Energy Prices

Item	Factory-Installed Insulation			Retrofitted Insulation		
50-Gallon Electric						
Electricity price (¢/kwh)	1	2	4	1	2	4
Minimum-cost insulation	4″	5″	7″	2″	2″	6″
Annual cost saving	$1.40	$4.00	$10.10	—	—	$4.90
Annual electricity saving (percent)	8.1	9.5	11.0	—	—	10.4
40-Gallon Natural Gas/LPG						
Natural gas price (¢/therm)	10	20	40	10	20	40
Minimum-cost insulation	3″	4″	5″	1″	3″	4″
Annual cost saving	$2.90	$7.90	$15.50	—	$2.50	$10.70
Annual gas saving (percent)	21.6	23.8	25.0	—	21.6	23.8
40-Gallon Fuel Oil						
Fuel oil price (¢/gallon)	20	30	40	20	30	40
Minimum-cost insulation	4″	4″	5″	3″	3″	4″
Annual cost saving	$4.80	$7.90	$11.20	$0.70	$3.50	$6.60
Annual oil saving (percent)	23.8	23.8	25.0	21.6	21.6	23.8

[a] Based on 10-year lifetime and an inflation-free, after-tax discount rate of 8 percent.

Source: *Residential Water Heating: Fuel Conservation, Economics, and Public Policy,* prepared for the National Science Foundation by James J. Mutch, (R-1498-NSF), Rand Corporation, Santa Monica, CA, May 1974.

provide near-boiling water almost instantly, for making tea, instant coffee, and instant soups and gravies. They may save water for those people who have a habit of bringing to a boil much more water than they actually need at a time. But because they keep water at 198°F (92.3°C) all the time, they consume a significant amount of electricity.)

Since we can't buy "point-of-use" heaters in this country yet, the best we can do to economize on energy used to heat water is to lower the thermostat setting on our heaters. Most are set at 160° to 140°F (71° to 60°C) but they can be set as low as 110°F (43.3°C). The Rand Corporation has calculated that you'll use about 15 percent less oil or gas fuel to heat water if you keep the thermostat at 110°F (43.3°C) instead of 140°F (60°C). 110°F (43.3°C) should be hot enough for most washing jobs and will eliminate the wasteful need to keep water very hot and then mix it with cold to achieve a comfortable washing temperature.

If you're in the market for a new hot water tank, be sure to get one that's well insulated. James Mutch of the Rand Corporation, in a study he did for the National Science Foundation, found that 21.2 to 35 percent of the heat in stored hot water is lost through the surface of the average hot water tank.

You may also be aware of the recent technological developments that make solar hot water heating practical today. There is much commercial hardware available now that enables individuals to heat water with solar collectors installed on house walls, roofs, and on adjacent ground areas and then store it in insulated water tanks.

Changing Habits to Save Water

Being just more aware of water use and waste around the house is going to mean that you'll probably unconsciously be taking shorter showers, adjusting the water level in your washing machine (if it has such an adjustment), and letting less water flow freely from a faucet when you're washing your hands or some dishes. Here are some more suggestions for saving water that you might not have thought of:

1. Don't thaw foods under running water. Take foods out of the freezer ahead of time.

2. Remove ice cube trays a few minutes before you need them so that they can thaw slightly, making running them under water to loosen them unnecessary.

3. Wash vegetables in a pan or pot of water rather than under free running water.

4. Don't use garbage grinders. Besides wasting water, these gadgets exist only to grind up valuable kitchen wastes and send them down the drain. Compost all your kitchen scraps (except meat, bones, and fats) instead.

5. Keep a container filled with water in the refrigerator so that cold drinking water will always be handy. This will prevent you from running the sink water for several seconds until it's good and cold every time you want a drink.

6. If you have trouble keeping your time in the shower down, remember that a conventional shower head uses between three and eight gallons of water a minute. The difference between a five- and ten-minute shower can mean as much as 40 gallons of water. A bath consumes about 25 gallons of water and an average-length shower 35 to 40 gallons.

7. If you're taking a bath, stop up the drain and turn the hot water on first. Let it run into the tub until it gets hot and then adjust the cold water. There's no sense wasting this less-than-hot water by letting it drain away.

8. Take a shower with someone else. Water use can be cut down by one-third or more, especially if one person is rinsing off while the other is lathering up.

9. Don't let the water run freely the whole time you're brushing your teeth, shaving, or washing your face or hands. For most of these needs, a pint of water should be enough, but several gallons will go down the drain if you leave the faucet on the whole time.

10. Flush the toilet only when you have to. Don't pull the handle just to flush away a tissue (which should go in the wastepaper basket anyway) or even necessarily to get rid of urine.

11. Sweep off the drive and sidewalks: don't hose them off.

12. Much water can be wasted outdoors by leaving a hose running unnecessarily while you're washing your car, garden furniture, windows, etc. Turn off the water between steps, or better yet, use a bucket of water for washing.

13. Make use of all the water you get "for free." Water your garden with rain water collected by placing barrels or other containers under gutter downspouts. Use the water that condenses from your dehumidifier or central air conditioner for watering house plants.

Hardware Listings

While by no means a comprehensive run-down of *all* the alternatives to flush toilets and septic tanks and of *all* the water-conserving hardware on the market today, these lists should give you a pretty good idea of what is available. For the most current information about such equipment don't neglect to check with plumbing supply houses and magazine classifieds.

Aerobic Systems

AQUAROBIC HOME SEWAGE TREATMENT SYSTEM
P.O. Box 1150
Penetanguishene, Ontario
Canada L0K 1P0

This system, designed for home use, decomposes household wastewater by pumping air into the underground holding and settling tank. An electric-mechanical unit, an aeration tank, and a settling tank are the major components.

BI-A-ROBI SYSTEMS
Box 133
Hamlin, Pennsylvania 18427

This aerobic sewage system is available for commercial and individual facilities. This unit does not require a special tank because each is custom-made in order to use as much of the present on-site system as possible. Units installed on an existing tank cost from $850 to $1,000, depending upon the labor rates in the area in which it is purchased.

BIODISC
Ames Croste Mills & Limited
105 Brisbane Rd.
Downsview, Ontario
Canada M3J 2K7

BioDisc is a complete above-ground aerobic sewage treatment system that can serve five to 500 people. The BioDisc principle is based on the slow rotation of discs alternating between atmosphere and sewage. The effluent may be discharged directly into a watercourse or onto land, depending upon local regulations. Chlorination of the effluent can be arranged.

CROMAGLASS SINGLE HOME AEROBIC WASTEWATER TREATMENT SYSTEM
Cromaglass Corporation
Williamsport, Pennsylvania 17701

This aerobic system is designed for single home use. It has a 1,000-gallon capacity tank at the water line and operates by using three chambers, hydraulic flows, and oxygen. The waste is broken up and partially oxidized in the primary comminution chamber. Continuous oxidation and aerobic decomposition of the sewage solids take place in the aeration chamber. The solids are settled in the effluent settling chamber, and then they are drawn into the aeration chamber as the settling chamber liquid level is equalized.

JET PLANT
Jet Aeration Co.
750 Alpha Drive
Cleveland, Ohio 44140

Jet Plant replaces the septic tank with an aerobic system like the others described here. Optional chlorination and tertiary treatment filters are available.

THE MINI-PLANT
Eastern Environmental Controls, Inc.
Box 475
Chestertown, Maryland 21620

This is an aerobic treatment plant for individual homes that includes air blower, air diffusers, submersible pump, mercury fluid level sensors, and control panel. The unit works on a fill and draw batch treatment process utilizing fine bubble diffused air. Units can treat between 250 and 1,500 gallons a day. Components of the unit can be adapted to most existing septic tanks. Maxi-Plants and Super Maxi-Plants can be custom built to process anywhere from 1,500 to 50,000 gallons per day.

Biological Toilets

BIO-FLO
Pure Way Corporation
301 42nd Ave.
East Moline, Illinois 61244

Bio-Flo is a self-contained recycling toilet in which wastes are converted into water by a biological process. This process is activated by the addition of digestive powder, and maintained by a weekly addition of this powder. The converted water is dual filtered and is poured into the clear water chamber by a gravity flow action. When the toilet is flushed, this water cleans the bowl, then drops back into the digestive chamber where the recirculating process starts again. Bio-Flo is available in a variety of models and is easy to install. For about $425 you can purchase a four-person unit using five gallons of water to start.

Composting Toilets

(For the distributor nearest you who handles the toilet that you're interested in, contact the following companies. For descriptions and specifications on all the toilets, see "A Comparison of Commercial Composting Toilets" on page 132.)

BIO LOO
Clivus Multrum USA, Inc.
14A Eliot St.
Cambridge, Massachusetts 02138

BIO TOILET (A, M, and 75)
Bio-Systems Toilets Corp. Ltd.
255 Gladstone St.
Hawkesbury, Ontario
Canada K6A 2G8

CLIVUS MULTRUM
Clivus Multrum USA, Inc.
14A Eliot St.
Cambridge, Massachusetts 02138

MULLBANK (ECOLET)
Recreation Ecology Conservation of United States, Inc.
9800 West Bluemound Rd.
Milwaukee, Wisconsin 53226

MULL-TOA (SODDY POTTY and BIU-LET)
Future Eco Systems Ltd.
680 Dennison Street
Markham, Ontario
Canada L3R IC1
or
ECOS Inc.
21 Imrie Road
Boston, Massachusetts 02134

SODDY POTTY #2
ASI Environmental Division
2 Industrial Parkway
Woburn, Massachusetts 01801

TOA THRONE
Enviroscope, Inc.
P.O. Box 752
Corona del Mar, California 92625

Low-Flush Toilets

CONSERVER
SILHOUETTE II CONSERVER
Briggs
P.O. Box 22622
Tampa, Florida 33622

Two water-saver toilets that flush with about 3½ gallons. The Conserver is a standard two-piece model, and the Silhouette II is a sleeker-looking one-piece water closet.

EMBLEM WATER-SAVING WATER CLOSET
Eljer Plumbingware
Wallace Murray Corporation
3 Gateway Center
Pittsburgh, Pennsylvania 15222

This water closet uses about 3½ gallons of water per flush. Its bowl has narrower sides than regular toilets and a trapway that allows a lower water level in the tank.

LF 210 LOW-FLUSH CERAMIC TOILET and LF 310 STAINLESS STEEL TOILET
Microphor
P.O. Box 490
Willits, California 95490

These toilets use compressed air and two quarts of water per flush. Their push buttons activate a flow of water into the bowl and open a valve in the toilet substructure. This valve automatically closes and a charge of air ejects waste material into the discharge line. You can purchase a Low-Flush Microphor toilet for about $400.

RADCLIFFE WATER MISER
Crane Co.
300 Park Avenue
New York, New York 10022

A conventional toilet with a smaller tank and bowl that uses 3½ gallons per flush.

WATER SAVER CADET TOILET
American Standard
P.O. Box 2003
New Brunswick, New Jersey 08403

The smaller tank and the new design of the bowl enable this otherwise conventional toilet to use only 3½ gallons per flush.

WELLWORTH WATER-GUARD TOILET
Kohler Co.
Kohler, Wisconsin 53044

Another conventional-looking and operating 3½-gallon flush toilet.

Greywater Treatment Systems

MINIPUR SYSTEM
Enviroscope, Inc.
P.O. Box 752
Corona del Mar, California 92625

This unit is designed to take the place of a septic tank when only greywater needs to be treated (such as when a composting toilet handles toilet wastes). The area of separation in the tank is increased because the separating chambers are placed one above the other. The 200-gallon capacity plastic tank can take care of the greywater from the kitchen, bathroom, and laundry for the average family and requires pump-out about once a year. The Minipur is available in two sizes: the smaller handles dishwater and laundry only, and the larger handles bath and shower water as well. The manufacturer claims it can be installed by the owner.

TRICKLE FILTER
Clivus Multrum USA, Inc.
14A Eliot Street
Cambridge, Massachusetts 02138

This unit treats greywater before it enters the leach field or other means of discharge by filtering out larger particles and fibers including hair, lint, and food wastes. The filter is contained in a four-foot-high cone-shaped, fiberglass tank that sits in the basement or utility room and can be installed by the owner. It has the capacity for maximum flow level of wastewater from a single family dwelling. The complete system including the cone, base, and cover, with internal pipes installed, pump, delay timer, and sump float switch (but not the leach lines) sells for about $450.

Incinerating Systems

A-C STORBURN
Lake Geneva A&C Corp.
Box 89 200 Elkhorn Road
Williams Bay, Wisconsin 53191
and in Canada:
Storburn Limited
Box 3368 Station "C"
Hamilton, Ontario
Canada L8H 7L4

This incinerating toilet is a large capacity unit that can be used 50 to 60 times in a row before incineration of the contents is necessary. Incineration is carried out by a 40,000 BTU propane gas burner. No electricity is required. The Storburn is intended for industrial use and can accommodate 15 to 20 workers in an average eight- to 10-hour workday. It can, however, be used in a residential application to meet the needs of 10 to 15 people in an average 16-hour day. Cost of the unit is about $900.

DESTROILET
LaMere Industries, Inc.
Walworth, Wisconsin 53184

Destroilet is an incinerating toilet that uses the processes of evaporation and oxidation to reduce human wastes to a nonpolluting vapor and ash. Heat to burn the waste is supplied by a gas burner or electric element. You can install this unit by connecting the gas supply pipe, attaching an outside standing vent, and connecting a source of electricity. The incinerating process begins when you close the lid after each use. The ashes must be removed weekly when the toilet is used regularly by four to six people. Destroilets are available for about $600.

INCINOLET
Research Products Blankenship
2639 Andjon
Dallas, Texas 75220

This electric incineration toilet requires that a wax paper liner be placed in the bowl prior to each and every use. Proper venting of the unit is essential both inside and outside the toilet enclosure. You can install this toilet unit wherever there is a source of electric power.

INCINOMODE
Incinomode Sales Company
P.O. Box 879
Sherman, Texas 75090

Incinomode is another incinerating toilet. This ash is collected in a pan located below the toilet seat, and should be emptied every 10 days to two weeks. A 120V or 220V source is needed to operate both heating elements and exhaust, and a flue to exit the by-products of the incineration. You operate this toilet by dropping in a sanitary liner before use, and then resetting the timer for the proper cycle.

THIOKOL CHEMICAL ZERO DISCHARGE WASTE TREATMENT SYSTEM
Thiokol/Wasatch Division
P.O. Box 524
Brigham City, Utah 84302

This system uses, in addition to incineration, a closed loop recirculating salt water treatment process. Liquid waste is chemically treated and recycled, while the solid waste is disposed through incineration. These systems are designed for ships, highway rest stops, recreation areas, etc.

Oil-Flush Toilets

AQUA-SANS
Space Division, Chrysler Corporation
P.O. Box 29200
New Orleans, Louisiana 70189

This automatic sewage treatment system operates by using permanent flush fluid and an incinerating method. The flush fluid is a mineral oil which carries the waste from the toilets to the separation tank where the waste settles to the bottom while the flush fluid rises to the top. The flush fluid is recirculated to the toilets, after being filtered. When sufficient waste is accumulated in the separation tank, a pump is activated to transport some of the waste to a preheated incinerator. This system is designed for public facilities.

"MAGIC FLUSH"
Monogram Industries, Inc.
P.O. Box 92545
Los Angeles, California 90009

"Magic Flush" is an oil-recirculating waterless toilet. It looks like a regular toilet except that the inside of the bowl is coated with Teflon. The clear flushing fluid is filtered through a purification system and used repeatedly. This fluid carries the waste to a small, sealed, underground separation and storage tank where the wastes are stored beneath the fluid. When the tank is full the wastes are removed for incineration, composting, or other treatment. The tank contains the electrical and recirculating equipment. "Magic Flush" systems are available for public facilities with various combinations of toilets, urinals, and purification systems. A combination of one toilet, one purification system, and one 400-gallon tank, sufficient for a single-family home, costs in the vicinity of $5,000.

SARMAX SYSTEM
Sar Industries, Inc.
2207 South Colby Ave.
Los Angeles, California 90064

This product is a self-contained, closed-loop flush toilet system. Each system has individual motors, pumps, controls and purification units. An oil-based clear flushing fluid is used instead of water. This flushing fluid performs the same function as the fluid used in the "Magic Flush" unit. The special tank, which holds 7,500 uses, should only need to be pumped out once a year under normal family use.

Recycling Systems

CYCLE-LET
Thetford Corporation
P.O. Box 1285
Ann Arbor, Michigan 48106

Cycle-Let is a self-contained recycling toilet system that operates by using gravity or vacuum flush. It recycles a low volume of clean water with each flush. A Cycle-Let consists of a waste transfer system, a waste treatment system, and a water recovery system. Within the waste treatment plant, the waste goes through anaerobic and aerobic digestion and a sedimentation process. From the waste treatment tank, treated water enters the water recovery system. Within this system the water passes through the membrane filter, the secondary filter, ultraviolet purifier, and is stored in the flush water storage tank and water pressure tank. Upon demand, the water enters the toilet, thus completing the water recycle. The Cycle-Let is slightly larger than a home refrigerator and can serve up to eight people continuously or 24 people for eight hours per day. The system should be pumped out and cleaned once every two years. This fairly expensive unit costs between $4,000 and $7,000, depending upon the installation site and the number of toilets. The system is in its final stages of product development and should be available by mid-1978.

MULTI-FLO
Multi-Flo, Inc.
500 Webster St.
Dayton, Ohio 45401

Wastewaters from bathroom, kitchen, and laundry are transported to an aerobic waste treatment unit. The clear effluent that leaves this aerobic unit is then disinfected by chlorine which is introduced in small tablets. From there the water is stored in a holding tank for reuse and any excess is discharged into a disposal field.

Vacuum Toilet Systems

AIRVAC
P.O. Box 508
Rochester, Indiana 46975

The Airvac vacuum sewage system is a central sewage transport and collection system. It is designed to serve from 50 to 100 homes. This system utilizes air moving at a high velocity, rather than water under gravity flow, to transport the sewage to the central collection site where the wastes are then pumped to an existing sewer hookup or a treatment facility.

ENVIROVAC
Colt Industries, Waster and Waste Management Operation
701 Lawton Ave.
Beloit, Wisconsin 53511

Envirovac is a vacuum sewage collection system that uses only three pints of water to flush the toilet. The vacuum central collection tanks are available in holding capacities of 200, 1,000, and 2,000 flushes. The toilet is connected to the discharge pipe that is under a constant vacuum, produced by a small vacuum pump. The waste is forced by the air through the pipe connected to the collecting tank. These systems are used on ships, in parks and for recreation facilities, and were recently used along the Alaskan pipeline.

VACU-FLUSH SYSTEM
Mansfield Sanitary, Inc.
150 First St.
Perrysville, Ohio 44864

This system operates on the vacuum principle and flushes wastes on less than one quart of water to the place of treatment and/or disposal. There is a small standard system for one toilet and a larger commercial unit that can flush six toilets at the same time.

Water-Conserving Faucets and Shower Heads

ANYFLOW SHOWER HEAD WITH AUTOFLO FLOW CONTROL
CONSERVAFLO
Speakman Company
Wilmington, Delaware 19899

Anyflow Shower Head has a built-in, concealed mechanism that works on water pressure alone, and this reduces the amount of water delivered. An optional thumb-operated volume control regulates the volume and allows the user to reduce the water flow while lathering up without losing the desired hot/cold mix.

Conservaflo is a water flow regulator and is sold as a separate unit for sink and lavatory faucets. It is designed for Speakman products but could be used in others. Conservaflo controls the water flow rate at one-half gallon per minute.

AQUAMIZER
American Standard
P.O. Box 2003
New Brunswick, New Jersey 88903

Aquamizer is a flow control device that is attached to a shower head and reduces the amount of water used to 2½ gallons per minute. You can install it easily by screwing it between the shower head and the shower water supply pipe.

BUBBLESTREAM ECOLOGY WATER SEWER KIT
Wrightway Manufacturing Co.
Distributed by G & S Supply Co.
5801 S. Halsted St.
Chicago, Illinois 60621

This kit includes a pushbutton volume control shower head and two water-saving aerators. The shower head makes it possible to adjust the spring and volume without affecting the temperature setting. You can turn the water on or off as you like during soap-down. The low-volume aerators, attached to faucets, use 40 percent less water at average pressures with good aeration.

DOLE AUTOMATIC VOLUME SHOWER CONTROLS
Eaton Corporation
Controls Division
191 East North Ave.
Carol Stream, Illinois 60187

This flexible rubber flow control, incorporated in the shower head, provides an automatic control of the volume of water that can flow through your shower head. It maintains a near constant flow of water at three gallons per minute. To install this device, you screw the Dole Control between your shower head and the shower water supply pipe. There are three types of these controls available, ranging from $2 to $5.

ECONO-FLO
The Chicago Faucet Co.
2100 South Nuclear Drive
Des Plaines, Illinois 60018

Econo-Flo is a water-saving device that can be installed on almost any faucet with a threaded spout. This mechanism limits the flow of water from a faucet to ¾ gallon per minute. You can install an Econo-Flo unit by screwing it onto the faucet outlet. There are two models of Econo-Flo available.

MERWIN 321 SHOWER HEAD
MERWIN SINK AERATOR
Merwin Manufacturing
136 East Fourth St.
Dunkirk, New York 14048

The Merwin shower head has a flow compensator that delivers less than two gallons of water per minute at any line pressure. It is available in four models and can be installed without special tools. The sink aerator fits most all faucets and cuts the water flow in half.

MINUSE SHOWER
Minuse Systems, Inc.
206 N. Main St., Suite 300
Jackson, California 95642

The Minuse shower mixes air with water as it leaves the shower head, thereby maintaining pressure and the feel of a vigorous shower without using great amounts of water. Two quarts of water are used per minute, as opposed to four gallons per minute in a regular shower. An enclosed stall is recommended since air movement is an important factor in the operation. This shower includes three major parts: a shower head, a shower valve, and a power unit. This unit can be easily installed. The Minuse shower sells for $260.

MOEN SINGLE-HANDLED FAUCETS
Moen
Elyria, Ohio 44035

These single-handled faucets contain a cartridge water control which reduces water consumption to about 50 percent of the amount used by ordinary two-handled faucets. The Moen Cartridge controls both temperature and volume. This cartridge needs to be replaced about every eight years on the average. These faucets cost about $17.

NOLAND SHOWER AND FAUCET FLOW CONTROLS
Noland Company
2700 Warwick Blvd.
Newport News, Virginia 23607

These simple Celcon cylinders with no moving parts regulate the flow of water at a predetermined rate and automatically compensate for varying pressures. They should cut water flow up to 50 percent.

NOVA CONTROLLED-FLOW SHOWER HEADS
Water Wizard, Inc.
P.O. Box 184
Croydon, Pennsylvania 19020

These flow-control shower heads use approximately one-quarter the amount of water used by a conventional shower head. An optional feature is a throttle valve control that enables one to cut off the water flow while lathering up and then restore the flow for rinsing without having to readjust the water temperature. A Nova Controlled-Flow Shower Head costs about $13.

NY-DEL SHOWER HEADS AND FLOW CONTROLS
Ny-Del Corporation
P.O. Box 155
740 E. Alosta Ave.
Glendora, California 91740

This shower head is equipped with a water-saving device that regulates the water flow to 2½ to three gallons per minute. The flow control can be installed on a present shower head and reduces water flow approximately 60 percent.

RADA 872
Richard Fife Inc.
140 Greenwood Ave.
Midland Park, New Jersey 07432

This restricter shower head reduces water flow to 1.5 to 2.4 gallons per minute and comes in both a fixed, wall-mounted model and a sliding bar shower that makes the shower head either hand-held or of any height.

ULTRAFLOW PUSH BUTTON ONE-LINE PLUMBING
Ultraflow Corporation
P.O. Box 2294
Sandusky, Ohio 44870

Ultraflow is a centralized water distribution system that eliminates the use of faucets to control water flow and temperature at sinks, tubs, and showers. It replaces the conventional hot and cold water supply system with a one-line system that is operated by buttons at each point of use that turn water off or on, fast or slow, cold, warm, or hot. Temperatures and flow rates are preselected so water temperature is always the same. The manufacturer claims that this system is cost-competitive with conventional plumbing and brings overall water and energy savings of about 30 percent.

WATER GATE SHOWER HEAD, FLOW CONTROL, and CONSERVARATOR
JKW 5000 LTD
10610 Culver Blvd.
Culver City, California 90230

The Water Gate Shower Head conserves as much as five gallons of water per minute during the average shower without loss of effectiveness. The shower head provides a forceful, adjustable flow of water and is engineered to prevent rust or corrosion or plug-up.

The Water Gate Shower Flow Control attachment can fit between any conventional shower head and the shower arm, which restricts the water flow while channeling the water to compensate for varying pressure. The result is a normal shower. The flow control conserves up to five gallons of water per minute.

The Water Gate Conservarator is an aerator that can be attached to kitchen sink and bathroom basin faucets, providing both steady flow benefits and water use reduction. The Conservarator, a dual-threaded attachment fitting either male or female faucets, supplies a consistent flow of aerated water while reducing water consumption by as much as 33 percent.

Water-Conserving Toilet Tank Devices

AQUA-MISER
Energy Recovery Systems, Inc.
P.O. Box 233
Lincroft, New Jersey 07738

This water-saving device saves up to 50 percent of the water used in flushing, while still allowing normal flushing action. The Aqua-Miser is made of thermoplastic rubber and fits conventional toilet tanks. It functions by forming a dam at the bottom of the tank, thereby retaining most of the water that is usually wasted. Aqua-Miser can be installed easily and costs about $5 per unit.

DUAL FLUSH SYSTEM
Savway Co., Inc.
930 Clarkson Ave.
Brooklyn, New York 11203

By using atmospheric pressure to control the escaping flow of water from the tank to the bowl, the Dual Flush can release a full tank or half tank of water. Operating much like the Duo-Flush, users either press and release the handle quickly or hold it down longer, depending upon how much water they want released. The unit can be installed in minutes and costs about $5.

DUO-FLUSH
Craig Ramsey and Associates
P.O. Box 2406, Dept. LS
Colorado Springs, Colorado 80901

This device saves water by giving you a choice between a half or full flush. To dispose of liquid, press the toilet lever about half way; this lets out about half the contents of the tank or about 2½ gallons of water. For solid disposal, press down all the way, opening the bottom valve and emptying the entire tank, about five to seven gallons. Chains are used to operate the valves and are connected to a brass wire compound lever, which in turn connects to the flush rod. Installation is easy, and the Duo-Flush costs about $13.

THE FLUIDMASTER FLUSHER FIXER KIT

Fluidmaster, Inc.
P.O. Box 4264, 1800 Via Burton
Anaheim, California 92803

This device is a fluid level control valve that controls the amount of water entering the tank. It also signals any leakage of water out of the tank. This fluid valve works with the fluid force of the toilet's water pressure instead of mechanical leverage to close the valve, providing a fast water shut-off. The toilet's regular ballcock, float ball, and rod arm are replaced with the corrosion-resistant, stainless steel ball chain, attached to the trap lever. You can install the Fluidmaster on any two-inch brass, china, or plastic valve seat and it can fit all tank sizes. This unit costs about $6.

THE FLUSHMATE

Water Control Products N.A., Inc.
110 Owendale, Suite E
Troy, Michigan 48084

Flushmate is a replacement for the traditional toilet tank. This unit uses air pressure to speed the flushing process, thereby using only two to 2½ gallons of water for a flush. This device is contained in a cylinder eight inches in diameter and is lower than the conventional tank-type toilet. It is easy to install on conventional toilets and retails for about $60.

LITTLE JOHN

North Shore Associates
Greenhurst, New York 14742

Little John is a toilet water-saver device consisting of two plastic pieces. Like the Aqua-Miser, these corrosion-free pieces are inserted in the toilet tank, creating a dam on each side of the drain hole. These dams retain water during flushing, saving two to three gallons in each flush. Little John does not need any tools to be installed and is available at $6.95 per unit.

SA-720 WATERSAVER

Ny-Del Corporation
740 E. Alosta Ave.
Glendora, California 91740

The water-saver reservoir dam claims to save from three to four gallons of water per flush.

"SUPER DRIP KIT"

Formulabs Inc.
529 W. 4th Ave. P.O. Box 1056
Escondido, California 92025

Not actually a kit, but an effervescent blue dye tablet mounted on a self-mailer and designed for water companies to distribute to their customers. Since a minimum order of $25 is necessary, these are not meant for individual customer sales.

The "kit" won't find all leaks, but it's a cheap and easy way to discover plunger-ball or overflow problems. The tablet should be dropped into the toilet tank. (Don't flush.) If the toilet leaks water, the dye will seep into the bowl within a matter of minutes.

THE WATER GATE
JKW 5000 LTD
10610 Culver Blvd.
Culver City, California 90230

This is another two-piece device that acts as a dam in the toilet tank. Water Gate saves approximately two gallons per flush and can be installed without tools. This mechanism costs around $5.

WATER WIZARD
Water Wizard, Inc.
P.O. Box 184
Croydon, Pennsylvania 19020

Water Wizard acts like the other toilet tank dams and saves approximately two gallons of water with each flush. It costs $3.49 and can be installed easily.

Bibliography

Bennett, E.R., Lindstedt, K.D., and Felton, J. "Comparison of Septic Tank and Aerobic Treatment Units: The Impact of Wastewater Variations on Those Units." *Water Pollution Control in Low Density Areas: Proceedings of the Rural Environmental Engineering Conference,* University of Vermont, Burlington, VT 05401.

Bernhart, Alfred P. *Treatment and Disposal of Wastewater from Homes by Soil Infiltration and Evapo-transpiration.* Toronto, Canada: University of Toronto Press, 1973. Available from Dr. Bernhart at 23 Cheriton Ave., Toronto, Ontario, Canada M4R 1S3 for $16.

A technical work on disposing of wastewater through evapo-transpiration and soil infiltration. Information on how living creatures clean wastewater is particularly good.

California Department of Water Resources. "Proceedings: An Urban Water Conservation Conference." Sacramento, CA: State of California Department of Water Resources, January 16–17, 1976.

At a meeting sponsored by the California Department of Water Resources, representatives from industry, government, agriculture, and conservation groups discussed the water situation in California past, present, and future.

Clark, Zandy, and Tibbetts, Steve. *Composting Toilets and Greywater Disposal: Building Your Own.* 1977. Alternative Waste Treatment Association, Star Route 3, Bath, ME 04530.

Includes a guide to composting toilets and greywater systems with recommendations for building and maintaining your own.

Clivus Multrum USA, Inc. 14A Eliot St., Cambridge, MA 02138.

The Clivus Multrum manufacturers offer publications intended to assist owners of their toilets. Among them is a three-page photocopied paper titled "Washwater Treatment Considerations for Houses with Multrums," which gives suggestions on handling wastewater. Margaret Fogel and Carl Lindstrom authored *The Treatment of Household Washwater in Homes Equipped with the Clivus Multrum Organic Waste Treatment System,* a brief summary of the scientific knowledge to date (1976) of greywater treatment. It includes a list of methods appropriate for homes where toilet wastes are excluded from the waterborne sewage.

Cohen, Sheldon, and Wallman, Harold. *Demonstration of Waste Flow Reduction from Households* (PB-236 904). General Dynamics Corporation, Groton, CT, 1974. Distributed by the National Technical Information Service, U.S. Department of Commerce, Springfield, VA 22157.

Prepared for the National Environmental Research Center, this reports on a two-year demonstration program conducted to evaluate water savings, cost, performance, and acceptability of various water-savings devices. Devices tested included reduced flow toilets, flow limiting shower heads, dual flush toilets, and systems in which wash water is recycled for use in toilet flushing and lawn sprinkling.

Cole, Charles A. "Sewage Flow Reduction by Improved Toilet Systems." *Water Pollution Control Association of Pennsylvania Magazine* (January-February 1975): 20.

A proposal to reduce the costs of municipal sewage treatment through the use of reduced-flow, dual flush, vacuum, recirculating, incinerating, and waterless toilets.

Compost Science: Journal of Waste Recycling. Emmaus, PA: Rodale Press. 1 year/six issues for $6.

This bimonthly magazine reports on the entire field of large-scale composting and recycling of organic solid wastes. It provides technical, scientific, and practical information for the conversion of municipal, agricultural, and industrial wastes into useful products.

Farallones Institute, 15290 Coleman Valley Road, Occidental, CA 95465.

The Farallones Institute, an innovative West Coast group researching and implementing alternative technology designs, has designed a safe compost privy which can be built by amateur builders using common materials and tools for less than $100. It is described in their "Technical Bulletin No. 1: Composting Privy." Homeowners interested in alternative toilets would also find helpful two reports by Max Kroschel in the Institute's 1977 annual report: "Waste Water Reclamation: Development of a Small Scale Greywater System for Agricultural Use" and "The Farallones Composting Privy."

"Final and Comprehensive Report: Washington Suburban Sanitary Commission's Water Conservation/Wastewater Reduction, Customer Education and Behavioral Change Program." Hyattsville, MD: Washington Suburban Sanitary Commission, November 1974.

The WSSC distributed about 900,000 plastic quart bottles for flush water displacement, 600,000 dye pills to reveal toilet leaks, and a half-million pamphlets with water conservation hints to residences in their service area. The report discusses the basic program and its results, among them an estimated five percent reduction of in-household water use.

Goldstein, Jerome. *Sensible Sludge.* Emmaus, PA: Rodale Press, 1977.

This book shows how sludge, through proper treatment and utilization, can help reclaim our depleted soils, save our finite energy resources, and reduce the large expenditure of tax dollars now spent on waste disposal.

Goldstein, Steven N., and Moberg, Walter J., Jr. *Wastewater Treatment Systems for Rural Communities.* Washington, D.C.: Commission on Rural Water, 1973.

A guide to systems and components available for treating wastewaters in rural areas, this book touches only briefly on composting toilets and privy construction. It presents basic concepts of domestic sewage and treatment processes appropriate to rural settings and describes the role and use of soils in wastewater treatment and disposal. It includes a review of many traditional and some new systems; one appendix reviews a representative collection of appropriate equipment.

Goldstein, S.N., Wenk, V.D., Fowler, M.C., and Poh, S.S. "A Study of Selected Economic and Environmental Aspects of Individual Home Wastewater Treatment Systems" (M72-45). Office of Water Programs Project No. 1320. MITRE Corporation, Washington, D.C., March 1972. Photocopied.

A study disclaiming the popularly held notion that individual home wastewater treatment systems are a last resort or a temporary expedient, this report indicates that the full potential of individual treatment is not being realized.

Golueke, Clarence G. *Biological Reclamation of Solid Wastes.* Emmaus, PA: Rodale Press, 1977.

One of the first comprehensive guides to the biological processes for reclaiming the resources found in solid waste, including municipal and industrial refuse, animal manures, and crop residues.

———. *Composting: A Study of the Process and Its Principles.* Emmaus, PA: Rodale Press, 1972.

A very readable and complete discussion of small- and large-scale composting with information on the process and scientific principles, the technology involved, and the health aspects of using compost as a soil amendment.

Gotaas, Harold B. *Composting: Sanitary Disposal and Reclamation of Organic Wastes.* World Health Organization Monograph Series No. 31. Geneva, Switzerland: World Health Organization, 1956.

This monograph presents methods and processes by which organic materials may be treated for sanitary disposal and utilization in agriculture. In addition to providing information on the municipal composting of urban refuse and sewage sludge, it delineates operations designed for villages, small towns, and individual farms and describes methods of methane recovery. Primary focus is promoting health through sanitary waste treatment and soil improvement with the use of compost.

Hershaft, A., Von Hasselin, R.H., and Roop, R.D. *Water Management Alternatives on Long Island.* October 1974. Available from Ms. Deborah Cooper, Booz, Allen & Hamilton Inc., 4733 Bethesda Ave., Bethesda, MD 20014 for $5.00 (payable to Booz, Allen & Hamilton Inc.). Copies of the summary report can be obtained by sending $1 to Environmental Technology Seminar, Inc., P.O. Box 391, Bethpage, NY 11714.

In response to the growing pollution of the ground water supplies in Long Island's Nassau and Suffolk Counties, a study of the available options in water management was undertaken. The resulting educational handbook considers these options in both the traditional economic and sanitary aspects as well as the environmental, ecological, and social impacts.

Kern, Ken, Kogon, Ted, and Thallon, Rob. *The Owner-Builder and the Code.* Oakhurst, CA: Owner-Builder Publications, 1976.

A discussion of the politics of constructing your own house is illustrated with case histories outlining how a number of owner-builders dealt with site design, economic problems, and the bureaucratic tangle of codes and standards. Since at least half the citations received by owner-builders are for building unapproved sewage systems, one section of the book addresses the problems of the typical waste removal system and the alternatives of the composting toilet and pit privy.

King, F.H. *Farmers of Forty Centuries: Permanent Agriculture in China, Korea and Japan.* Emmaus, PA: Rodale Press, 1973.

How the Chinese, Korean, and Japanese systems of agriculture and their recycling of all wastes have maintained soil fertility through forty centuries of production.

Kira, Alexander. *The Bathroom.* 2d rev. ed. New York: The Viking Press, 1976.

Though it does not concern itself with water conservation, waste treatment, or composting toilets, the book questions the complacency with which we accept standard bathroom fixtures and encourages their redesign with an eye to providing the most practical, safe, and comfortable means for human cleansing and elimination purposes.

Laak, R. "Home Plumbing Fixture Waste Flows and Pollutants." Unpublished report. University of Connecticut, Storrs, CT 06268, 1972.

This reports actual measurements made of the wastewater generated at each home plumbing fixture, and the character of the pollutants in the wastes.

Leborg, Sonja, ed. "21 Biological Toilets: Decomposition Toilets for Cabins and Holiday Homes." (Translation of an extract from Consumer Report No. 10.) Norway: Microbiological Institute, Agricultural College of Norway.

Results of a test of 21 decomposition toilets which are available in Norway.

Ligman, K. "Rural Wastewater Simulation." M.S. Independent Study Report, University of Wisconsin, Madison, WI 53706, 1972.

Lindstrom, Rikard. "A Simple Process for Composting Small Quantities of Community Waste." *Compost Science: Journal of Waste Recycling* (Spring 1965): 30–32.

A detailed explanation of the operation of the Clivus Multrum by its inventor.

Love, Sam. "An Idea in Need of Rethinking: the Flush Toilet." *Smithsonian* (May 1975): 61–66.

A brief history of human waste treatment systems and an overview of the present spate of alternatives, concentrating on the Clivus Multrum composting toilet.

Mann, H.T., and Williamson, D. *Water Treatment and Sanitation: A Handbook of Simple Methods for Rural Areas in Developing Countries.* Rev. ed. January 1976. Intermediate Technology Publications, 9 King St., London, England WC2E 8HN.

This booklet (90 pages) describes low-cost methods of selecting a water source, testing water quality, and handling wastes and waste water. Although it was written for poor communities in Third World countries, the methods recommended are applicable everywhere.

McClelland, Nina I., ed. *Individual On-Site Wastewater Systems: Proceedings of the Third National Conference 1976.* Ann Arbor: Science Publishers, Inc., 1977.

Sponsored by the National Sanitation Foundation and the U.S. Environmental Protection Agency Technology Transfer Program, the conference presented the state-of-the-art of methods of on-site treatment as seen by government and engineering. E.P.A.'s response to the rural wastewater problem was discussed without considering radical changes in technology, and without mention of composting toilets.

Milne, Murray. *Residential Water Conservation.* California Water Resources Center, University of California/Davis. Report #35 (1976).

The best nontechnical report on ways to reduce residential water consumption available. Includes a review of many plumbing fixtures now available.

Minimum Cost Housing Group. *Stop the Five Gallon Flush: A Survey of Alternative Waste Disposal Systems.* Montreal: McGill University, 1976.

A careful catalog and explanation of different alternative waste treatment systems ranging from an incinerating toilet to modifications of the basic pit privy.

Nesbitt, Patricia M., and Smythe, Robert B. "Wastewater Treatment Systems: An Assessment Guide for Citizens." Washington, D.C.: Environmental Impact Assessment Project of the Institute of Ecology, January 4, 1974. Available from Patricia M. Nesbitt, Rt. 1, Box 106A, Strasburg, VA 22657.

A detailed analysis of the different ways waste is treated at municipal levels.

Nicholas, H. Wayne. *Analysis of Bacterial Populations in the Final Product of the Clivus Multrum.* Center for the Biology of Natural Systems, Washington University, Saint Louis, MO 63130, 1976.

Compost samples were taken from Clivus Multrum composting toilets in the United States and in Sweden and were tested for residual pathogenic bacteria. Results showed that the bacterial composition and pathogenic potential of the final Multrum product were similar to those of soil itself.

Otis, Richard J., and Boyle, William C. "Performance of Single Household Treatment Units." *Journal of the Environmental Engineering Division,* ASCE, Vol. 102, No. EE 1, Proc. Paper 11895, February 1976, pp. 175–88.

An evaluation of the present state-of-the-art in household waste treatment with recommendations for improvement through design modification.

Rybczynski, Witold. "Small Is Beautiful . . . but Sometimes Bigger Is Better: New Developments in Moldering and Composting Toilets." *Solar Age* 1, No. 5:8–11.

An overview of current designs in moldering and composting toilets, focusing on those which require no energy input.

Small Scale Waste Management Project. 1 Agriculture Hall, University of Wisconsin, Madison, WI 53706.

This federally funded project has produced a number of publications relating to their research on alternatives and improvements to leach fields and sewage treatment plants in small towns and rural areas. Though they do concentrate on treatment for combined flow wastes, several of their publications will be of some interest to homeowners designing their own greywater systems, among them "On-Site Disposal of Small Wastewater Flows." For an overview of their work ask for their "Publications List," which is revised periodically.

Stewart, James M., ed. *North Carolina Conference on Water Conservation: Proceedings Sept. 3–4, 1975.* Available from Water Resources Research Institute, 124 Riddick Building, North Carolina State University, Raleigh, NC 27607, for $8 per copy if prepaid, $10 per copy if billed.

Papers presented at the conference addressed water conservation on the industrial, municipal, and personal levels. Flow reduction toilets were considered, though composting types were not discussed.

United Stand Privy Booklet (and Greywater Systems). United Stand, P.O. Box 191, Potter Valley, CA 95469.

United Stand, a group of concerned build-it-yourselfers working for the legalization of alternative waste systems, has produced a brief practical booklet aimed at helping people build their own compost privies and greywater systems.

Warshall, Peter. *Septic Tank Practices.* Bolinas, CA: Mesa Press, 1976.

A very readable primer on the conservation and reuse of household wastewaters, it defines the sewage problem and the contribution of soil to water purification and offers guidelines for the design, construction, and maintenance of a septic tank and drainfield.

Windblad, Uno. *Compost Latrines: A Review of Existing Systems.* 2d rev. ed. Alternative Waste Disposal Systems, P.O. Box 1588, Dar Es Salaam, Tanzania. 1975. Photocopied.

Description of ten different compost latrines used in different parts of the world.

Witt, Michael D. "Water Use in Rural Homes." Small Scale Waste Management Project, University of Wisconsin, Madison, WI 53706, 1974.

Another report that examines the amount of water used in rural homes, according to how much wastewater is generated at each fixture and the character of the pollutants in these wastes.

Wright, Lawrence. *Clean and Decent.* Toronto, Canada: University of Toronto Press, 1967.

A history of the bathroom and water closet from ancient times up to recent years.

Index

A

ABC process, of waste disposal, 14-15
Absorption trenches, for greywater disposal, 174, 177-79
Acidity/alkalinity. *See* pH
Activated sludge, 16
Active composting toilets, 108-12
See also specific names
Aeration, of compost, 46
in owner-built toilets, 141-42
of septic tanks, 169
in waste treatment, 32, 38
See also Ventilation
Aerators, faucet, for water conservation, 247
Aerobic composting, 43
Aerobic tanks, for waste treatment, 32-34
Air pollution. *See* Pollution, air
Air staircase, in Toa Throne, 117
Algae, in greywater treatment, 185, 187
in pollution from sewage, 18
in sewage treatment, 39-40
Algal Regenerative System, of waste disposal, 39-41
Anaerobic composting, 43
Asia, waste recycling in, 2-3
See also China

B

Bacteria, for waste disposal, 16
See also Composting
Ballcock, leak-signalling, 243-44
Bio-gas. *See* Methane
Biological toilets, 29
Bio Loo, 108-12, 130-31, 138-39
air-pull of, 102
pasteurizing hotplate, 109, 111, 130, 218
Bio-Mat. *See* Mull-Toa
Bio Toilet A (and M), 108-12, 128-30, 136-37
Bio Toilet 75 (and 75B), 108-12, 127-28, 136-37
Biu-Let. *See* Mull-Toa
BOD_5, 164
Building codes. *See* Government
Building materials, for owner-built privies, 61, 75, 77, 83

C

Cadmium, in sludge, 26-27
CANWEL System, of waste treatment, 36-38
Carbon dioxide, from composting, 67
Carbon/nitrogen ratio, importance in composting, 46-49, 67
    maintaining in toilets, 79, 103
    in Mullbank compost, 225, 226
Cesspits, for waste disposal, 2, 175
Children, composting toilets and, 208
China, waste disposal in, 2-3, 60
    See also Asia
Chlorine, as waste treatment, 24-25, 38
Cholera, 54
Clivus Minimus, owner-built toilet, 157-60
Clivus Multrum, 97-107, 112-17, 132-33, 159
    air-pull of, 102
    cost, 132, 231-32
    installation, 113-14
    spiral conveyor, 114
    temperatures in, 218
    trouble shooting manual, 106-7
Compaction, in owner-built toilet, 143, 144
    See also Stirring arms
Compost, from toilets, use of, 103, 111, 117, 217-30
    See also Gardens, Land disposal
Composting, 43-49, 59-60, 90-91
    of raw sewage, 65-67
    of septage, 25
    of sewage sludge, 24, 26
Composting privies, 51-95
    backcountry bin composter, 91-95
    Farallones Drum Privy, 83-90
    Farallones Two-Holer, 68-79
    Ken Kern's, 58-68
Composting toilets, 41, 97-160
    commercial, 97-139
    owner-built, 140-160
    See also specific names
Conservation, water, 64, 206, 235-55
Conveyor, spiral, in Clivus Multrum, 114
Copper, in sludge, 26-27
Costs, of composting toilets, 132-38, 207, 230-33
    of sewage treatment plants, 21

D

Detergents, in greywater, 189, 191
Disease, 54-55
    sewage and, 8-10, 12, 24
    urine and, 77
    See also Pathogens
Dishwashers, water conservation in, 252
Domestic Sewage-Methane Cycle, 38-39
Drum, rotating, in Bio Toilet A, 128
    in Evans's Coprophage #3, 150-51
Drum privy, owner-built, 83-90
Drywell. *See* Seepage pits
Dual-flush toilets, for water conservation, 240-42
Duckweed, for greywater treatment, 186

E

Earth closet, for waste disposal, 11
Ecolet. *See* Mullbank
Energy, used by sewage treatment plants, 21-22, 23
England, early waste disposal in, 2
    sewage farming in, 15
Eutrophication, of rivers, 18-19
Evacuators, odorless, for sewage disposal, 7, 8
Evans's Coprophage #3, owner-built toilet, 147-51
Evapo-transpiration beds, for greywater disposal, 174, 179-82

F

Fans, in composting toilets, 101, 109, 115, 117, 121
    in owner-built toilets, 146
    See also Ventilation
Farallones composting toilets, 68-91
Farming, use of sewage in, 4-10, 15
    use of sludge in, 19, 24
    See also Gardens, Land disposal
Faucets, water conservation in, 245-49
Filtration, of drinking water, 17-18
    of greywater, 169-74, 188
    of sewage, 16
Flanders, night soil use in, 2
Flies. *See* Insects
Flow restricters, for water conservation, 245-47
Flush toilet, water conservation in, 237-44
Fly paper, in composting toilets, 104
Franchino, Peter, greywater system, 194-203
Fruit flies. *See* Insects

G

Garbage, in composting toilets, 79, 88
    See also Kitchen wastes
Gardens, greywater use in, 190-91
    sludge use in, 26, 103
    See also Compost, Land disposal
"Good Garden Dirt"—song, 234
Government, composting toilets and, 209-17
Gravel, as greywater filter, 172
Grease, in greywater, 165, 169
Greenhouses, for greywater disposal, 183-84, 185
Greywater, 161-203
    definition, 161
    disposal of, 174-92
    filtration of, 169-74
    irrigation with, 187-89, 190
    kitchen wastes in, 163-65
    owner-built system, 194-203
    septic tank and, 166-69
Ground water, 22
    pollution by sewage, 23
Gypsum, in gardens, 189

H

Heat, from composting toilets, 104

Heater, in composting toilets, 109, 121, 124
pasteurizing hotplate, 109, 111, 130
Heat exchanger, on ventilation pipes, 102, 104
Heavy metals, in Mullbank compost, 228-30
in sludge, 23, 24, 26-27
Hot water heating, 252-54
Humus Toilet H5. *See* Mull-Toa
Hygrometer, in composting toilets, 110, 147

I

Incinerating toilets, 29
Incineration, of sludge, 19, 23, 38
Industrial wastes, in sludge, 24
Insecticides, in composting toilets, 104, 107, 111
Insects, in compost, 59
in composting toilets, 103-4, 107, 111, 142
owner reaction to, 208
in privies, 57, 68, 83
Installation, of large composting toilets, 100-1, 113-14, 119-20
of small composting toilets, 108
Insulation, of composting chamber, 66-67, 113
of vent pipe, 114, 115, 144-45
Interceptor tank, 34
Intermittent filtration, of sludge, 16, 17

K

Ken Kern, composting privy, 58-68
Kitchen chute, on Clivus Multrum, 114
on composting toilets, 103-4
Kitchen wastes, in composting toilets, 102, 111
See also Garbage

L

Lagoon, for greywater disposal, 185-87
Land disposal, of compost from toilets, 79-80, 103
of sewage, 4-10, 16
of sludge, 19, 23
See also Compost, Farming, Gardens
Leach fields, clogging of, 165
for irrigation, 189
size of, 163
See also Absorption trenches
Leaching chambers, 175, 182-83
Legislation, governing sewage disposal, 17
See also Government
Lime, in composting toilet, 120
Linings, in greywater lagoons, 186
Liquid. *See* Moisture
Low-flush toilets, 239-40

M

Maine tank, owner-built toilet, 160

Maintenance, of large compost toilets, 103, 115
of Mullbank, 122-24
Medical science, and waste disposal, 8-10, 12
Methane, from sewage sludge, 39-40
Milorganite, 24
Moisture, in Clivus Multrum, 115
in composting toilets, 106-7, 110-11
importance in composting, 46
in owner-built toilets, 151
in privy, 65-66, 78-79, 88
Mounds, for greywater disposal, 175, 182
Mullbank, 108-11, 120-24, 134-35
air-pull of, 102
content of finished compost, 224-230
costs, 134, 231
temperatures, 218
Mull-Toa, 108-12, 124-27, 134-35

N

Nickel, in sludge, 26-27
Night soil, 2
in China, 2-3
use in farming in U.S., 4-10
Nitrogen, content of Mullbank compost, 226
Nutrient elements, in Mullbank compost, 226-28

O

Oak leaves, as priming material, 120
Ocean disposal, of sludge, 19, 22, 23
Odor, in composting toilets, 106, 142, 207
in owner-built toilets, 145
in privies, 57, 60
Odorless evacuators, 7, 8
Oil-flush toilets, 29-32
Owner/builder, greywater system, 194-203
privies, 58-90
toilets,140-60
Owner opinion, on composting toilets, 205-9
Oxygen, importance in composting, 46
Ozone, as waste purifier, 38
Ozone layer, chlorine and, 25

P

Pail system, use in England, 2
Paris, early waste disposal in, 2
Passive composting toilets, 98-104
Pasteurizing hotplate, in Bio Loo, 109, 111, 130, 218
Pathogens, in composting toilets, 218
destruction of, 45, 117
in sludge, 26
PBB (polybrominated biphenyls), in sludge, 27
PCB (polychlorinated biphenyls), in sludge, 27
Peat moss, as priming material, 102, 115, 120, 121
pH, importance in composting, 49
in Mullbank compost, 224-26
Philorganic, 24
Pit privy, 51, 157

Pollution, from rural household, 164
Pollution, air, by incinerating sludge, 23
Pollution, water, by sewage, 10, 16-17, 23, 54
Ponds, for greywater treatment, 185-87
Pressure sewers, 34-36
Pressurized toilets, 240
Prices. *See* Costs
Priming, of Clivus Multrum, 114-15
  of composting toilets, 102, 112, 121
  of Toa Throne, 120
Public health departments. *See* Government
Pyrethrum, for insect control, 104

R

Raincap, over vent pipe, 145
Romans, use of human wastes by, 2

S

Sand, as greywater filter, 172-74
Sanitation, in composting, 59, 68
Sawdust, in composting privies, 88
  as insect control, 104
Screening, insect, 57, 83
Seepage beds, for greywater disposal, 174, 179
Seepage pits, for greywater disposal, 174, 175
Septic tanks, 16, 25-28
  for greywater treatment, 166-69
  with pressured sewers, 34-36
Settling tank, for greywater systems, 188
Sewage treatment plants, 21-25
Showering, for water conservation, 64
Showers, water conservation in, 245-47
Size, of composting pile, 44
  of privy compartments, 73
Sludge, as commercial fertilizer, 24
  disposal of, 19
  use on land, 19, 24
    in gardens, 26-27
  See also Farming, Gardens
Soakaway. *See* Seepage pits
Soaps, in greywater, 189, 191
Soddy Potty. *See* Mull-Toa
Soddy Potty #2, 108-10, 125, 136-37
Soil, as greywater treatment, 166
Solar heating, of composting compartments, 89-90, 100, 125, 152-57
Spiral conveyor, in Clivus Multrum, 114
Spray taps, for water conservation, 248
Squat plate, 65, 68, 79
Stabilization pond. *See* Lagoon
Starter bed. *See* Priming
Sterilization chamber. *See* Bio Loo
Stirring arms, in composting toilets, 109, 121

T

Tank ball, adjustments for water conservation, 243
Tannic acid, in composting toilet, 120
Temperature, importance in composting, 44-46, 66-67
  in privies, 79

in toilets, 99, 117, 145
Toa Throne, 98-105, 117-20, 132-33
air-pull of, 102
air staircase, 117
Toilets, biological, 29
Toilets, composting. *See* Composting toilets
Toilets, incinerating, 29
Toilets, oil-flush, 29-32
Travis's Solar Composting Toilet, 152-57

U

United States, early waste disposal in, 3-11
sewage treatment plants in, 21-25
See also Government
Urine, 76-77
in compost, 78-79
See also Moisture

V

Ventilation, in Clivus Multrum, 114
in composting toilets, 98, 101, 108-9, 121, 124
in owner-built toilets, 145-46, 153-57
in privies, 68, 74-75, 83, 86
in Toa Throne, 117-19
See also Fans
Volume reducers, for flush toilets, 238

W

Washing machines, water conservation in, 249-51
Waste disposal, early methods of, 1-11
water carriage for, 11-27
Wastewater Recycling Systems, 242-43
Water, filtration of, 17
use by sewage treatment plants, 22-23
use in average household, 162, 239
Water closet, 11
Water conservation. *See* Conservation
Water heating, 252-54
Water hyacinths, for greywater treatment, 186
Water pollution. *See* Pollution, water
Waterproofing, owner-built toilets, 145
Wind turbine, as ventilator and raincap, 145
Wood chips, as greywater filter, 172

Z

Zinc, in sludge, 26-27